Edited by
Rudolf Eggers

Industrial High Pressure Applications

Related Titles

Zaichik, L.I., Alipchenkov, V. M., Sinaiski, E.G.

Particles in Turbulent Flows

2008

Hardcover

ISBN: 978-3-527-40739-2

Bonem, J. M.

Process Engineering Problem Solving

Avoiding "The Problem Went Away, but it Came Back" Syndrome

Hardcover

ISBN: 978-0-470-16928-5

Koltuniewicz, A., Drioli, E.

Membranes in Clean Technologies

Theory and Practice

2008

Hardcover

ISBN: 978-3-527-32007-3

Doona, C. J., Feeherry, F. E. (eds.)

High Pressure Processing of Foods

2008

E-Book

ISBN: 978-0-470-37631-7

Helmus, F. P.

Process Plant Design

Project Management from Inquiry to Acceptance

2008

Hardcover

ISBN: 978-3-527-31313-6

Häring, H.-W. (ed.)

Industrial Gases Processing

2008

Hardcover

ISBN: 978-3-527-31685-4

Aehle, W. (ed.)

Enzymes in Industry

Production and Applications

2007

Hardcover

ISBN: 978-3-527-31689-2

Asua, J. (ed.)

Polymer Reaction Engineering

2007

Hardcover

ISBN: 978-1-4051-4442-1

Meyer, R., Köhler, J., Homburg, A.

Explosives

2007

Plastic

ISBN: 978-3-527-31656-4

Ingham, J., Dunn, I. J., Heinzle, E., Prenosil, J. E., Snape, J. B.

Chemical Engineering Dynamics

An Introduction to Modelling and Computer Simulation

2007

Hardcover

ISBN: 978-3-527-31678-6

Edited by Rudolf Eggers

Industrial High Pressure Applications

Processes, Equipment and Safety

WILEY-VCH Verlag GmbH & Co. KGaA

The Editor

Prof. Dr.-Ing. Rudolf Eggers
TU Hamburg-Harburg, Inst. für
Thermische Verfahrenstechnik
Eißendorfer Str. 38
21073 Hamburg

■ All books published by **Wiley-VCH** are carefully produced. Nevertheless, authors, editors, and publisher do not warrant the information contained in these books, including this book, to be free of errors. Readers are advised to keep in mind that statements, data, illustrations, procedural details or other items may inadvertently be inaccurate.

Library of Congress Card No.: applied for

British Library Cataloguing-in-Publication Data
A catalogue record for this book is available from the British Library.

Bibliographic information published by the Deutsche Nationalbibliothek
The Deutsche Nationalbibliothek lists this publication in the Deutsche Nationalbibliografie; detailed bibliographic data are available on the Internet at http://dnb.d-nb.de.

© 2012 Wiley-VCH Verlag & Co. KGaA, Boschstr. 12, 69469 Weinheim, Germany

All rights reserved (including those of translation into other languages). No part of this book may be reproduced in any form – by photoprinting, microfilm, or any other means – nor transmitted or translated into a machine language without written permission from the publishers. Registered names, trademarks, etc. used in this book, even when not specifically marked as such, are not to be considered unprotected by law.

Print ISBN: 978-3-527-32586-3
ePDF ISBN: 978-3-527-65268-6
ePub ISBN: 978-3-527-65267-9
mobi ISBN: 978-3-527-65266-2
oBook ISBN: 978-3-527-65265-5

Cover Design Formgeber, Eppelheim, Germany
Typesetting Thomson Digital, Noida, India
Printing and Binding   Markono Print Media Pte Ltd, Singapore

Printed on acid-free paper

Contents

Preface *XIII*
List of Contributors *XV*

Part One Introduction *1*

1 Historical Retrospect on High-Pressure Processes *3*
Rudolf Eggers
References *6*

2 Basic Engineering Aspects *7*
Rudolf Eggers
2.1 What are the Specifics of High-Pressure Processes? *7*
2.2 Thermodynamic Aspects: Phase Equilibrium *9*
2.3 Software and Data Collection *10*
2.4 Phase Equilibrium: Experimental Methods and Measuring Devices *10*
2.5 Interfacial Phenomena and Data *12*
2.6 Material Properties and Transport Data for Heat and Mass Transfer *20*
2.7 Evaporation and Condensation at High Pressures *37*
2.7.1 Evaporation *37*
2.8 Condensation *43*
References *46*

Part Two Processes *49*

3 Catalytic and Noncatalytic Chemical Synthesis *51*
Joachim Rüther, Ivo Müller, and Reinhard Michel
3.1 Thermodynamics as Driver for Selection of High Pressure *51*
3.1.1 Chemical Equilibrium: Law of Mass Action *51*
3.1.2 Reaction Kinetics *53*
3.1.3 Phase Equilibria and Transport Phenomena *55*
3.2 Ammonia Synthesis Process *55*
3.2.1 Basics and Principles *56*
3.2.2 History of the Ammonia Process *57*
3.2.3 Development of Process and Pressure *58*

3.2.4	Special Aspects 63
3.3	Urea Process 64
3.3.1	Basics and Principles 65
3.3.2	History of Urea Process 67
3.3.3	Integration of Ammonia and Urea Processes 71
3.3.4	Special Construction Materials 71
3.4	General Aspects of HP Equipment 72
3.4.1	Multilayered Vessels 73
3.4.2	Recommendations to Vessel Design 73
3.4.3	Gaskets and Bolting 74
	References 75

4	**Low-Density Polyethylene High-Pressure Process** 77
	Dieter Littmann, Giulia Mei, Diego Mauricio Castaneda-Zuniga, Christian-Ulrich Schmidt, and Gerd Mannebach
4.1	Introduction 77
4.1.1	Historical Background 77
4.1.2	Properties and Markets 77
4.1.3	Polyethylene High-Pressure Processes 78
4.1.4	Latest Developments 78
4.2	Reaction Kinetics and Thermodynamics 78
4.2.1	Initiation 79
4.2.2	Propagation 79
4.2.3	Chain Transfer 80
4.2.4	Termination 81
4.2.5	Reaction Kinetics 81
4.3	Process 82
4.3.1	General Process Description 82
4.3.2	Autoclave Reactor 84
4.3.3	Tubular Reactor 85
4.3.4	Safety 88
4.4	Products and Properties 89
4.4.1	Blown Film 89
4.4.2	Extrusion Coating 90
4.4.3	Injection Molding 90
4.4.4	Wire and Cable 90
4.4.5	Blow Molding 90
4.4.6	Copolymers 91
4.5	Simulation Tools and Advanced Process Control 91
4.5.1	Introduction 91
4.5.2	Off-Line Applications 91
4.5.2.1	Flow Sheet Simulations 91
4.5.2.2	Steady-State Simulation of the Tubular Reactor 92
4.5.2.3	Dynamic Simulation of the Process 92
4.5.3	Online Application 93

4.5.3.1	Soft Sensors	93
4.5.3.2	Advanced Process Control	94
	References	96

5 High-Pressure Homogenization for the Production of Emulsions 97
Heike P. Schuchmann, Née Karbstein, Lena L. Hecht, Marion Gedrat, and Karsten Köhler

5.1	Motivation: Why High-Pressure Homogenization for Emulsification Processes?	97
5.2	Equipment: High-Pressure Homogenizers	98
5.2.1	Principal Design	98
5.2.2	Disruption Systems for High-Pressure Homogenization	98
5.2.2.1	Valves	98
5.2.2.2	Orifices and Nozzles	99
5.2.3	Flow Conditions	100
5.2.3.1	Flow Conditions in the Disruption System	100
5.2.3.2	Effect of Flow Conditions in Homogenization Valves on Emulsion Droplets	101
5.2.4	Simultaneous Emulsification and Mixing (SEM) Systems	101
5.3	Processes: Emulsification and Process Functions	103
5.3.1	Droplet Disruption in High-Pressure Valves	103
5.3.2	Droplet Coalescence in Homogenization Valves	104
5.3.3	Droplet Agglomeration in Homogenization Valves	107
5.4	Homogenization Processes Using SEM-Type Valves	107
5.4.1	Dairy Processes	107
5.4.2	Pickering Emulsions	109
5.4.3	Melt Homogenization	111
5.4.4	Emulsion Droplets as Templates for Hybrid (Core–Shell) Nanoparticle Production	112
5.4.5	Submicron Emulsion Droplets as Nanoreactors	114
5.4.6	Nanoparticle Deagglomeration and Formulation of Nanoporous Carriers for Bioactives	116
5.5	Summary and Outlook	117
	References	118

6 Power Plant Processes: High-Pressure–High-Temperature Plants 123
Alfons Kather and Christian Mehrkens

6.1	Introduction	123
6.2	Coal-Fired Steam Power Plants	125
6.2.1	Thermodynamics and Power Plant Efficiency	125
6.2.2	Configuration of Modern Steam Power Plants	127
6.3	Steam Generator	130
6.3.1	Steam Generator Design	130
6.3.2	Membrane Wall	134
6.3.3	Final Superheater Heating Surface	135

6.3.4	Final Superheater Outlet Header and Live Steam Piping	*136*
6.4	High-Pressure Steam Turbines	*138*
6.4.1	Configuration of Modern Steam Turbines	*138*
6.4.2	Design Features of High-Pressure Steam Turbines	*139*
6.5	Summary and Outlook	*142*
	References	*142*

7	**High-Pressure Application in Enhanced Crude Oil Recovery**	**145**
	Philip T. Jaeger, Mohammed B. Alotaibi, and Hisham A. Nasr-El-Din	
7.1	Introduction	*145*
7.1.1	Principal Phenomena in Oil and Gas Reservoirs	*145*
7.1.2	Reservoir Conditions	*145*
7.2	Fundamentals	*147*
7.2.1	Miscibility at Elevated Pressures	*147*
7.2.2	Physical Chemical Properties of Reservoir Systems at Elevated Pressures	*148*
7.2.2.1	Density	*148*
7.2.2.2	Rheology	*150*
7.2.2.3	Interfacial Tension	*151*
7.2.2.4	Wetting	*151*
7.2.2.5	Diffusivity	*153*
7.2.2.6	Permeability	*154*
7.3	Enhanced Oil Recovery	*155*
7.3.1	Water Flooding	*157*
7.3.2	Chemical Injection	*158*
7.3.3	Thermal Recovery	*158*
7.3.4	Gas Injection	*159*
7.3.5	Carbon Dioxide Capture and Storage (CCS) in EOR	*160*
7.3.6	Combustion	*160*
7.4	Oil Reservoir Stimulation	*161*
7.5	Heavy Oil Recovery	*161*
7.6	Hydrates in Oil Recovery	*162*
7.7	Equipment	*163*
7.7.1	Pumps	*163*
7.7.2	Pipes	*164*
7.7.3	Seals	*164*
7.7.4	Separators	*165*
	References	*166*

8	**Supercritical Processes**	**169**
	Rudolf Eggers and Eduard Lack	
8.1	Introduction	*169*
8.2	Processing of Solid Material	*172*
8.2.1	Isobaric Process	*174*

8.2.2	Single or Cascade Operation with Multistep Separation	*174*
8.2.3	Cascade Operation and Multistep Separation	*175*
8.2.4	Extractable Substances	*175*
8.2.4.1	Selective Extraction	*176*
8.2.4.2	Total Extraction	*176*
8.2.5	Pretreatment of Raw Materials	*176*
8.2.6	Design Criteria	*177*
8.2.7	Design with the Use of Basket	*178*
8.2.8	Thermodynamic Conditions	*179*
8.2.9	Mass Transfer	*179*
8.2.10	Hydrodynamics	*182*
8.2.11	Energy Optimization	*182*
8.2.12	Pump Process	*182*
8.2.13	Compressor Process	*183*
8.2.14	Some Applications of Supercritical Extraction of Solids	*184*
8.2.14.1	Decaffeination of Green Coffee Beans	*184*
8.2.14.2	Production of Hops Extract	*184*
8.2.14.3	Extraction of Spices and Herbs	*186*
8.2.14.4	Extraction of Essential Oils	*186*
8.2.14.5	Production of Natural Antioxidants	*188*
8.2.14.6	Production of High-Value Fatty Oils	*189*
8.2.14.7	Extraction of γ-Linolenic Acid	*189*
8.2.14.8	Cleaning and Decontamination of Cereals Like Rice	*189*
8.2.14.9	Impregnation of Wood and Polymers	*190*
8.2.14.10	Cleaning of Cork	*193*
8.2.14.11	Economics – Especially Investment Cost for Multipurpose Plants	*193*
8.3	Processing of Liquids	*194*
8.4	Future Trends	*202*
8.4.1	Drying of Aerogels	*202*
8.4.2	Treating of Microorganisms	*203*
8.4.3	Use of Supercritical Fluids for the Generation of Renewable Energy	*204*
8.4.4	Gas-Assisted High-Pressure Processes	*205*
	References	*206*
9	**Impact of High-Pressure on Enzymes**	***211***
	Leszek Kulisiewicz, Andreas Wierschem, Cornelia Rauh, and Antonio Delgado	
9.1	Introduction	*211*
9.2	Influence of Pressure on Biomatter	*212*
9.3	Influence of Pressure on the Kinetics of Enzyme Inactivation	*215*
9.4	Technological Aspects	*218*
9.5	Summary	*226*
	References	*227*

10	**High Pressure in Renewable Energy Processes** *235*
	Nicolaus Dahmen and Andrea Kruse
10.1	Introduction *235*
10.2	Thermochemical Processes *236*
10.2.1	Pyrolysis *237*
10.2.2	Liquefaction *238*
10.2.3	Gasification *240*
10.2.3.1	Fixed Bed Gasifier *242*
10.2.3.2	Fluidized Bed Gasifiers *243*
10.2.3.3	Entrained Flow Gasifiers *244*
10.3	Hydrothermal Processes *248*
10.3.1	Hydrothermal Carbonization *250*
10.3.2	Hydrothermal Liquefaction *251*
10.3.3	Hydrothermal Gasification *253*
10.3.3.1	Catalytic Hydrothermal Gasification *253*
10.3.3.2	Supercritical Hydrothermal Gasification *254*
	References *256*

11	**Manufacturing Processes** *257*
	Andrzej Karpinski and Rolf Wink
11.1	Autofrettage: A High-Pressure Process to Improve Fatigue Lifetime *260*
11.2	Waterjet Cutting Technology *265*
11.2.1	Generation of Waterjets *265*
11.2.2	Cutting Process and Parameters *267*
11.2.3	High-Pressure Pumps *269*
11.2.4	Waterjet Cutting with 6000 bar *272*
11.2.5	Cutting Devices *273*
11.2.6	New Trends in the Waterjet Cutting *276*
11.2.6.1	Abrasive Water Suspension Jet *276*
11.2.6.2	Microcutting *276*
11.2.6.3	Medical Applications *277*
	References *278*

Part Three Process Equipment and Safety *283*

12	**High-Pressure Components** *285*
	Waldemar Hiller and Matthias Zeiger
12.1	Materials for High-Pressure Components *285*
12.1.1	Steel Selection Criteria *286*
12.1.2	High-Strength Low-Alloy Steel *287*
12.1.3	Weldable Fine-Grain and High-Temperature Structural Steels *287*
12.1.4	High-Strength High-Alloy Steels *287*
12.1.5	Austenitic Stainless Steels *288*

12.1.6	Austenitic–Ferritic Duplex Steels	288
12.1.7	Chromium–Molybdenum Hydrogen-Resistant Steels	288
12.1.8	Fatigue and Fracture Properties of High-Strength Steels	289
12.2	Pressure Vessels	290
12.2.1	Leak Before Burst	292
12.2.2	Welded Pressure Vessels	292
12.2.3	Nonwelded Pressure Vessels	294
12.2.4	Prestressing Techniques	298
12.2.5	Sealing Systems	300
12.3	Heat Exchangers	301
12.4	Valves	303
12.5	Piping	304
	References	309
13	**High-Pressure Pumps and Compressors**	**311**
	Eberhard Schluecker	
13.1	Selection of Machinery	311
13.2	Influence of the Fluid on Selection and Design of the Machinery	313
13.3	Design Standards for High-Pressure Machines	314
13.4	Materials and Materials Testing	316
13.5	High-Pressure Centrifugal Pumps and High-Pressure Turbocompressors	317
13.6	Rotating Positive Displacement Machines	319
13.6.1	Discharge Rate	319
13.6.2	Gear Pumps	320
13.6.3	Screw Pumps	321
13.6.4	Progressing Cavity Pump	323
13.7	Reciprocating Positive Displacement Machines	323
13.7.1	Drive Technology for Reciprocating Positive Displacement Machines	324
13.7.2	Flow Behavior of Reciprocating Positive Displacement Machines	325
13.7.3	Pulsation Damping	327
13.7.4	Design Versions	328
13.7.4.1	Vertical Pump Head for 70 MPa	328
13.7.4.2	Horizontal Pump Head with Y-Piece for 300 MPa	329
13.7.4.3	Diaphragm Pump Heads	329
13.7.4.4	Piston Compressor for 30 MPa at the Maximum	330
13.7.4.5	Compressor for 300 MPa	332
13.7.4.6	Piston Compressor for 1400 MPa	333
	References	334
14	**High-Pressure Measuring Devices and Test Equipment**	**335**
	Arne Pietsch	
14.1	Introduction	335
14.2	Process Data Measuring – Online	336

14.2.1	Sensor Choice and Installation	*337*
14.2.2	Pressure and Differential Pressure	*338*
14.2.3	Temperature	*341*
14.2.4	Flow	*343*
14.2.5	Fluid Level	*350*
14.2.6	Density	*351*
14.2.7	Viscosity	*351*
14.2.8	Concentration – Solute in High-Pressure Gases and Fluids	*352*
14.2.9	Concentration – Gas Traces Dissolved in Liquids	*358*
14.3	Lab Determination – Additional Offline Test Equipment	*359*
14.3.1	Phase Equilibrium	*359*
14.3.2	Magnetic Sorption Balance	*362*
14.3.3	Interfacial Tension and Wetting	*362*
14.3.4	Gas Hydrates	*363*
14.3.5	Other Properties Online	*364*
14.4	Safety Aspects	*364*
14.5	Future	*366*
	References	*367*

15 Sizing of High-Pressure Safety Valves for Gas Service *369*
Jürgen Schmidt

15.1	Standard Valve Sizing Procedure	*369*
15.2	Limits of the Standard Valve Sizing Procedure	*371*
15.3	Development of a Sizing Method for Real Gas Applications	*372*
15.3.1	Equation of State and Real Gas Factor	*375*
15.3.2	Isentropic Exponent	*378*
15.3.3	Critical Pressure Ratio	*379*
15.4	Sizing of Safety Valves for Real Gas Flow	*380*
15.5	Summary	*382*

Appendix 15.A Calculation of Sizing Coefficient According to EN-ISO 4126-1 and a Real Gas Nozzle Flow Model *383*

15.A.1	Inlet Stagnation Conditions	*383*
15.A.2	Property Data and Coefficients for Ethylene	*383*
15.A.3	Calculation of Flow Coefficient According to EN-ISO 4126-1	*384*
15.A.4	Calculation of Flow Coefficient Accounting for Real Gas Effects	*385*
15.A.5	Approximation of Mass Flux by an Analytical Method (Averaging Method)	*386*

Appendix 15.B List of Symbols *387*
Subscripts *388*
References *389*

Appendix: International Codes and Standards for High-Pressure Vessels *391*

Index *397*

Preface

In 2010, when Wiley-VCH Verlag GmbH asked me to edit a new book on high-pressure applications, the first thought that came to my mind was whether there was really a requirement for compiling such a reference book. In fact, numerous conference proceedings and even some textbooks were available that illustrated the state of the art and special applications of high-pressure processes in detail, offering support for production of innovative products. However, the application of high pressure covers many different industries – from basic material production, mechanical engineering, energy management, chemical engineering to bioprocessing and food processing. In engineering education, these applications even postulate different courses of study.

Based on this background, it is not surprising that a general and comprehensive description of industrial high-pressure processes is hardly possible. Next to basic knowledge, the aim was now to especially include overall aspects such as the need for applying high pressure, desirable and undesirable effects, and prospects and risks of high-pressure processes. In this respect, my activities on high-pressure engineering in industry and university since 1977 facilitated access to experts from various different fields of industrial applications and scientific research who were willing to contribute with their knowledge to special high-pressure applications.

The book is structured in three main parts. Part One is an introductory section dealing with the history and the engineering basics of high-pressure techniques. Part Two demonstrates classical and more recent high-pressure applications from chemical engineering, energy management and technology, bioengineering and food engineering, and manufacturing techniques. Part Three concentrated on equipment, measurement, and safety devices in high-pressure processes. The book concludes with a short survey and an evaluation of international rules that are valid for the calculation and design of high-pressure vessels.

It is my pleasure to thank all the authors for their commitment and their highly valuable and professional contributions. I also thank Wiley-VCH Verlag GmbH for consistent assistance and patience.

Hamburg, June 2012 *Rudolf Eggers*

List of Contributors

M. Alotaibi

Diego Mauricio Castaneda-Zuniga
Lyondell Basell GmbH
Gebäude B 852
65926 Frankfurt
Germany

Nicolaus Dahmen
Karlsruhe Institute of Technology (KIT)
Institute for Technical Chemistry
Hermann-von-Helmholtz-Platz 1
76344 Eggenstein-Leopoldshafen
Germany

Antonio Delgado
Universität Erlangen-Nürnberg
Lehrstuhl für Strömungsmechanik
Cauerstr. 4
91058 Erlangen
Germany

Rudolf Eggers
Technische Universität Hamburg-Harburg
Institut für Thermische Verfahrenstechnik/Wärme und Stofftransport
Eißendorfer Str. 38
21073 Hamburg
Germany

Marion Gedrat
Uni Karlsruhe
Institut für Bio- und Lebensmitteltechnik (BLT)
Fritz-Haber-Weg 2, Geb. 30.44
76131 Karlsruhe
Germany

Lena L. Hecht
Uni Karlsruhe
Institut für Bio- und Lebensmitteltechnik (BLT)
Fritz-Haber-Weg 2, Geb. 30.44
76131 Karlsruhe
Germany

Klaus Heinrich
Uhde GmbH
Friedrich-Uhde-Str. 15
44141 Dortmund
Germany

Waldemar Hiller
Uhde High Pressure Technologies GmbH
Buschmuehlenstr. 20
58093 Hagen
Germany

Philip Jaeger
TU Hamburg–Harburg
Institut für Thermische
Verfahrenstechnik/Wärme und
Stofftransport
Eißendorfer Str. 38
21073 Hamburg
Germany

Née Karbstein
Uni Karlsruhe
Institut für Bio- und
Lebensmitteltechnik (BLT)
Fritz-Haber-Weg 2, Geb. 30.44
76131 Karlsruhe
Germany

Andrzej Karpinski
Uhde High Pressure Technologies
GmbH
Buschmuehlenstr. 20
58093 Hagen
Germany

Alfons Kather
TU Hamburg–Harburg
Institut für Energietechnik
Denickestr. 15
21073 Hamburg
Germany

Karsten Köhler
Uni Karlsruhe
Institut für Bio- und
Lebensmitteltechnik (BLT)
Fritz-Haber-Weg 2, Geb. 30.44
76131 Karlsruhe
Germany

Andrea Kruse
Karlsruhe Institute of Technology (KIT)
Institute for Technical Chemistry
Hermann-von-Helmholtz-Platz 1
76344 Eggenstein-Leopoldshafen
Germany

Leszek Kulisiewicz
Universität Erlangen-Nürnberg
Lehrstuhl für Strömungsmechanik
Cauerstr. 4
91058 Erlangen
Germany

Eduard Lack
NATEX Prozesstechnologie GesmbH
Werkstr. 7
Ternitz
2630 Österreich
Germany

Dieter Littmann
Lyondell Basell GmbH
Gebäude B 852
65926 Frankfurt
Germany

Gerd Mannebach
Lyondell Basell GmbH
Gebäude B 852
65926 Frankfurt
Germany

Christian Mehrkens
TU Hamburg–Harburg
Institut für Energietechnik
Denickestr. 15
21073 Hamburg
Germany

Giulia Mei
Lyondell Basell GmbH
Gebäude B 852
65926 Frankfurt
Germany

Ivo Müller
Uhde GmbH
Friedrich-Uhde-Str. 15
44141 Dortmund
Germany

Hisham Nasr-El-Din
Texas A&M University
Dwight Look College of Engineering
401L Richardson Building, 3116 TAMU
College Station, TX 77843-3116
USA

Arne Pietsch
Eurotechnica GmbH
An den Stuecken 55
22941 Bargteheide
Germany

Cornelia Rauh
Universität Erlangen-Nürnberg
Lehrstuhl für Strömungsmechanik
Cauerstr. 4
91058 Erlangen
Germany

Joachim Rüther
Uhde GmbH
Friedrich-Uhde-Str. 15
44141 Dortmund
Germany

Eberhard Schluecker
Universität Erlangen-Nürnberg
Lehrstuhl für Prozessmaschinen und Anlagentechnik
Cauerstr. 4, Haus 5
91058 Erlangen
Germany

Christian-Ulrich Schmidt
Lyondell Basell GmbH
Gebäude B 852
65926 Frankfurt
Germany

Jürgen Schmidt
BASF SE
Safety & Fluid Flow Technology
GCT/S-L511
67056 Ludwigshafen
Germany

Heike P. Schuchmann
Uni Karlsruhe
Institut für Bio- und Lebensmitteltechnik (BLT)
Fritz-Haber-Weg 2, Geb. 30.44
76131 Karlsruhe
Germany

Andreas Wierschem
Universität Erlangen-Nürnberg
Lehrstuhl für Strömungsmechanik
Cauerstr. 4
91058 Erlangen
Germany

Rolf Wink
Uhde High Pressure Technologies GmbH
Buschmuehlenstr. 20
58093 Hagen
Germany

Matthias Zeiger
Uhde High Pressure Technologies GmbH
Buschmuehlenstr. 20
58093 Hagen
Germany

Part One
Introduction

1
Historical Retrospect on High-Pressure Processes
Rudolf Eggers

The historical development of high-pressure processes since the beginning of the industrial period is based on two concepts: first, the transfer of the inner energy of water vapor at elevated pressures into kinetic energy by the invention of the steam engine; second, the movement of gas-phase reaction equilibrium at high pressures enabling the production of synthetic products like ammonia. Thus, the industrial use of high-pressure processes goes back to both mechanical and chemical engineering. Beginning in the second half of the eighteenth century, the need of safe and gas-tight steam vessels up to few megapascals became essential because that time many accidents happened by bursting of pressure vessels. Chemical industry started high-pressure synthesis processes in the early twentieth century. Compared to moderate working pressures of steam engines, the pressure range now was extremely high between 10 and 70 MPa. As a consequence, a fast growing requirement for high-pressure components like high-pressure pumps, compressors, heat transfer devices, tubes and fittings, reliable sealing systems, and in particular new pressure vessel constructions developed.

Besides, mechanical and chemical engineering material science has promoted the development of new high-pressure processes by creating high ductile steels with suitable strength parameter.

Finally, the safety of high-pressure plants is of outstanding importance. Thus, in the course of development, national safety rules for vessels, pipes, and valves have been introduced by special organizations. For example, in 1884, the American Society of Mechanical Engineering (ASME) launched its first standard for the uniformity of testing methods of boilers. The German society TÜV was founded in 1869 in order to avoid the devastating explosions of steam vessels.

The following list of year dates shows essential milestones of high-pressure processes concerning their development and technical design:

> **1680**: Papins construction of the first autoclave for evaporating water. The design shows the idea of an early safety valve working on an adjustable counterbalance.
> **1769**: James Watt introduced the steam engine transferring thermal energy in motive power.

Industrial High Pressure Applications: Processes, Equipment and Safety, First Edition. Edited by Rudolf Eggers.
© 2012 Wiley-VCH Verlag GmbH & Co. KGaA. Published 2012 by Wiley-VCH Verlag GmbH & Co. KGaA.

1826: Jacob Perkins demonstrated the compressibility of water by experiments above 10 Mpa. Caused by the increasing application of steam engines, the boiling curve of different media became of interest. It was observed that boiling temperatures increase with rising pressure. That time one assumes a remaining coexistence of liquid and gas phase up to any high pressure. It was the Irish physicist and chemist Thomas Andrews who in 1860 disproved this assumption. On the basis of experiments with carbon dioxide, he was able to demonstrate a thermodynamic state with no difference between liquid and gas phase characterized by a distinct value of temperature, pressure, and density. This point has been called the "critical point."

1852: J.P. Joule and W. Thompson discovered the cooling effect caused by the expansion of gases during pressure release.

1873: J.D. van der Waals gives a plausible explanation for the behavior of fluids at supercritical condition.

1900: W. Ostwald claimed a patent on the generation of ammonia by the combination of free nitrogen and hydrogen in the presence of contacting substances.

1913: F. Haber and C. Bosch: First commercial plant synthesizing ammonia from nitrogen and hydrogen at 20 Mpa and 550 °C. The reactors were sized at an inner diameter of 300 mm and a length of 8 m. The productivity of one reactor was 5 ton/day [1]. The pressure vessel was equipped with an in-line tube made from soft iron and degassing holes in order to protect the pressure-resistant walls against hydrogen embrittlement. This process was the forerunner of many others that have been developed into commercial processes [2].

1920: First application of methanol synthesis as a conversion of carbon monoxide and hydrogen at a pressure of 31 MPa and temperatures between 300 and 340 °C.

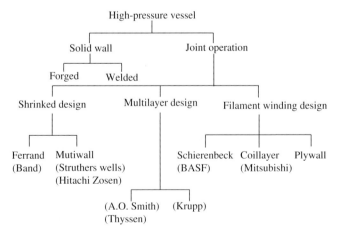

Figure 1.1 Survey on high-pressure vessel design [3].

1924: First industrial plant for direct hydrolysis of fuel from coal at 70 MPa based on the Bergius process, which was claimed at 1913.
1953: Initiation of a polyethylene production at about 250 MPa.
1978: First commercial decaffeination plant using supercritical carbon dioxide as a solvent.

The development of high-pressure vessel design is characterized by the initiation of seamless and forged cylindrical components. The two versions are the forged solid wall construction and a group of different layered wall constructions. Among these, the BASF Schierenbeck vessel plays an important role, because these vessels are manufactured without welding joints. Figure 1.1 presents an overview.

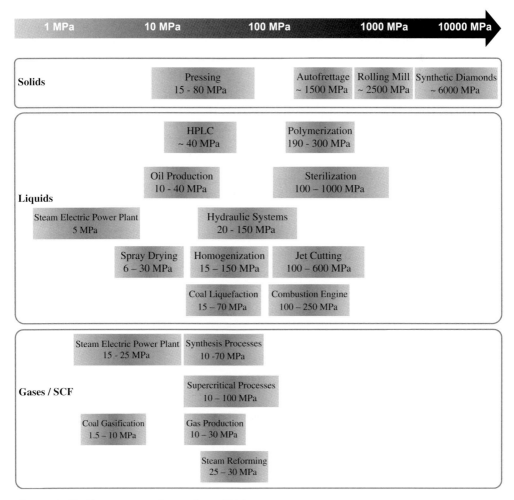

Figure 1.2 Working pressures of currently used high-pressure processes.

Special high-pressure closures have been developed equipped with single or double tapered sealing areas. A breakthrough toward leak-tight high-pressure devices was without doubt the "principle of the unsupported area" from Bridgman [2]. His idea extended the accessibility of pressures up to 10 000 MPa. Another concept is that the metallic lens ring enabled safe connections of high-pressure tubes and fittings.

Up to now new high-pressure processes have been introduced constantly. Materials like ceramics, polymers, or crystals having special properties are generated and formed in high-pressure processes. The current increase in liquid natural gas (LNG) plants is not possible with safe high-pressure systems. Also, the enhanced recovery of oil and gas by fluid injection at very high pressures requires qualified compressors, tubes, and safety valves. High-pressure fuel injection decreases the efficiency of combustion engines.

An example of current development is the investigation of processes aiming homogenization and even sterilization in industrial scale at high pressures up to 1000 MPa. Figure 1.2 illustrates the pressure regimes of currently operated high-pressure processes.

References

1 Witschakowski, W. (1974) Hochdrucktechnik, Schriftenreihe des Archivs der BASF AG, Nr. 12.
2 Spain, I.L. and Paauwe, J. (eds.) (1977) *High Pressure Technology*, vol. I, Marcel Dekker Inc., New York.
3 Tschiersch, R. (1976) Der Mehrlagenbehälter. *Der Stahlbau*, **45**, 108–119.

2
Basic Engineering Aspects
Rudolf Eggers

2.1
What are the Specifics of High-Pressure Processes?

It is obvious that with increasing process pressure, the distances between molecules of solid, liquid, or gaseous systems become smaller. Generally, such diminishing of distances results in alterations of both the phase behavior of the system and the transport effects of the considered process. Consequently, in designing the high-pressure processes, not only the knowledge of phase equilibrium data for pure and heterogeneous systems is needed from thermodynamics but also the reliable data for material and transport properties at high pressures are of high importance, because these can fluctuate strongly especially in the near-critical region of a medium.

In Figure 2.1, an easily interpreted image illustrates the molecule distances depending on pressure and temperature. The three phases – solid, liquid, and gas – are differentiated by the phase transition lines for melting, evaporation, and sublimation. At the critical point, the processes of condensation and evaporation merge.

Besides the decreasing molecule distances at enhanced pressures, the diagram reveals the continuous transfer from the gas phase into the liquid region by passing the so-called supercritical region without any crossing of a phase change line. Because this region is connecting the low-density region of gas and the high-density region of liquid state, it is evident that the corresponding density gradients without phase change are highest in the near-critical region. As a consequence, high pressure enables the use of fluid phases as solvents with liquid-like densities and gas-like diffusivities. Table 2.1 exemplifies that the basic engineering aspects of high-pressure processes are based on phase equilibrium data and material properties for both single and multicomponent systems and further they will be influenced by relevant transport data.

Of course, plant engineering and vessel design are also basic aspects of high-pressure processes. Due to their significance in industrial applications of high-pressure processes, these aspects are discussed in Chapter 12. Nevertheless,

Industrial High Pressure Applications: Processes, Equipment and Safety, First Edition. Edited by Rudolf Eggers.
© 2012 Wiley-VCH Verlag GmbH & Co. KGaA. Published 2012 by Wiley-VCH Verlag GmbH & Co. KGaA.

Figure 2.1 Molecule distances dependent on pressure and temperature.

this chapter focuses on the thermodynamic aspects of high-pressure phase equilibrium and the influence of pressure on material and transport data for heat and mass transfer at high pressures, including some information on basic measuring principles, which are given in detail in Chapter 14.

Table 2.1 High-pressure phase equilibrium: material properties and transport data in corresponding phase state.

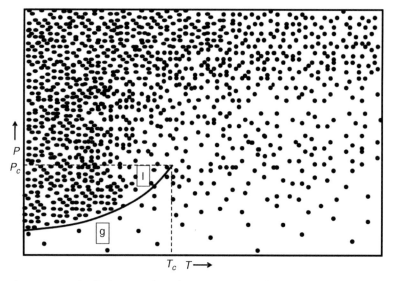

	Density	Viscosity	Diffusivity	Interfacial tension
g	liquid like	gas like	between gas and liquid	reduced in case of partial miscibility
l	liquid	strongly reduced	enhanced	impact on wettability

2.2
Thermodynamic Aspects: Phase Equilibrium

In many industrial high-pressure processes, the involved mass flows are getting in direct contact in order to enable heat and mass transport. The well-known examples are extraction processes using supercritical fluids (see Chapter 8) or liquefying processes of gas mixtures under pressure in combination with transport and storage of natural gas [1]. Further examples are the carbon capture and storage technology (CCS) (see Chapter 6), enhanced oil recovery processes (see Chapter 7), refrigeration cycles, and renewable energy processes (see Chapter 10).

Transport processes across phase boundaries of contacting phases are controlled by driving gradients of pressure, temperature, and chemical potential of each component inside a phase as long as phase equilibrium is not established and these gradients are existing. A phase is defined as a homogeneous region without discontinuities in pressure, temperature, and concentration. Thus, phase equilibrium is accomplished when the corresponding phases are of the same pressure (mechanical equilibrium), of the same temperature (thermal equilibrium), and of the same chemical potential (material equilibrium) for each component the system contains [2]. The chemical potential of a single component represents the change of internal energy of a system when the molar mass of this component varies. Instead of using the relative inaccessible chemical potential, it is possible to equalize the fugacities of the different phases. As the fugacity demonstrates an adjusted pressure considering the forces of interaction between the molecules in a real system, this quantity is of high importance for phase equilibrium especially in heterogeneous high-pressure systems [3]. The Gibbs phase rule predicts the number of degrees of freedom F for a mixture of K coexisting phases and n components:

$$F = 2 - K + n \tag{2.1}$$

The phase equilibrium constitutes a thermodynamic limitation of transfer processes. Therefore, the knowledge of phase equilibrium is an essential precondition for specification and calculation of high-pressure processes.

High-pressure processes need the knowledge of phase equilibrium for pure substances, binary systems, and multicomponent system. Nowadays data of high precision are available for pure components like water [4] and for numerous gases [5, 6] up to very high pressures. These data are computable by empirical equations. So far, the calculation of phase equilibrium for mixtures is recommended by use of equations of state. As such there are modified Redlich–Kwong and Peng–Robinson equations that have been proven for high-pressure systems [3]. Recently, the perturbation theory has attracted increasing research interest [7]. Thus, the so-called PCSAFT equation is established for polymeric systems and further application in high-pressure processing [8].

2.3
Software and Data Collection

For modern industrial engineering, an increasing number of capable software tools have been developed and are commercially available. Some of these are well-known examples that have been proved of value for calculation of high-pressure phase equilibrium: ASPEN PLUS (www.aspentech.com), Simulis Thermodynamics (www.prosim.net), and PE 2000 Phase Equilibrium (www.sciencecentral.com). Furthermore, there are data banks with experimental data for pure components and even multicomponent systems at high pressures (www.ddbst.com). Also, data on material properties are available, for instance, at www.dechema.de or webbook.nist.gov/chemistry. Finally, the well-experienced companies offer experimental determination of unknown data for high-pressure processes.

As an example for high-pressure system properties, Figure 2.2 demonstrates the phase behavior of CO_2 and Figure 2.3 illustrates the different phases of the binary system CO_2–water [9].

2.4
Phase Equilibrium: Experimental Methods and Measuring Devices

Although the direct measurement of equilibrium data for mixtures at high pressures requires detailed experimental experience and expensive equipment, it is still an essential and reliable way in order to obtain the data needed for the evaluation of high-pressure processes. Recently, Dohrn et al. [10] presented a classification of experimental methods for high-pressure phase equilibria. Figure 2.4 illustrates the two main groups: analytical methods and synthetic methods. In case of analytical

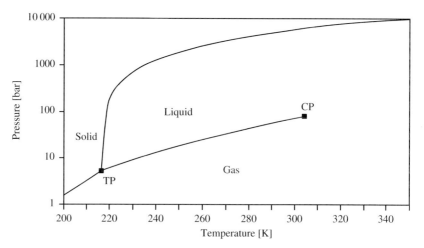

Figure 2.2 p, T diagram for CO_2.

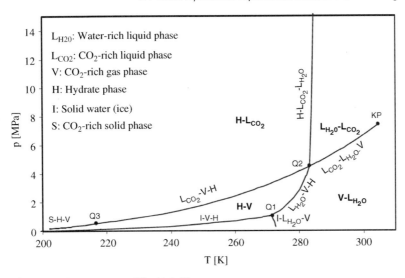

Figure 2.3 Binary system: CO_2–H_2O [7].

methods, the composition of each of two phases have to be analyzed without need to know precisely the composition of the overall mixture after achieving phase equilibrium of the system under investigation within a high-pressure cell. In contrast, synthetic methods require the precise knowledge of the composition of the overall mixture and no analysis of the equilibrium phases is required. The subgroups in Figure 2.4 indicate the different possibilities of analyzing with

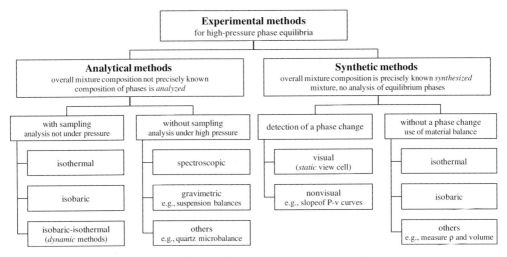

Figure 2.4 Classification of experimental methods for high-pressure phase equilibria. Adapted from Ref. [10].

and without taking samples alternatively by detection of a phase change or direct measuring of pressure, temperature, or volume.

Because there is no ideal method for all types of phase equilibrium, some hints on the applicability of the different measuring systems for high-pressure systems may be obvious:

- Synthetic measuring elucidates the advantages of applicability for material systems with coexisting phases of similar densities.
- Synthetic measuring does not need sampling. As a consequence, this method is applicable with test cells of small volumes and it can be used up to extremely high pressures. Moreover, the synthetic measurement normally is cheaper compared to analytical measurements due to a small number of high-pressure components.
- For multicomponent systems, the analytical method offers the advantages of a precise information on the composition of the phases, whereas the synthetic method reveals only the position of the phase transfer region.

2.5
Interfacial Phenomena and Data

In high-pressure process engineering, interfacial phenomena play an important role especially in case of processes that exhibit high specific areas for heat and mass transfer. It is well known that the efficiency of transport processes is related directly to the size of the phase boundaries. Also, reaction rates have a role in high-pressure processes. Furthermore, the activity of ingredients in pharmaceutical, cosmetic, and food products is increased by processing high levels of interfacial area. Interfaces can be formed between volume phases in solid, liquid, or gaseous state. Thus, except for gas/gas interfaces, five types are possible, which are listed in Table 2.2 together with related high-pressure processes.

Interfacial phenomena like formation of bubbles and droplets, foaming, falling films, coalescing, dispersing, wetting of surfaces, and interfacial convection are often present in moving boundaries, which are processed in high-pressure columns, nozzles, heat exchangers, and pump systems.

Table 2.2 Interfaces in high-pressure processes.

Type		Example of high-pressure process
Solid/solid	(s/s)	Drilling
Solid/liquid	(s/l)	Jet cutting
Solid/gas	(s/g)	Supercritical extraction
Liquid/liquid	(l/l)	Homogenizing
Liquid/gas	(l/g)	Evaporation

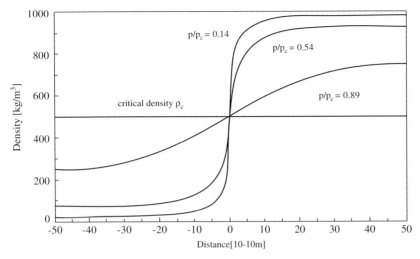

Figure 2.5 Density profile of CO_2 within the phase boundary region for different reduced pressures as a function of the distance coordinate [11].

A microscopic view down to molecular scale reveals the nature of fluid-phase boundaries as finite volume phases between two contacting and corresponding phases. As a consequence, properties are changing continuously from one phase to the other across the finite interfacial distance, as demonstrated in Figure 2.5, for the density profile of CO_2 [11]. At high pressures, the interfacial region increases, which also happens for interfaces between different fluids being partial miscible.

An illustrative explanation of the interfacial tension σ is the effort of reversible energy that is needed in order to create new interface between two phases.

$$\sigma = \frac{dE_{rev}}{dA} \tag{2.2}$$

A usual interpretation from mechanics is the work of creating new surface area by moving molecules from the bulk phase to the interface, whereas the thermodynamic interpretation is a change of inner energy when the surface area is altered.

Including the finite interface as a further phase into the fundamental equation of thermodynamics for the free energy F and the free enthalpy G [12],

$$dF = -Sdt - pdV + \sigma dA + \mu_i dn_i \tag{2.3}$$

$$dG = -Sdt + Vdp + \sigma dA + \mu_i dn_i \tag{2.4}$$

opens the possibility of deriving the dependency of the interfacial tension from the pressure and the temperature of a process:

$$\left(\frac{\partial \sigma}{\partial T}\right)_{V,A,n_i} = -\left(\frac{\partial S}{\partial A}\right)_{T,p,n_i} \tag{2.5}$$

$$\left(\frac{\partial \sigma}{\partial p}\right)_{T,A,n_i} = \left(\frac{\partial V}{\partial A}\right)_{T,p,n_i} \tag{2.6}$$

Equation (2.5) explains the temperature dependency of the interfacial tension as the change in entropy during the formation of surface. Increasing temperatures are combined with rising mobility of molecules. Thus, the work of forming new surface area becomes lower and the interfacial tension is decreased at elevated process temperatures.

On the contrary, the interfacial tension between fluid phases usually decreases at higher process pressures. This reveals an attractive feature for spray processes at high pressures [13]. Following Eq. (2.6), this means an enhancement of molecule concentration within the interfacial volume (adsorption) and in summary a volume decrease of the total system at constant pressure appears.

As an example, Figure 2.6 illustrates the surface tension of water as a function of the vapor pressure [14].

A further example of a two-component system demonstrates the interfacial tension of water and ethanol in contact with carbon dioxide (Figure 2.7) [15].

The reverse is rarely possible: a decrease in molecule concentration of the interfacial volume leading to an isobaric increase of total volume and thus the interfacial tension increases at higher process pressures. A known example is represented by the interfacial tension between helium and water [16].

Although most data of interfacial tension are experimentally determined, the prediction of interfacial tension by application of the Cahn–Hillliard theory [17] is of increasing interest. This theory describes the thermodynamic properties of a system where an interface exists between two phases being in equilibrium. Recently, the calculation of the interfacial tension for some binary systems at high pressures has approved the applicability of the Cahn–Hillliard theory [18, 19].

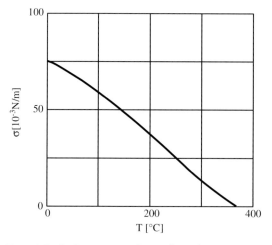

Figure 2.6 Surface tension of water dependent on temperature at saturation pressure. Adapted from Ref. [14].

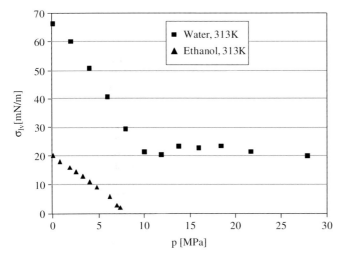

Figure 2.7 Interfacial tension of the system CO_2–H_2O [15].

In high-pressure processes, often droplets and bubbles are important owing to their high surface area for heat and mass transport. Whereas plane fluid surfaces without external forces are pressure equilibrated, curved interfaces are balanced by the size keeping surface forces and the forces that are acting by the inner pressure in a spherical geometry like a bubble or a droplet with diameter D in equilibrium.

$$\sigma \pi D = \Delta p \frac{\pi}{4} D^2 \tag{2.7}$$

The pressure difference is called capillary pressure.

$$\Delta p = \frac{4\sigma}{D} = \frac{2\sigma}{r} \tag{2.8}$$

As an example, high hydrostatic pressures have to be applied for recovery of oil from reservoirs having capillary structures (see Chapter 7):

$$\Delta \varrho g h = \frac{2\sigma}{r} \tag{2.9}$$

Arbitrary curved interfaces are to be calculated by the two orthogonal radii of curvature r_1 and r_2. The balance of forces [12] leads to the Young–Laplace equation that enables the calculation of the inner pressure or the interfacial tension.

$$\Delta \varrho = \sigma \left(\frac{1}{r_1} + \frac{1}{r_2} \right) \tag{2.10}$$

The Young–Laplace equation incorporates the special cases for spherical, cylindrical, and plane surfaces.

Sphere:

$$\Delta p = \frac{2\sigma}{r}, \quad r_1 = r_2 = r \tag{2.11}$$

Cylinder:

$$\Delta p = \frac{\sigma}{r_1}, \quad r_2 = \infty \tag{2.12}$$

Plane layer:

$$\Delta p = 0, \quad r_1 = r_2 = \infty \tag{2.13}$$

These equations are useful for measuring principles that enable the determination of interfacial tension even for high-pressure applications (see Chapter 7).

Furthermore, the wetting behavior of liquids on solid surfaces is of high interest for many high-pressure processes. As such the design of separation processes that are carried out in high-pressure columns requires knowledge on the degree of wetted surfaces in structured packing or falling films [20]. Generally, the Young equation describes the static wetting angle of liquids on solid surfaces by a balance of the three acting forces corresponding to the surface tensions $\sigma_{l,g}$, $\sigma_{s,l}$, $\sigma_{s,g}$ (Figure 2.8) [12].

$$\sigma_{s,g} = \sigma_{s,l} + \sigma_{l,g} \cos\theta \tag{2.14}$$

Liquids in capillary structures revealing high wetting corresponding to capillary rising of convex fluid interfaces show a lowering of vapor pressure. This is in accordance with a positive hydrostatic pressure that has to be equilibrated by an equivalent decrease of vapor pressure (Eq. (2.9)). On the other hand, nonwetting liquids are resulting in concave fluid interfaces combined with a capillary depression that has to be balanced by an increased vapor pressure. Thus, strong convex curved liquid-phase boundaries being present in very small bubbles are characterized by vapor pressure lowering and the concave liquid surfaces represented by droplets are of enhanced vapor pressure. As a consequence, small droplets are quick evaporating and small bubbles are often very stable. These phenomena may be useful in high-pressure processes like spraying and drying of liquids [14], particle generation of supersaturated solutions [21], foaming by bubble generation in polymers [22], and even in processing high-pressure vapor in power plants (see Chapter 6).

The change of vapor pressure at very low radii in droplets and bubbles is given by the Kelvin equation that combines the capillary pressure (Eq. (2.8)) and the capillary

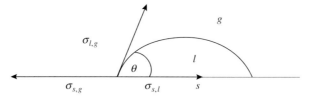

Figure 2.8 Wetting angle on solid surfaces.

height h (Eq. (2.9)) with $h > 0$ for convex liquid-phase boundaries like bubbles and $h < 0$ for concave liquid-phase boundaries like droplet surfaces. In order to use the Kelvin equation (2.15) for high-pressure applications, one has to consider that this basic law is derived by expressing the vapor density via the law of ideal gases and further introducing the pressure altitude.

$$\frac{p'}{p} = \exp\left(\pm\frac{2Mg\sigma}{RTr\varrho'}\right) \quad (2.15)$$

Nevertheless, the exponent of the Kelvin equation is of interest for high-pressure processes because both figures, the interfacial tension in most cases and the enthalpy of evaporation, decrease at elevated pressures.

From the Young equation (2.14), it is evident that in case of decreasing interfacial tension between the liquid and the gas phase, the wetting angle θ continuously increases as long as it does not pass 90°. However, the other two surface energies $\sigma_{s,l}$ and $\sigma_{s,g}$ may also change with pressure. This will happen when the pressurized gas phase or even components from a dense gas mixture become partially miscible in the liquid phase. Thus, without the knowledge of $\sigma_{s,l}$ and $\sigma_{s,g}$, no reliable prediction of the wetting angle is possible. Unfortunately, a direct measurement of the interfacial energies of solid surfaces is not possible. There are theoretical approaches for the calculation of these data, for instance, the method of Wenn, Girifalco, and Folkes [23, 24], but these theories are not proven yet for the application at high pressures.

As an example, Figure 2.9 illustrates experimental results of the static wetting angle when a droplet of water or oil is deposited on a clean steel surface [23].

In fact, in case of water, the wetting angle rises with increasing pressure until a constant remaining value at high pressures. Thus, a falling aqueous film may detach in contact with dense carbon dioxide operated in a desorption column at high

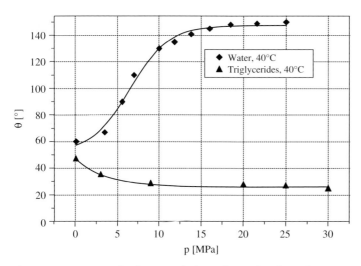

Figure 2.9 Wetting angle of water and oil on a flat steel surface [18].

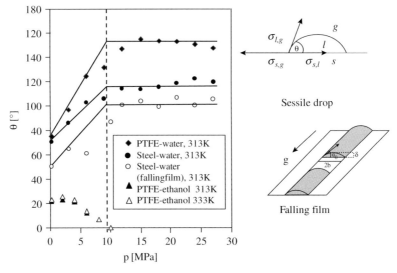

Figure 2.10 Pressure dependence of the wetting angle [25].

pressures [24]. In contrast, no risk of a detaching film will exist when an oily system is processed in a high-pressure column [24]. An improved knowledge of the wetting behavior is available by measuring both the static and the dynamic wetting angle. This is demonstrated for the aqueous and oily system in Figure 2.10 [25].

Although the comparison elucidates the lowered wetting angle in case of a moved aqueous liquid due to the additional acting inertial force, the risk of film detaching from the solid surface is enhanced at high pressures. Furthermore, the mass transfer in a high-pressure liquid/gas column is influenced by high densities of the gas phase. The decreasing density difference between both phases results in an enhanced thickness of the falling film in combination with decreasing flow velocities. Thus, the mass transfer resistance is increased. Finally, the shear stress τ at the interface is no longer negligible and instabilities may support the formation of a wavy film. Often heat and mass transport across these moved phase boundaries lead to gradients of the interfacial tension along the falling film surface. In consequence, interfacial convection phenomena (called Marangoni convection) [26] occur that superpose the convective transport normal to the phase boundary.

The general equations for the calculation of the film thickness δ, the film velocity w, and shear stress [27] read as follows:

Film thickness
Laminar:

$$\delta_l = \left(3\frac{\eta_l^2}{\varrho_l^2 g}\right)^{1/3} Re^{1/3} \tag{2.16a}$$

Turbulent:

$$\delta_t = 0.302 \delta_l \, Re^{1/5}, \quad Re > 400 \tag{2.16b}$$

Film velocity

$$w_{(y)} = \frac{\Delta\varrho g}{\eta_l}\left(-\frac{y^2}{2} + \delta y\right) + \frac{\tau}{\eta_l} y \tag{2.17}$$

$$\tau = \pm\eta_l\left(\frac{\partial w}{\partial y}\right) = \delta \tag{2.18}$$

High-pressure processes like absorption or condensation of gases show significant reduction of density differences and interfacial tension. Often the viscosity data of the film phase becomes low at higher pressures by dissolving gases [25].

Also, the formation of droplets in spray processes is controlled by interfacial phenomena revealing at high pressures. Falling droplets are characterized by the Weber number.

$$We = \frac{\varrho_g w^2 d}{\sigma} \tag{2.19}$$

The Weber number is derived by a balance of the inertial and the surface forces [14]. High-pressure spraying processes lead to very small droplet diameters because the gas density ϱ_g and the velocity w_l at the nozzle outlet are high, while the interfacial tension normally diminishes at elevated pressures.

Figure 2.11 illustrates the extended spray regime as a function of the modified Ohnesorge number Oh and the Reynolds number. Spray formation is significantly

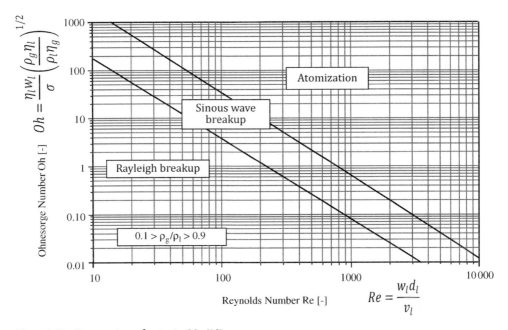

Figure 2.11 Spray regime of water in CO_2 [28].

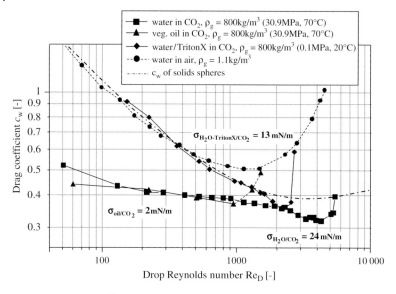

Figure 2.12 Drag coefficients of droplets in high-pressure atmosphere [29].

supported by high pressures. Ongoing applications are fuel atomization in combustion engines or high-pressure spray columns [28, 29].

In contrast to solid spheres, liquid droplets develop mobile-phase boundaries exerting a circulating flow due to friction forces when falling down in a dense gas atmosphere. Thus, the drag coefficients are reduced and free falling velocities are slightly enhanced. Figure 2.12 demonstrates a comparison of drag coefficients that are valid for solid spheres, droplets at ambient pressures, and droplets at high pressures [29].

2.6
Material Properties and Transport Data for Heat and Mass Transfer

With increasing pressure, the density of a process medium will rise. Due to the compressibility of the medium, the effect of enhanced densities will be highest for gases. Thus, heat and mass transfer processes with gases have to consider strong variations in material properties and transport data.

Generally, transport processes are based on the conservation laws of momentum (flow field), energy (temperature field), and mass (concentration field) taking into account the conditions like gradients of flow velocity, temperature, and concentration at the phase boundaries of the system under investigation usually by the application of transfer coefficients. In order to describe the interaction between fluid flow, heat flow, and mass flow, at least the behavior of the following data has to be clarified:

Material properties: Density ϱ, specific heat capacity c_p corresponding to the specific enthalpy h of the material and latent energies like melting Δh_m, condensation Δh_c, and evaporation Δh_v.

Transport data: Cinematic viscosity ν (related to the density as dynamic viscosity η), thermal conductivity λ, and diffusivity D.

Transport coefficients: Heat transfer coefficient h and mass transfer coefficient k.

At first, Figure 2.2 transfers the phenomenological character of Figure 2.1 to the p, T phase diagram of carbon dioxide as an example for a pure medium. Above the critical temperature ($T_c = 304.8$ K), there is no difference between liquid and gaseous state. Thus, liquefaction of CO_2 gas is not possible higher than 31.6 °C. In Figure 2.13, the density dependence of CO_2 depicted as a function of pressure is at four temperatures: lower T_c, T_c, near-T_c, and higher T_c.

The resulting profiles illustrate the nonsteady density difference between the gas and liquid phases at $T < T_c$, which diminishes down and vanishes at the critical point ($p_c = 7.18$ MPa, $T_c = 304.8$ K). The following continuous density profile in the near-critical region at $T > T_c$ also shows strong density change that decrease at increasing temperatures while forming a slightly increasing plateau at higher pressures.

Generally, the strong density variations in the vicinity of the critical point appear together, simultaneously with a sharp decrease of the specific enthalpy (Figure 2.14), as the specific heat capacity further approaches infinity (Figure 2.15). Regarding the transport properties, the thermal conductivity takes ostentatious high values (Figure 2.16).

As the pressure is further increased, there is still an attenuating maximum of c_p at the so-called pseudocritical temperature, whereas the heat conductivity slightly

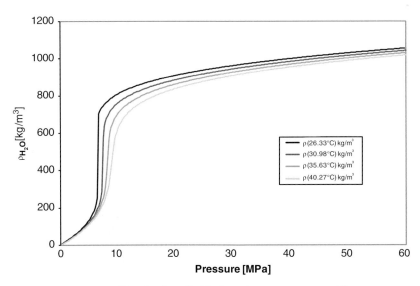

Figure 2.13 Density profiles of carbon dioxide dependent on pressure at different temperatures $T_1 < T_c < T_2 < T_3$.

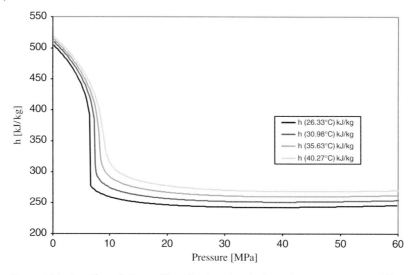

Figure 2.14 Specific enthalpy profiles of carbon dioxide dependent on pressure at different temperatures $T_1 < T_c < T_2 < T_3$.

increases with pressure, because the probability of molecule collisions is rising due to shorter distances between the gaseous molecules. The comparison of data for carbon dioxide and water in Figures 2.15 and 2.16 illustrates similar thermophysical anomalies of near-critical profiles as a basic principle. This statement is valid too for the specific behavior of the rheological data: dynamic and cinematic viscosities that are demonstrated in Figures 2.17 and 2.18. The relatively low cinematic viscosities for high pressures at supercritical temperatures are noticeable, because this is a reason for low friction forces at high mass flows and enhanced heat and mass transport by natural convection at high pressures (see Chapter 8).

Data used for Figures 2.13–2.18 are calculated by the Gerg [5] equation.

Table 2.3 lists some approved correlations for predicting basic engineering data in high-pressure processes.

Similar to heat transport by conduction, the transport mechanism by diffusion depends on the distance between the molecules in a given system. But in contrast, the higher the mean free path length, the lower the diffusion resistance by collisions with other molecules or even by wall surfaces, for instance in porous media. In general, the diffusion coefficient decreases at enhanced pressures due to reduced distances between the molecules, whereas the diffusivity at higher temperatures increases based on the higher mobility of the molecules. Furthermore, often there is a dependency on the concentration of diffusing components in a system. Related to fluids, the diffusivity behaves inversely proportional to the dynamic viscosity, which is the product of cinematic viscosity and density. This is stated by the basic equation of Stokes–Einstein [32]. Even in multicomponent system, the diffusion of a single component is normally calculated as a binary diffusion coefficient (Table 2.4). The corresponding data range from $10^{-4}\,m^2/s$ down to $10^{-14}\,m^2/s$ considering gaseous, liquid, and solid matter [33].

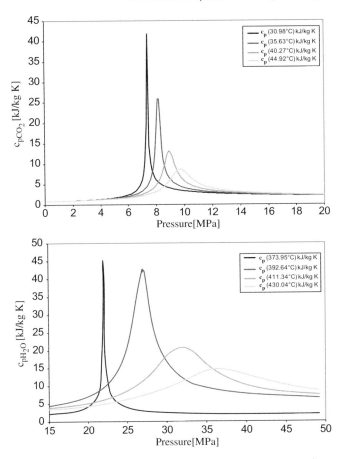

Figure 2.15 Comparison of near-critical heat capacities for CO_2 and H_2O.

The design of heat and mass transfer operations in chemical engineering is based on the well-known correlations that use the dimensionless numbers Nu (Nusselt) for heat transfer and Sh (Sherwood) for mass transfer. By balancing the acting forces, energies, and mass flows within the boundary layers of velocity, temperature, and concentration, the theoretical derivation of general relations for Nu and Sh is given in fundamental work [35].

$$Nu = \frac{\alpha L}{k} = f(Re, Pr) = f\left(\frac{wL}{\nu}, \frac{\nu \varrho c_p}{k}\right) \quad (2.20)$$

$$Sh = \frac{\beta L}{D} = f(Re, Sc) = f\left(\frac{wL}{\nu}, \frac{\nu}{D}\right) \quad (2.21)$$

These correlations are valid for steady-state processes of forced convection. In case of free convection, the influence of gravity force is to be considered as an additional independent parameter. Thus, the balance of forces leads to the Grashof number

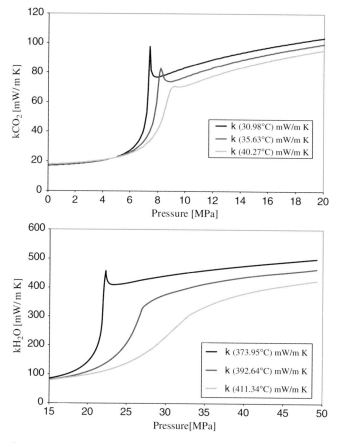

Figure 2.16 Comparison of near-critical heat conductivities for CO_2 and H_2O.

and the general equations for heat and mass transfer by free convection read as follows:

$$Nu = f(Gr, Pr) = f\left(\frac{g\Delta\varrho L^3}{\varrho v^2}, \frac{v\varrho c_p}{k}\right) \tag{2.22}$$

$$Sh = f(Gr, Sc) = f\left(\frac{g\Delta\varrho}{\varrho}\frac{L^3}{v^2}, \frac{v}{D}\right) \tag{2.23}$$

A descriptive interpretation of these dimensionless groups of parameters may be helpful to better understand the specifics under conditions of high pressures.

The Nusselt number represents the relation between the forced or free convective transport mechanism and the conductive heat transport within the boundary layer of the fluid flow.

The Sherwood number compares the convective mass transport and the transport by diffusion within the fluid flow boundary layer.

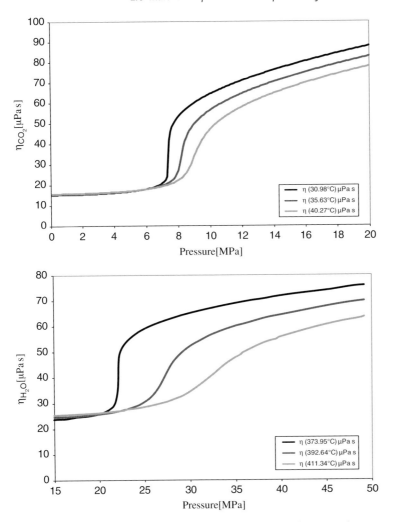

Figure 2.17 Comparison of near-critical dynamic viscosity data for CO_2 and H_2O.

The Reynolds number characterizes the relation between the inertial forces and the friction forces of the fluid flow.

The Prandtl number relates the dimensions of the boundary layers for fluid velocity and temperature.

The Schmidt number relates the dimensions of the boundary layers for fluid velocity and concentration.

The Grashof number represents a combination of the three acting forces for movement: inertia, friction, and gravity.

A great number of semiempirical and full empirical equations have been derived by experimental work in certain limits of process data, flow conditions, and

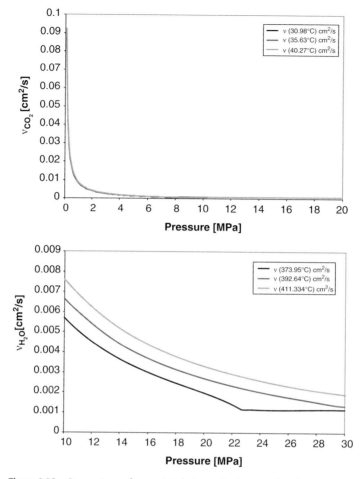

Figure 2.18 Comparison of near-critical cinematic viscosity data for CO_2 and H_2O.

geometries of the device under investigation. Thus, the user has to account for the following:

- Range of pressure and temperature given by limits of Prandtl and Schmidt number.
- Laminar or turbulent flow conditions expressed by the Reynolds number.
- External flow over flat plates and cylindrical and spherical geometries.
- Internal flow through tubes or channels regarding inlet flow or fully developed flow.

With respect to high-pressure application, some specifics on heat and mass transfer are of importance. Especially for gases in near-critical state, the described thermophysical anomalies (see Figures 2.15–2.18) result in maxima of heat capacity

Table 2.3 Correlations for thermophysical data at high pressures.

	Correlations	Source
Viscosity	Gases: $\eta = (\eta \cdot \xi)^r \cdot f_P \cdot f_Q \cdot \dfrac{1}{\xi}$ with $\xi = \dfrac{[T_C/K]^{1/6} \left[\bar{R}/(\text{J}/\text{kmol K})\right]^{1/6} \left[N_A/(1/\text{kmol})\right]^{1/3}}{\left[\tilde{M}/(\text{kg}/\text{kmol})\right]^{-1/2} [p_c/(\text{N}/\text{m}^2)]^{2/3}}$ $(\eta \cdot \xi)^r = (\eta \cdot \xi)^{r,o} \left[1 + \dfrac{Ap_r^E}{Bp_r^F + (1 + Cp_r^D)^{-1}}\right]$ $(\eta \cdot \xi)^{r,o} = 0.807 T_r^{0.618} - 0.357 e^{-0.449 T_r}$ $+ 0.340 e^{-4.058 T_r} + 0.018$ $\tilde{M}$: Molecular weight (kg/kmol) $\bar{R}$: General gas constant (J/kmol K) N_A: Avogadro constant (1/kmol) p_c: Critical pressure $p_r = p/p_c$: Reduced pressure $0 \leq Pr \leq 100$ $T_r = T/T_c$: Reduced temperature $1 \leq T_r \leq 40$ The subfunctions for (polarity factor) f_P, f_Q (Quantal factor), and the coefficients A, B, C, D, E, F are given in Ref [27], Section DA	Section DA 27 in Ref. [27]
	Liquids: $\eta = \eta_{bl} \left(\dfrac{\eta}{\eta_{bl}}\right)^0 / (1 + F)$ η_{bc}: Dynamic viscosity of saturated liquid (boiling line) (Ns/m²) The subfunctions $(\eta/\eta_{bc})^0$ (pressure influence) and F (temperature influence) are given by diagrams in Ref. [27], Section DA	Section DA in Ref. [27]
Heat conductivity	Gases: $(k-k_0) \cdot Z_c^5 \cdot \Gamma = 0.0122[\exp(0.535\varrho_r) - 1]$, $\varrho_r = \varrho/\varrho_c < 0.5$ $(k-k_0) \cdot Z_c^5 \cdot \Gamma = 0.0155[\exp(0.67\varrho_r) - 1.069]$, $0.1 < \varrho_r < 2.0$ $(k-k_0) \cdot Z_c^5 \cdot \Gamma = 0.0262[\exp(1.155\varrho_r) + 2.016]$, $2.0 < \varrho_r < 2.8$ k_0: Heat conductivity at low pressures and temperatures Z_c: Real gas factor at p_c, T_c $\Gamma = \dfrac{(T_c/K)^{1/6} (\bar{M}/(\text{kg}/\text{kmol}))^{1/2} (N_A/(1/\text{kmol}))^{1/3}}{(p_c/(\text{N}/\text{m}^2))^{2/3} (\bar{R}/(\text{J}/\text{kmol}))^{5/6}}$ Γ (km/W)	[30]
	Liquids: For $T_R > 0.8$, use above equations for gases For $T_R < 0.8$ approximation via enhancement factor	[27], Section DA in Ref. [34]
Heat capacity	Gases and liquids: $c_p = c_p^{id}(T) + \Delta c_p(T, p)$ $\Delta c_p(T, p)$: Derivation from ideal gas tabulated in Ref. [27], Section DA 23–26	[31] [24–27]

Table 2.4 Range of binary diffusion coefficients in gases, liquids, and solids [34].

Diffusion	D (m²/s)
Gases $p < p_c$	10^{-4}
Supercritical fluids $p > p_c$; $T > T_c$	10^{-2}
Liquids	10^{-9}
Solids	10^{-10}–10^{-14}

in the pseudocritical region. Thus, the Prandtl number also reveals maximum values like Figure 2.19 demonstrates for carbon dioxide.

As a consequence, heat transfer to gases in high-pressure processes is enhanced when the process conditions become pseudocritical. A successful method is to calculate the improved heat transfer when designing a heat transfer apparatus for options of heating and cooling as w is the relation of the thermophysical data at wall and bulk temperature. Polyakov [36] and Krasnoshchekov [37] developed the following equations:

$$Nu = Nu_0 \left(\frac{\bar{c}_p}{c_{p_b}}\right)^p \left(\frac{\eta_w}{\eta_b}\right)^q \left(\frac{k_w}{k_b}\right)^r \left(\frac{\varrho_w}{\varrho_b}\right)^s \qquad (2.24)$$

With Nu_0 valid in the nonpseudocritical pressure range for turbulent flow [38],

$$Nu_0 = 0.0183 \, Re^{0.82} \, Pr^{0.5} \qquad (2.25)$$

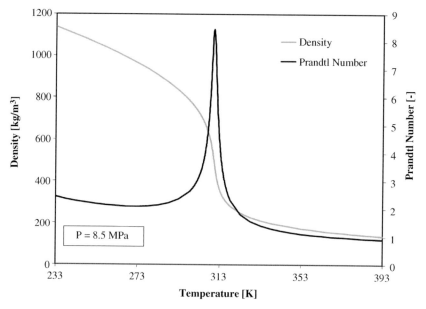

Figure 2.19 Prandtl numbers for CO_2.

2.6 Material Properties and Transport Data for Heat and Mass Transfer | 29

$$Nu = 0.0183 \cdot Re^{0.82} \cdot Pr^{0.5} \left(\frac{\bar{c}_p}{c_{p,b}}\right)^p \left(\frac{\varrho_w}{\varrho_b}\right)^{0.3} \quad (2.26)$$

$$\bar{c}_p = \frac{H_w - H_b}{T_w - T_b}$$

$$p = 0.4 \quad \text{at } T_b < T_w < T_{pc} \quad \text{and} \quad 1.2 T_{pc} < T_b < T_w$$

$$p = 0.4 + 0.2 \left[\frac{T_w}{T_{pc}} - 1\right] \quad \text{at} \quad T_b < T_{pc} < T_w$$

$$p = 0.4 + 0.2 \left[\frac{T_w}{T_{pc}} - 1\right] \left[1 - 5\left(\frac{T_b}{T_{pc}} - 1\right)\right]$$

$$\text{at} \quad T_{pc} < T_b < 1.2 T_{pc} \quad \text{and} \quad T_b < T_w$$

As an example, the length of a gas cooler in a refrigeration plant was designed. The heat exchanger is to be operated with the natural refrigerant carbon dioxide at a working pressure of 9 MPa.

Figure 2.20 shows calculated temperature profiles considering the strongly variable local heat transfer coefficient within the near-critical temperature range of CO_2 [39].

Nevertheless, at higher pressures far away from the critical pressure, the thermophysical anomalies for gases shown in Figures 2.15–2.18 vanish. Consequently, high-

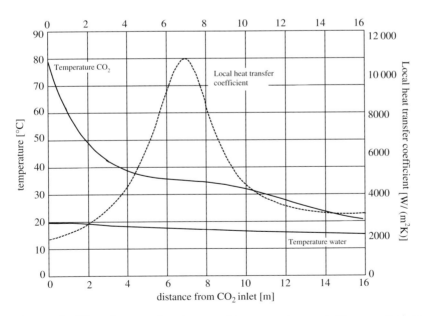

Figure 2.20 Effect of heat transfer enhancement in a gas cooler operated in near-critical state.

pressure transfer processes beyond the pseudocritical region are well calculable for gases by the application of usual Nusselt correlations. Although the Prandtl number slightly decreases (Figure 2.19), the Reynolds number may increase because the cinematic viscosity of gases diminishes at high pressures (Figure 2.18).

Heat transfer coefficients for liquid flow are well predictable by common Nusselt correlations even for high-pressure processes. The Prandtl number increases with pressure slightly improving the mechanism of heat transfer.

In Table 2.5, some approved heat transfer correlations are summarized for liquid and gas flow taking into consideration the specifics of high-pressure effects.

For heat transfer in near-critical pressure, use Eqs. (2.24)–(2.26).

Free convection: Use equations for heat transfer (Table 2.6) with Sh, Gr, and Sc numbers from Eq. (2.23).

In case of heat transfer by free convection, again high pressure affects the transport mechanism for gases in a special way. As the data for the cinematic viscosity of gases decrease to very low values (see Figure 2.18), the influence of the gravity force becomes high on gas flows having local differences in density. This is expressed in the denominator of the Grashof number:

$$Gr = \frac{g \Delta \varrho}{\varrho} \frac{L^3}{\nu^2} = \frac{g \beta_T \Delta T L^3}{\nu^2}$$

$$\beta_T = -\frac{1}{\varrho} \left(\frac{\partial \varrho}{\partial T} \right)_P$$

(2.27)

Furthermore, the coefficient of thermal expansion β_T reveals a strong maximum at near-critical conditions caused by high-density gradients. In addition, high-pressure processes are normally operated with low gas velocities, often ranging around some millimeter/seconds. In consequence, free convection may become the dominant effect in high-pressure gas flow. The superposition of free convective transfer is of high impact to processes like high-pressure gas transport in pipelines as recently discussed in carbon capture storage technology (CCS) [40]. The horizontal forced gas flow will be overlapped by secondary vertical eddy flow induced by free convection. Even in high-pressure vessels, the effect of free convective gas flow causes differences in mass transfer operations depending on the direction of fluid flow (up or down) and heat flow (heating or cooling). A heated gas flow in vertical upward direction is supported by buoyancy, whereas it will be impeded in down flow.

The importance of free convective transfer in high-pressure gas flow has been reported by Lenoir and Comings [41] and Grassmann [14]. The results given by Lenoir and Comings for free convection of CO_2 and N_2 were recalculated using the Gerg equation [5] in order to get precise material properties at high pressures, especially the thermal expansion coefficient.

As a result, Figure 2.21 shows the product of Grashof and Prandtl numbers that are relevant for the effect of free convection. The assumed data for the driving temperature difference and the height are given in Figure 2.21. It is clearly to be seen that free convection increases strongly with pressure due to the dropping

Table 2.5 Heat transfer correlations applicable for high-pressure processes.

Forced convection	Tube flow	Heat transfer	Limits of validity
$Re = \omega \cdot d/\nu$	Laminar	Average heat transfer: $Nu_m = 0.664 \cdot \sqrt{Re_d \cdot (d/L)} \cdot \sqrt[3]{Pr}$ Local heat transfer: $Nu_x = 0.332 \cdot \sqrt{Re_d \cdot (d/L)} \cdot \sqrt[3]{Pr}$	$0.1 < (d/L), Pr > 0.1$
Material properties at average flow temperature $Nu = \dfrac{\alpha \cdot L}{\lambda}, Re = \omega \cdot d/\nu$	Turbulent full developed flow friction factor: $\xi = (1.8 \cdot \log_{10} Re - 1.5)^{-2}$	$Nu_m = \dfrac{(\xi/8) Re\, Pr}{1 + 12.7\sqrt{(\xi/8)}(Pr^{2/3}-1)} \cdot \left(1 + (d/L)^{2/3}\right)$	$10^4 < Re_d < 10^6, 0.6 < Pr < 1000, (d/L) < 1$
	External flow Flat plate parallel flow Laminar	Average heat transfer: $Nu_L = 0.664 \cdot Re_L^{0.5} \cdot Pr^{1/3}$ Local heat transfer: $Nu_x = 0.332 \cdot Re_x^{0.5} \cdot Pr^{1/3}$	$Re < 10^5, 0.6 < Pr < 1000$
	Turbulent full developed flow	$Nu = \dfrac{0.037 \cdot Re_L^{0.8} \cdot Pr}{1 + 2.443 \cdot Re^{-0.1} \cdot (Pr^{2/3}-1)}$	$5 \cdot 10^5 < Re < 10^7, 0.6 < Pr < 1000$
$Pr = \nu/a$	Cylinder cross-flow Laminar–turbulent	$Nu_L = \sqrt{Nu_{L,\text{lam}}^2 + Nu_{L,\text{turb}}^2}$	$10^1 < Re < 10^7, 0.6 < Pr < 1000$

(Continued)

Table 2.5 (Continued)

Forced convection	Tube flow	Heat transfer	Limits of validity
Material properties at average flow temperature	Laminar	Like plate with $L = (\pi \cdot d)/2$	
Forced convection	Turbulent		
	Sphere cross-flow		
	Laminar	Like plate with $L = d$	
	Turbulent		
Natural convection $Gr = \dfrac{g \cdot \beta_T \cdot \Delta T \cdot L^3}{\nu^2}$, β_T = coefficient of thermal expansion	Vertical plate		
	Laminar	$Nu_L = 0.667 \cdot \left(\dfrac{Pr^2 \cdot Gr_L}{0.952 + Pr} \right)^{1/4}$	$Gr \cdot Pr < 10^9$ $0.6 < Pr < 1000$
	Turbulent	$Nu_L = 0.13 \cdot (Gr \cdot Pr)^{1/3}$	$10^9 < Gr_L \cdot Pr$ $0.6 < Pr < 10$
	Overall	$Nu_L = \left[0.825 + 0.387 (Gr_L \cdot Pr \cdot f(Pr))^{1/6} \right]^2$ $f(Pr) = \left[1 + (0.492/Pr)^{9/16} \right]^{-16/9}$	$10^{-1} < Gr_L \cdot Pr < 10^{12}$ $0.001 < Pr$

Table 2.6 Mass transfer correlations relevant for free and forced convection and valid for internal and external flow regimes.

Convective mass transfer	Tube flow	Mass transfer	Limits of validity
$Sh = (\beta \cdot d)/D$	Laminar	$Sh = 0.664 \cdot \sqrt{Re \cdot (d/L)} \cdot Sc^{1/3}$	$Re < 2300$
$Sh = v/D$	Turbulent	$Sh = 0.037 \cdot (Re^{0.75} - 180) \cdot Sc^{0.42} \cdot \left[1 + (d/L)^{2/3}\right]$	$Re < 2300$
	External flow		
	Plate		
$Sh = (\beta \cdot d)/D$	Laminar	$Sh = 0.664 \cdot \sqrt{Re} \cdot Sc^{1/3}$	$Re < 5 \cdot 10^5$, $0.6 < Sc < 1000$
$Sh = \dfrac{v}{D}$	Turbulent boundary layer	$Sh_L = \dfrac{0.037 \cdot Re^{0.8} \cdot Sc}{1 + 2.443 \cdot Re^{-0.1} \cdot (Sc^{2/3} - 1)}$	$5 \cdot 10^5 < Re < 10^7$, $0.6 < Sc < 1000$
	Cylinder cross-flow		
	Laminar	Like plate with $L = \dfrac{\pi \cdot d}{2}$	
	Turbulent		
	Transition region	$Sh_L = 0.3 + (Sh_{L,\text{lam}}^2 + Sh_{L,\text{turb}}^2)^{1/2}$	
	Sphere cross-flow		
	Laminar	Like plate with $L = d$	
	Turbulent		
	Transition region	$Sh_L = 2 + (Sh_{L,\text{lam}}^2 + Sh_{L,\text{turb}}^2)^{1/2}$	$0.1 < Re < 10^7$ $0.6 < Sc < 10^4$

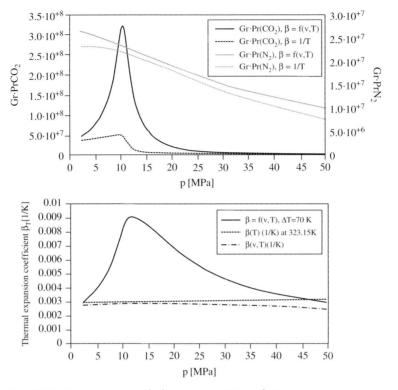

Figure 2.21 Free convection at high pressures in CO_2 and N_2.

cinematic viscosities. In addition, in the near-critical region of CO_2, the maximum of the thermal expansion coefficient leads to a maximum of the free convective flow. A calculation using the ideal gas law in order to determine the thermal expansion coefficient ($\beta_T = 1/T$) does not consider the discussed anomalies and thus fails.

The Archimedes number Ar expresses the relation between forced convection and natural convection:

$$Ar = \frac{Gr}{Re^2} \qquad (2.28)$$

The Archimedes number quotes the ratio of the gravity force and the inertial force in a fluid flow. Thus, high values $Ar \gg 1$ indicate the applicability of correlations valid for free convection, whereas $Ar \ll 1$ permits the calculation of the problem by using equations for forced convection. The intermediate region is characterized by forced convection with superposed free convection. This type of mixed convection often occurs in high-pressure processes, particularly in gas flow due to high Grashof numbers at low cinematic viscosities and enhanced gas densities. The calculation of the summarized heat transfer is possible by the established equation [42]:

$$Nu = \left(Nu_{fc}^3 + Nu_{nc}^3\right)^{1/3} \tag{2.29}$$

It is worth mentioning that mass transfer may be additionally supported in case of free convection by density differences that are induced by concentration differences. As an example, absorption of gas in liquid film flow at high pressures may even increase the liquid density and thus make influence on mass transport [43]. Nevertheless, free convective heat and mass transfer in liquids are of minor importance because the driving density differences decrease at high pressures. Thus, common correlations are applicable for free convective transfer in liquids.

Like free convective heat transfer, the mass transport in gases is supported by free convection due to the very low cinematic viscosity data at high pressures.

Forced convective mass transport in high-pressure processes is governed by the pressure dependency of both the dynamic viscosity and the diffusion coefficient. Whereas gases show dropping values of cinematic viscosity, their dynamic viscosity increases strongly with pressure due to rising densities. Liquids reveal marginal change of dynamic viscosity data at higher pressures. However, as stated above, often gas diffuses at high pressures in liquids or even in solids – like polymers – and decreases the dynamic viscosity. Thus, in gas/liquid or gas/solid systems with partial miscibility of gas or single gas components in the corresponding liquid or solid phase, the transport coefficients of diffusion (D) and mass transfer β may increase. As an example, Figure 2.22 shows experimental data on diffusion coefficients of CO_2 in oil [44]. At high pressures, CO_2 dissolves in the liquid reducing the dynamic viscosity and thus the diffusion increases, although the pressure is increased. However, in general, mass transfer is hindered in high-pressure processes.

For a rough estimation of diffusion coefficients in gaseous systems, the well-known Stokes–Einstein equation is a help even for high pressures.

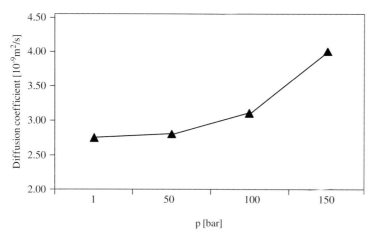

Figure 2.22 Diffusion coefficient of CO_2 in oil [44].

$$D = \frac{kT}{C\pi\sigma\eta} \qquad (2.30)$$

In high-pressure gas systems, a correlation is proposed that is based on a data collection of diffusion coefficients in dense gases [45]:

$$D = 5.152 D_s T_r \left(\varrho^{-2/3} - 0.5410\right) \frac{R}{x} \qquad (2.31)$$

with D_s being the self-diffusion coefficient (see Table 2.7).

In liquid systems, equations exist that enable the calculation of diffusion coefficients via the knowledge of the course of dynamic viscosity. As an example, Mc Manamay and Woolen stated for the diffusivity of CO_2 in some organic liquids [48]:

$$D(m^2/s) = 1.41 \cdot 10^{-10} \eta^{0.47} (Pa \cdot s)^{0.47} \qquad (2.32)$$

Another way to predict diffusion coefficients is to make use of a reference data point from experiment or even from literature. Cussler [47] stated for diffusion in fluids:

$$D(p, t) = D(P_1, T_1) \frac{T + 273.15 \cdot \eta(P_1, T_1)}{T + 273.15 \cdot \eta(P, T)} \qquad (2.33)$$

Table 2.7 summarizes equations that are valid for the prediction of diffusion coefficients at high pressures in gaseous and liquid systems.

Table 2.7 Calculation of diffusion coefficients.

Diffusion coefficient

Gases:
$D = 5.152 D_s T_r \left(\varrho^{-2/3} - 0.4510\right) R/X, \quad 1 < \varrho_r < 0.25$ [34, 35]

with $\begin{aligned} R &= 1.0 \pm 0.1, \quad \text{for } 2 < X \\ R &= 0.664 X^{0.17} \pm 0.1, \quad \text{for } 2 < X < 10 \end{aligned}$

where X is the size-to-mass parameter that is calculated from
$X = \left[1 + (V_{c2}/V_{c1})^{1/3}\right] / (1 + M_1/M_2)^{1/2} X$
where 1 is the solvent and 2 is the solute
D_s is the self-diffusion coefficient, correlated using critical quantitates as follows:
$$D_s = \frac{4.30 \times 10^{-7} M^{1/2} T_c^{0.75}}{\left(\sum V\right)^{2/3} \varrho_c}$$
where
M is the molecular weight (g/mol)
T_c is the critical temperature (K)
ϱ_c is the critical density (kg/m^3)
$\sum V$ is the atomic diffusion volumes (Fuller) [30, 39]
Liquids:
No reliable equations
Approximative calculation:
$D(p, T) = D(p_1, T_1) \dfrac{T + 273.15 \eta(T_1, p_1)}{T_1 + 273.15 \eta(T, p)}$ [46]

$D(m^2/s) = 1.41 \cdot 10^{-10} \cdot \eta (Pa \cdot s)^{-0.47}$ [47, 48]

Examples for application of heat and mass transfer in high-pressure processes are given in

- Ref. [1]: Refrigeration cycles
- Chapter 6: Evaporator in power plants
- Chapter 8: Free convection in extraction vessels
- Chapter 8: Diffusion of dense gas in liquids
- Chapter 8: Condenser in supercritical extraction plants

2.7
Evaporation and Condensation at High Pressures

2.7.1
Evaporation

The process of evaporation is basically classified in pool boiling accordant to a free convective transfer or flow boiling being a forced convective transfer. The fundamental question is how the bubbles are generated in interaction with different heating surfaces. Although the evaporation depends on numerous parameters, the general impact of high pressures becomes plausible by the well-known equation for the radius r_b of a bubble being in mechanical equilibrium between the pressure difference inside and outside the bubble and the surface tension, tending to maintain a spherical shape according to Eq. (2.8) [42]

$$r_b = \frac{2\sigma}{\Delta p} \qquad (2.34)$$

The vapor pressure curve of a liquid shows rising temperatures at increasing pressure. Moreover, the gradient dp/dT along the temperature curve increases with pressure. However, the surface tension whose physical meaning is the energy needed for creating new surface between liquid and gaseous states drops with temperature. As an example, Figure 2.6 demonstrates the decreasing surface tension along the temperature at saturation pressure of water from ambient up to the critical conditions: $p_c = 22.12$ MPa and $T_c = 374.15\,°C$ [14]. Furthermore, thermal equilibrium requires the same temperature inside and outside the bubble. However, the liquid ambient contacting the bubble is superheated at the process pressure, whereas the vapor inside the bubble, being held at the right elevated pressure by surface tension, is just in saturated state. It becomes clear that bubbles, whose radii are described by the mechanical equilibrium given in Eq. (2.41), are unstable [42]. Thus, smaller bubbles are generated with high frequency at high pressures for evaporation processes.

Based on Nukiyama [49], the well-known boiling curves show the heat flux q of a heating device depending on the wall superheating $T_w - T_s$ that is needed for evaporation. Increasing superheating leads to different regimes for pool boiling. First, natural convection starts induced by density changes of the superheated liquid. First bubbles are created at nucleation sites that are caves or grooves from the roughness of the heating surface. These first bubbles collapse when they start to rise and come in contact with liquid layers at temperatures below the saturation temperature. The regime of nucleate boiling represents wall superheating sufficient for bubble rise from the nucleation sites up to the liquid surface of the pool. The movement of the increasing amount of bubbles provokes a strong increase in heat transfer. The nucleation regime is the boiling section of high technical relevance. However, if bubble density becomes very high, vapor cushions will appear on the heating surfaces leading to decreasing heat transfer by just partial wetting of the transferring surface. The maximum value is called the critical heat flux q_{max}. The heat transfer regime of partial wetting is an unstable region. Thus, a possible change of heat flux beyond the maximum q_{max} – often called the burn out point – may damage the heat transfer device because the temperature difference jumps to very high values that are located in the transfer regime of film boiling, where the total heat transfer surface is covered by a closed vapor film.

Even forced convective boiling elucidates the described regimes, but the superheating of the wall and the wetting character become additionally dependent on the mass flow of the liquid to be evaporated mostly in tubes. Of course, for technical application, the nucleate boiling is decisive. The heat transfer in case of flow boiling is generally enhanced because the latent heat containing bubbles are transported away from the heating surface. However, the profiles of the heat transfer regimes are not that pronounced like in pool boiling.

Figure 2.23 visualizes the qualitative course of the boiling curve. Regarding high-pressure processes, the general profile of the nucleation boiling is shifted to lower superheating at constant heat flux or enhanced heat flux at constant superheating of the wall. This is due to Eq. (2.41) because at high pressures, the saturation temperatures increase strongly and simultaneously the surface tension drops resulting in a higher number of smaller bubbles.

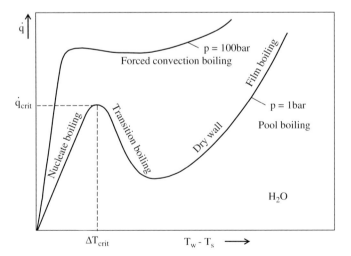

Figure 2.23 Heat transfer regimes for boiling: influence of pressure [50].

2.7 Evaporation and Condensation at High Pressures

The well-known equation of Kutateladze reveals precise data for $\dot{q}_{crit}$ in pool boiling valid even for high pressures [51, 52]:

$$\dot{q}_{crit} = K_1 \Delta h_v \varrho_g^{0.5} \left[\sigma_{g,l} (\varrho_l - \varrho_g) g \right]^{0.25} \tag{2.35}$$

with

$K = 0.14$ for flat heating plates and 0.16 for tubes [42]
Δh_v: Enthalpy of evaporation
$\sigma_{g,l}$: Surface tension liquid/vapor depending on pressure (Figure 2.6)
ϱ_g: Density of vapor
ϱ_l: Density of liquid

However, the enhanced heat transfer regime of nucleate boiling at high pressures does not imply an increase of the critical heat flux $\dot{q}_{crit}$. As with the increasing pressure, the density differences between vapor and liquid phases decrease the buoyancy force of bubbles drop rapidly when the pressure approaches the critical pressure p_c. Thus, an increase of $\dot{q}_{crit}$ is only to be observed at reduced pressures up to about 0.3. Figure 2.24 illustrates the data for water.

There are also correlations for the heat transfer coefficient in nucleate pool boiling that take into account the influence of process pressure. The equations give data related to a basis heat transfer coefficient h_0 dependant on the properties of the heating surface (C_w), the reduced pressure $p^* = p/p_c$, and the related heat flux $\dot{q}/\dot{q}_0$ [53, 54].

$$\frac{h}{h_0} = C_w F(p^*) \left(\frac{\dot{q}}{\dot{q}_0} \right)^n \tag{2.36}$$

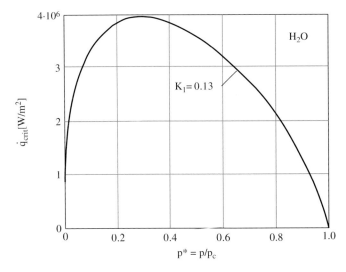

Figure 2.24 Critical heat flux $\dot{q}_{crit}$ for pool boiling of water dependent on pressure [27].

The function $F(p^*)$ considers the influence of pressure for a wide range of liquids.

$$F(p^*) = 1.2 p^{*0.27} + \left(2.5 + \frac{1}{1-p^*}\right) p^* \tag{2.37}$$

In particular, for water,

$$F(p^*) = 1.73 p^{*0.27} + \left(6.1 + \frac{0.68}{1-p^{*2}}\right) p^{*2} \tag{2.38}$$

Forced convection boiling is a much more complex evaporation process than pool boiling, especially because of the different flow regimes. For instance, the development of an evaporating two-phase flow in vertical tubes heated with a uniform heat flux starts as single-phase liquid flow that is followed by bubbly flow, slug flow, annular flow, annular flow with entrainment of liquid drops, drop flow, and finally single-phase vapor flow [50]. The flow regimes correspond to different regimes of heat transfer: single-phase forced convection, subcooled boiling, saturated nucleate boiling, forced convective heat transfer through liquid film, liquid-deficient region, and finally convective heat transfer to vapor. A further complication is the appearance of two critical aspects of boiling: the generation of a vapor film on the tube surface; this is possible when the critical heat flux – similar to pool boiling – is exceeded. The second critical point in forced convection boiling is called the dry out point that is characterized as the change from annular flow to drop flow. Both critical values depend not only on mass flow but also on the steam quality of the two-phase flow.

The variety of regimes during the forced convection boiling in tubes or ducts requires different correlations in order to determine the heat transfer coefficient related to the respective boiling mechanisms. The well-established correlations have been developed for nucleate boiling controlled heat transfer – when evaporation occurs at the inner tube surface – and convective boiling heat transfer – when evaporation occurs at the liquid film interface.

While convective boiling is strongly affected by the vapor content x of the flow, nucleate boiling is governed by the heat flux $\dot{q}$. With respect to high-pressure processes, it is important to know that nucleate boiling reveals increasing heat transfer coefficients with pressure. This is caused by smaller bubbles due to decreasing surface tensions at high pressures. Furthermore, at high pressures, the nucleate boiling regime is expanded almost over the entire range of heat flux. This behavior matches the theory that at higher pressures less energy is needed for activating new bubble sites. On the contrary, forced convective boiling shows slightly decreased data for heat transfer as the velocity of the dense vapor flow drops.

Recently, Schael [55] demonstrated these characteristics of flow boiling by measurements of evaporating CO_2 at elevated pressures. Figure 2.25 elucidates the influence of pressure on the heat transfer coefficient within the regime of nucleate boiling.

Approved correlations for the determination of heat transfer coefficients for vertical tube flow of pure components are as follows [27]:

2.7 Evaporation and Condensation at High Pressures

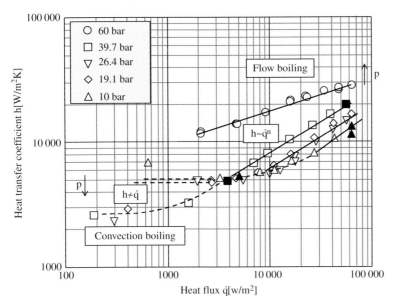

Figure 2.25 Influence of pressure on the evaporation of CO_2. Flow boiling $\dot{m} = 3000$ kg/m^2 and vapor content $\dot{x} = 0.7$ [55].

Regime of forced convective boiling with complete wetting of the inner tube surface:

$$\frac{h_{(2)}}{h_{l,o}} = \left[(1-\dot{x})^{0.01} \left[(1-\dot{x})^{1.5} + 1.9\dot{x}^{0.6} \left(\frac{\varrho_l}{\varrho_v} \right)^{0.35} \right]^{-2.2} \right.$$

$$\left. + \dot{x}^{0.01} \left[\frac{h_{g,o}}{h_{l,o}} \left(1 + 8(1-\dot{x})^{0.7} \left(\frac{\varrho_l}{\varrho_v} \right)^{0.67} \right) \right]^{-2} \right]^{-0.5} \quad (2.39)$$

$h_{(2)}$: Local heat transfer coefficient for convective boiling in vertical tubes
$\dot{x}$: Vapor content of the flow

$h_{g,o}$ and $h_{l,o}$ are the local heat transfer coefficients at the position z, to be calculated with full mass flow density $\dot{m}$ in the gaseous state ($h_{g,o}$) and in the liquid state ($h_{l,o}$).t

Regime of nucleate boiling:

$$\frac{h}{h_0} = C_F \left(\frac{\dot{q}}{\dot{q}_0} \right)^n F_{(p^*)} F_{(d)} F_{(w)} F_{(\dot{m},\dot{x})} \quad (2.40)$$

Different factors are to be considered by distinct coefficients [27]:C_F
Representing the function $F_{(p^*)}$ as the influence of the related pressure p^*/p_c

$F_{(d)}$ As the tube geometry
$F_{(w)}$ As the properties of the heating wall surface
$F_{(\dot{m},\dot{x})}$ As the influence of vapor flow content $\dot{x}$ and mass flow density $\dot{m}$
h_0 Representing the heat transfer coefficient that is valid in case of pool boiling

With respect to high-pressure processes, it is of interest that the heat transfer coefficient h for nucleate boiling also increases with both $\dot{q}$ and p^*. The impact factors of Eq. (2.40) are available from Ref. [27]. The dependency on the pressure has been evaluated by numerous data from literature [27].

$$F_{(p^*)} = 2.816 p^{*0.45} + \left(3.4 + \frac{1.7}{1-p^{*7}}\right) p^{*3.7} \tag{2.41}$$

In order to determine the critical heat flux for flow boiling, one has to distinguish the data for film boiling (low vapor content, liquid forms the continuous phase) and the dry out point (high vapor forms the continuous phase in annular flow).

Film boiling (upflow of water in a tube) [56]:

$$\dot{q}_{cr} = 10^3 \left[10.3 - 17.5\left(\frac{p}{p_c}\right) + 8\left(\frac{p}{p_c}\right)^2\right] \left(\frac{0.008}{d}\right)^{0.5} \left(\frac{\dot{m}}{1000}\right)^{0.68\left(\frac{p}{p_c}\right)-1.2\dot{x}-0.3} e^{-1.5\dot{x}}$$

$$\tag{2.42}$$

Limits of validity:

$2.9\,\text{MPa} \leq p \leq 20\,\text{MPa}$	Process pressure
$500\,\text{kg/m}^2\text{s} \leq \dot{m} \leq 5000\,\text{kg/m}^2\text{s}$	Mass flow density
$0.004\,\text{m} \leq d \leq 0.0025\,\text{m}$	Tube diameter

The critical heat flow density $\dot{q}_{cr}$ at the dry out point of the tube surface strongly depends on the pressure range [57].

Pressure range p

$$\dot{q}_{cr} = 1.8447 \cdot 10^8 \dot{x}^{-8} \dot{m}^{-2.664} \cdot (d \cdot 1000)^{-0.56} e^{0.1372 p} \quad 4.9\,\text{MPa} < p < 2.94\,\text{MPa}$$

$$\dot{q}_{cr} = 2.0048 \cdot 10^{10} \dot{x}^{-8} \dot{m}^{-2.664} \cdot (d \cdot 1000)^{-0.56} e^{-0.0204 p} \quad 2.94\,\text{MPa} < p < 9.8\,\text{Mpa}$$

$$\dot{q}_{cr} = 1.1853 \cdot 10^{12} \dot{x}^{-8} \dot{m}^{-2.664} \cdot (d \cdot 1000)^{-0.56} e^{-0.0636 p} \quad 9.8\,\text{MPa} < p < 20.0\,\text{Mpa}$$

$$\tag{2.43}$$

Limits of validity: $200\,\text{kg/m}^2\text{s} \leq \dot{m} \leq 5000\,\text{kg/m}^2\text{s}$
$4\,\text{mm} \leq d \leq 32\,\text{mm}$

In the film boiling forms, a vapor film is formed between the inner tube surface and the liquid bulk flow. The heat transfer drops with increasing vapor content of the flow. This phenomenon is called "departure from nucleate boiling" (DNB). In case of higher contents of vapor have been generated along the tube, an annular film flow of liquid is contacting the heating surface. The liquid film dries out at the critical heat flux $\dot{q}_{cr}$, which is called dry out point.

Regarding the Eqs. (2.49) and (2.50), Figure 2.26 reveals the strong decrease of $\dot{q}_{cr}$ at dry out conditions. The diagram is valid for upflow of water in a vertical tube [27]. For high pressures, it is of interest that the $\dot{q}_{cr}$ drops and the transfer from film boiling to dry out conditions moves to lower critical contents of vapor.

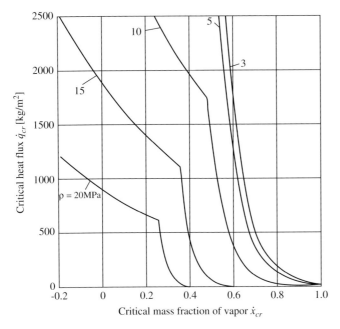

Figure 2.26 Critical heat flux dependent on critical vapor content along film boiling and dry out conditions ($\dot{m} = 1000\,\text{kg/m}^2\text{s}$; tube diameter $d = 10\,\text{mm}$).

2.8
Condensation

There are two essential factors of impact of high pressures on condensation processes. Based on the increased vapor density ϱ_g at elevated pressures, at first the density difference $\varrho_l - \varrho_g$ between liquid and vapor becomes smaller. The second effect of the increasing vapor density is a rising shear stress at the moving interface of the condensing liquid that potentially is no longer neglectable. The decreased density difference causes slower falling velocities of a condensate film or of droplets in case of mist condensation. As a consequence, the film thickness accumulates and the heat transfer diminishes.

Increased shear stress at the liquid–vapor phase boundary may generate a wavy film or even tear the film. In case of a forced convective vapor flow, the effect of the significant shear stress at the interface depends on the flow direction. Downward flow of the vapor – which is normal in practice – will accelerate the film surface and result in a thinner film. Thus, the heat transfer is supported. However, vapor flow opposing gravity will have the effect of thickening the film thickness that reduces the heat transfer.

Nusselt himself derived the development of the film thickness and the heat transfer coefficient in case of laminar flow and neglected shear stress at the film surface [27]. Regarding a finite shear stress τ_δ, the film thickness of a condensating pure vapor phase at a distinct vertical position x reads as follows [42, 58]:

$$\frac{4\eta k(T_s-T_w)x}{g\Delta h_c \varrho_1\left(\varrho_1-\varrho_g\right)} = \delta^4 \pm \frac{4}{3}\left[\frac{\tau_\delta \delta^3}{\left(\varrho_1-\varrho_g\right)g}\right] \qquad (2.44)$$

The minus sign is valid for upflow conditions of the vapor flow.

The evaluation of the shear stress τ_δ requires knowledge of the gradient of the vapor velocity at the phase boundary. Another way of regarding the shear stress is the modification of approved correlations by experimental values [27].

The influence of an elevated gas density at high pressures on the average velocity of a falling film in case of a quiescent CO_2 phase [24] has been investigated.

Figure 2.27 shows the deviation of the real velocity profile compared to the theoretical profile according to Eq. (2.44) with $\tau_\delta = 0$ and to a model with definite wall shear stress at the film surface (zero velocity at the film surface). The results elucidate the approach of the experimental results to the wall model at high pressures.

The following equations are of fundamental importance for film condensation in laminar and turbulent flow on vertical and horizontal devices.

The average heat transfer coefficient $\bar{a}$ for film condensation of a pure unmoved vapor [59, 60] is given by the following:

Vertical plate – laminar film flow.

$$\bar{a} = 0.943\left[\frac{\varrho_1\left(\varrho_1-\varrho_g\right)g\Delta h_c k_1^3}{\eta_1(T_s-T_m)H}\right]^{1/4} \qquad (2.45)$$

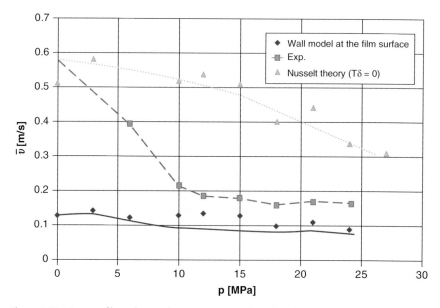

Figure 2.27 Average film velocity of water on a vertical steal surface in a gas atmosphere (CO_2 unmoved) at high pressures [24].

Horizontal tube:

$$\bar{a} = 0.728 \left[\frac{\varrho_1 (\varrho_1 - \varrho_g) g \Delta h_c k_1^3}{\eta_1 (T_s - T_m) d} \right]^{1/4} \quad (2.46)$$

Vertical plate – turbulent flow:

$$Nu = \frac{\bar{a}(v_1^2/g)^{1/3}}{k_1} = \frac{0.02 \, Re_1^{7/24} Pr^{1/3}}{1 + 20.52 \, Re_1^{-3/8} Pr_1^{-1/6}} \quad (2.47)$$

$$Re = \frac{\dot{M}_1}{b \eta_1}, \quad 400 < Re < 10000 \quad (2.48)$$

$$Pr = \frac{v_1 \varrho_1 c_{p_1}}{k_1}, \quad 0.5 < Pr < 500 \quad (2.49)$$

v_1: Kinematic viscosity of the liquid
$\dot{M}_1$: Liquid mass flow
η_1: Dynamic viscosity of the liquid
ϱ_1: Density of the liquid
c_{p_1}: Warm capacity of the liquid
k_1: Heat conductivity of the liquid

In general, heat transfer is calculated based on quiescent vapor. In case of vapor flow cocurrent to the vertical film flow, the heat transfer is calculated by an enhancement factor relative to the heat transfer to unmoved vapor. The enhancement factor depends on the shear stress at the film boundary [61]. Thus, for high-pressure processes, the increase in heat transfer depends not only on the relative velocity between the liquid and gas phase but also on the increased gas density.

The calculation considering laminar and turbulent flow is described in detail in Ref. [27].

In case of inert gases in vapor, the heat transfer of film condensation is attenuated by an additional resistance at the vapor side of the phase boundary because the condensation temperature of the vapor phase is related to the partial pressure of the vapor within the gas mixture. Due to the continuous decrease of the vapor content along the length of condensation, the condensation temperature and thus the driving temperature difference drops. Precise calculation of vapor condensation with inert gases requires the knowledge of the temperature of the film surface. Furthermore, a mass transport is superposed to the heat transport due to the inversed movement of the inertial gas at the phase boundary. Low contents of inertial gases lead to heat transfer-controlled condensation, whercas condensation of vapor with high contents of inert gases is mass transfer controlled.

Exceeding certain limits in vapor flow results in the risk of an entrainment of droplets or even may cause dispatching of the condensate film [62].

References

1 Haering, H.W. (2008) *Industrial Gases Processing*, Wiley-VCH, Weinheim
2 Gmehling, J. and Kolbe, B. (1988) *Thermodynamic*, Thieme Verlag, Stuttgart.
3 Dohrn, R. (1994) *Berechnung von Phasengleichgewichten*, Vieweg Verlag, Braunschweig.
4 Wagner, W. and Overhoff, U. (2006) *Thermofluids*, springeronline.com.
5 Kunz, O., Klimeck, R., Wagner, W., and Jaeschke, M. (2007) *The GERG-2004 Wide-Range Equation of State for Natural Gases and Other Mixtures*, VDI Verlag GmbH.
6 Span, R. and Wagner, W. (1996) A new equation of state for carbon dioxide covering the fluid region from the triple point temperature to 1100K at pressures up to 800 MPa. *Journal of Physical and Chemical Reference Data*, **25** (6), 1509–1596.
7 Sadowski, G. (2005) Phase behaviour of polymer systems in high pressure carbon dioxide, in *Supercritical Carbon Dioxide in Polymer Reaction Engineering* (eds F.K. Kemmere and T. Meyer), Wiley–VCH Verlag GmbH, Weinheim.
8 Enders, S., Amézquita, O.G., Jaeger, P., and Eggers, R. (2010) Interfacial properties of mixtures containing supercritical gases. *Journal of Supercritical Fluids*, **5** (2), 724–734.
9 Gmelin, L. (1973) *Gmelin Handbuch der Anorganischen Chemie, 8. Auflage Kohlenstoff, Teil C3 Verbindungen*, vol. **3**, Part 3, Verlag Chemie, Weinheim.
10 Dohrn, R., Peper, S., and Fonseca, J.M.S. (2010) High-pressure fluid phase equilibria: experimental methods and systems investigated (2000–2004). *Fluid Phase Equilibria*, **288** (1–2), 1–54.
11 Lockemann, C. (1994) *Flüssigseitiger Stofftransport in Hochdrucksystemen Mit Einer Überkritischen und Einer Flüssigen Phase*, Karlsruhe Institut für Technology.
12 Schwuger, M.J. (1996) *Lehrbuch Grenzflächenchemie*, Thieme Verlag, Stuttgart, Germany.
13 Eggers, R., Wagner, H., and Jaeger, P. (1996) *High Pressure Chemical Engineering*, Elsevier, pp. 247–252.
14 Grassmann, P. (1983) *Physikalische Grundlagen der Verfahrenstechnik*, Verlag Sauerländer, Frankfurt, pp. 350.
15 Jaeger, P. (1998) *Grenzflächen und Stofftransport in verfahrenstechnischen Prozessen am Beispiel der Hochdruck-Gegenstromfraktionierung mit überkritischem Kohlendioxid*. Technische Universität Hamburg-Harburg.
16 Wiegand, G. (1993) *Messung der Grenzflächenspannung binärer wäßriger Systeme bei hohen Drücken und Temperaturen*, Universität Karlsruhe.
17 Cornelisse, P.M.W. (1997) *The Gradient Theory Applied*, Delft University of Technology.
18 Amezquita, O.G.N., Enders, S., Jaeger, P., and Eggers, R. (2010) Interfacial properties of mixtures containing supercritical gases. *Journal of Supercritical Fluids*, **55** (2), 724–734.
19 Holzknecht, C., Kabelac, S., Klank, D., and Eggers, R. (2002) Simulation of the wetting behavior of liquids on a teflon surface in dense nitrogen. *Forschung Ingenieurwesen*, **67** (2), 54–79.
20 Marckmann, H. (2005) *Überkritische Extraktion von aufkonzentrierten Kaffeelösungen in Hochdruckkolonnen*, Technische Universität Hamburg-Harburg.
21 Weidner, E., Steiner, R., and Knez, Z. (1996) *High Pressure Chemical Engineering*, Elsevier Science B.V., pp. 223–228.
22 Duarte, R.C., Lindsley, E.A., Duarte, C., and Kazarian, G. (2005) A comparison between gravimetric and *in situ* spectroscopic methods to measure the sorption of CO_2 in a biocompatible polymer. *The Journal of Supercritical Fluids*, **36** (6), 160–165.
23 Girifalco, L.A. and Good, R.J. (1960) A theory for estimation of surface and interfacial energies III: estimation of surface energies of solids from contact angle data. *Journal of Physical Chemistry*, **64** (5), 561–565.
24 Sutjiad-Sia, Y., Marckmann, H., Eggers, R., Holzknecht, C., and Kabelac, S. (2007) Zum Einfluss von in Flüssigkeiten unter Druck gelösten Gasen auf

Grenzflächenspannungen und Benetzungseigenschaften. *Forschung Ingenieurwesen*, **71** (1), 29–45.
25 Sutjiadi-Sia, Y. (2005) *Interfacial Phenomena of Liquids in Contact with Dense CO_2*, Technische Universität Hamburg-Harburg.
26 Arendt, B. and Eggers, R. (2007) Interaction of interfacial convection and mass transfer effects in liquid/liquid systems. Proceedings of the Conference on Transport Phenomena with Moving Boundaries, October 11–12, Berlin, Germany.
27 VDI, Verein Deutscher Ingenieure (2006) Wärmeatlas Springer.
28 Czerwonatis, N. (2002) *Zerfall flüssiger Strahlen und Widerstand von Tropfen in verdichteten Gasen am Beispiel des Verfahrens der Hochdruck-Sprühextraktion*, Technische Universität Hamburg, Harburg, Germany.
29 Hobbie, M. (2005) *Bildung von Tropfen in verdichteten Gasen und stationäre Umströmung fluider Partikel bei Drücken bis zu 50 Mpa*, Technische Universität Hamburg, Harburg, Germany.
30 Stiel, L.J. and Thodos, G. (1964) The thermal conductivity of nonpolar substances in the dense gaseous and liquid regions. *AICHE Journal*, **10**, 26–30.
31 Lee, B.J. and Kesler, M.G. (1975) A generalized thermodynamic correlation based on three-parameter corresponding states. *AICHE Journal*, **21**, 510.
32 Brunner, G. (1992) *Gas Extraction*, Steinkopf Verlag.
33 Middelman, S. (1998) *An Introduction to Mass and Heat Transfer*, John Wiley & Sons, Inc., New York.
34 Bertucco, A. and Vetter, G. (2001) *High Pressure Process Technology: Fundamentals and Applications*, Elsevier, p. 101.
35 Baehr, H.D. and Stephan, K. (2006) *Heat and Mass Transfer*, Springer, New York.
36 Polyakov, A.F. (1991) Heat transfer under supercritical conditions, in *Advances in Heat Transfer*, vol. **21** (eds J.P. Hartnett and T.F. Irvine), Academic Press, San Diego.
37 Krasnoshchekov, A. and Protopopov, V.S. (1959) Heat exchange in the supercritical region during the flow of carbon dioxide and water. *Teploenergetika*, **6** (12), 26130.

38 Jackson, J.D. and Hall, W.B. (1979) Forced convection heat transfer to fluids at supercritical pressure, in *Turbulent Forced Convection in Channels and Bundles*, vol. **2** (eds S. Kakac and D.B. Spaldin), Hemisphere, Washington.
39 Eggers, R. and Sievers, U. (2001) Near critical heat transfer in CO_2 process cycle. *IHT Congress Thessaloniki*, III-2335.
40 Carrol, J.J. (2010) *Acid Gas Injection and Carbon Dioxide Sequestration*, John Wiley & Sons, Inc., New York.
41 Lenoir, J.M. and Comings, E.W. (1951) Thermal conductivity of gases measurement at high pressure. *Chemical Engineering Progress*, **47** (5), 223–231.
42 Lienhard, J. (2005) *A Heat Transfer Textbook*, Phlogiston Press, Cambridge, MA, p. 739.
43 Tegetmeier, A., Dittmar, D., Fredenhagen, A., and Eggers, R. (2000) Density and volume of water and triglyceride mixtures in contact with carbon dioxide. *Chemical Engineering and Processing*, **39** (5), 399–405.
44 Dittmar, D. (2007) *Untersuchungen zum Stofftransport über Fluid/Flüssig: Phasengrenzen in Systemen unter erhöhten Drücke*, Technische Universität Hamburg, Harburg, Germany.
45 Catchpole, O.J. and King, M.B. (1994) Measurement and correlation of binary diffusion coefficients in near critical fluids. *Industrial & Engineering Chemical Research*, **33** (7), 1828–1837.
46 Perry, R.H. and Green, D.W. (2008) *Chemical Engineering Handbook*, 8th edn, McGraw Hill, New York.
47 Cussler, E.J. (1975) *Diffusion Mass Transfer in Fluid Systems*, Cambridge University Press, Oxford.
48 McManamey, W.J. and Woolen, J.M. (1973) The diffusivity of carbon dioxide in some organic liquids at 25 and 50 °C. *AICHE Journal*, **19** (3), 667–669.
49 Nukiyama, S. (1966) The maximum and minimum values of the heat Q transmitted from metal to boiling water under atmospheric pressure. *International Journal of Heat and Mass Transfer*, **9**, 1419–1433 (translation from *Journal of the Japan Society of Mechanical Engineers*, 37, 367–374).

50 Mayinger, F. (1982) *Strömung und Wärmeübergang in Gas-Flüssigkeits-Gemischen*, Springer, New York.

51 Kutateladze, S.S. (1959) Kritische Wärmestromdichte bei einer unerkühlten Flüssigkeitsströmung. *Energetika*, **7** (1951), 229–239 (also *Izvestia Akademia Nauk Otdelenie Tekhnicheski Nauk*, 4, 529).

52 Zuber, N. and Tribus, M. (1958) *Further Remarks on the Stability of Boiling Heat Transfer*, UCLA Report No. 58–5, University of California, Los Angeles.

53 Stephan, K. and Abdelsalam, M. (1980) Heat transfer correlations for natural convection boiling. *International Journal of Heat and Mass Transfer*, **23**, 73–87.

54 Luke, A., Gorenflo, D., and Danger, E. (1998) Interactions between heat transfer and bubble formation in nucleate bowling, in *Heat Transfer*, vol. **I** (ed. J.S. Lee), Taylor and Francis, Levittown, pp. 149–174.

55 Schael, A.E. (2009) *Über das Strömungsverdampfen von CO_2 im glatten und innen berippten Rohr-Hydrodynamik*, Universität Karlsruhe, Wärmeübergang.

56 Doroshchuk, V.E., Levitan, L.L., and Lantsmann, F.P. (1975) Recommendations for calculating burnout in a round tube with uniform heat release. *Teploenergika*, **22** (12), 66–70.

57 Końkov, A.S. (1965) Experimental study of the conditions under which heat exchange deteriorates when a steam-water mixture flows in heated tubes. *Teploenergika*, **13** (12), 77.

58 Rohsenow, W.M., Webber, J.H., and Ling, A.H. (1956) Effect of vapor velocity on laminar and turbulent film condensation. *Journal of Heat Transfer-Transactions of the ASME*, **78**, 1637–1643.

59 Stephan, K. (1988) *Wärmeübergang beim Kondensieren und beim Sieden*, Springer.

60 Müller, J. (1992) Wärmeübergang bei der Filmkondensation und seine Einordnung in Wärme- und Stoffübergangsvorgänge bei Filmstörungen. Fortschritt-Berichte VDI. *Reihe*, **3**, 270.

61 Numrich, R. (1990) Influence of gas flow on heat transfer in film condensation. *Chemical and Engineering Technology*, **13** (1), 136–143.

62 Andreussi, P. (1980) The onset of droplet entrainment in annular downward flows. *Canadian Chemical Engineering*, **58** (4), 267–270.

Part Two
Processes

3
Catalytic and Noncatalytic Chemical Synthesis

Joachim Rüther, Ivo Müller, and Reinhard Michel

3.1
Thermodynamics as Driver for Selection of High Pressure

This chapter intends to summarize the general thermodynamic basics for selecting the pressure of a chemical reaction. The most important point may be the understanding of chemical equilibrium. Particularly for gas-phase reactions, the reaction kinetics can be a second driver for choosing high pressures. The influence of phase equilibria and transport phenomena is a widespread and complicated field, which for reasons of space can only be touched on some exemplary aspects.

3.1.1
Chemical Equilibrium: Law of Mass Action

The *Law of chemical affinity*, found by the Norwegian chemists C.M. Guldberg and P. Waage in 1864, is considered to be the first formulation of the *Law of mass action*. The law of mass action allows to theoretically predict the equilibrium of any reversible chemical reaction and thus is recognized to be one of the most important influences of thermodynamics on modern chemistry. Based on a kinetic approach at the time of Guldberg and Waage, it was later on derived from equilibrium thermodynamics [1] and gives an interesting insight into the key parameters of chemical equilibrium.

Given that the chemical equilibrium of a reacting system at constant pressure and temperature is characterized by a minimum of the *Gibbs energy G*, it follows that

$$dG = -SdT + Vdp + \sum_i \mu_i dn_i = \sum_i \mu_i dn_i \stackrel{!}{=} 0 \qquad (3.1)$$

Introducing the extent of reaction

$$d\xi = \frac{1}{\nu_i} dn_i \qquad (3.2)$$

where ν_i is the stoichiometric coefficient of component i, and the following expression for the chemical potential of component i in a liquid mixture,

---

Industrial High Pressure Applications: Processes, Equipment and Safety, First Edition. Edited by Rudolf Eggers.
© 2012 Wiley-VCH Verlag GmbH & Co. KGaA. Published 2012 by Wiley-VCH Verlag GmbH & Co. KGaA.

$$\mu_i = \mu_{0i} + RT \ln a_i = g_{0i} + RT \ln a_i \tag{3.3}$$

where

g_{0i} is the Gibbs energy of the pure component i, and
a_i is the activity of component i within the mixture,

Eq. (3.1) can be rewritten to

$$\frac{dG}{d\xi} = \sum_i \nu_i \mu_i = \sum_i \nu_i g_{0i} + RT \sum_i \nu_i \ln a_i = 0 \tag{3.4}$$

Finally, the activity-based formulation of the law of mass action follows:

$$K_a(p, T) = \prod_i a_i^{\nu_i} = \prod_i (\gamma_i x_i)^{\nu_i} = e^{-\frac{\Delta^R g^0(p,T)}{RT}} \tag{3.5}$$

where

K_a represents the activity-based equilibrium constant of the reaction,
γ_i is the activity coefficient,
x_i is the molar fraction of component i, and
$\Delta^R g^0$ is the standard Gibbs energy of the reaction, that is, $\sum_i \nu_i g_{0i}$.

Similarly, the law of mass action can be derived for gas phases, that is, based on the component fugacity f_i,

$$K_f(T) = \prod_i \left(\frac{f_i}{p^+}\right)^{\nu_i} = \prod_i \left(\frac{\varphi_i y_i p}{p^+}\right)^{\nu_i} = e^{-\frac{\Delta^R g^0(p^+,T)}{RT}} \tag{3.6}$$

where

K_f is the fugacity-based equilibrium constant,
p^+ is the reference pressure,
φ_i is the fugacity coefficient, and
y_i is the molar fraction of component i.

From Eqs. (3.5) and (3.6), it can be seen that for a given system, the equilibrium constants K_a and K_f both depend on the temperature. This correlation is also described by the *van't Hoff isochore*:

$$\frac{d \ln K}{dT} = \frac{\Delta^R h^0}{RT^2} \tag{3.7}$$

where $\Delta^R h^0$ is the standard enthalpy of the reaction.

As also postulated by the *le Chatelier–Braun principle*, it is obvious that for an exothermic reaction, the equilibrium is shifted to the products by a lowered system temperature and vice versa. The opposite is true for an endothermic reaction.

From the pressure point of view, it needs to be considered whether the physical state of the system is liquid or gaseous. In case of a liquid system, the Gibbs energy of a reaction is a function of pressure and so is K_a. It can be written that

$$\frac{-RT\,d\ln K_a}{dp} = \frac{d\Delta^R g}{dp} = \Delta^R v \tag{3.8}$$

with $\Delta^R v$ as the specific volume change caused by the reaction.

For many liquid-phase reactions, however, the reaction-induced volume change is small and might be neglected.

In case of a gas-phase reaction, K_f is calculated at a reference pressure p^+ and thus is not a function of the actual system pressure. Consequently, the right-hand side of Eq. (3.6) is nonvarying with respect to pressure:

$$K_f(T) = \prod_i \varphi_i^{\nu_i} \cdot \prod_i y_i^{\nu_i} \cdot \prod_i \left(\frac{p}{p^+}\right)^{\nu_i} = K_\varphi \cdot K_y \cdot \left(\frac{p}{p^+}\right)^{\sum_i \nu_i} \tag{3.9}$$

Assuming an ideal gas mixture and thus neglecting the influence of K_φ, it is therefore clear that the equilibrium composition of a gas-phase reaction is affected by the system pressure for all reactions with $\sum \nu_i \neq 0$, that is, where the reaction causes a change in the overall number of molecules and thus in volume.

Also, this result matches the le Chatelier–Braun principle, as, for example, the equilibrium composition of a gas-phase reaction with reduction of substance quantity can be forced to the product side by raising the pressure and vice versa. Consequently, it has to be noted that shifting the equilibrium to the side of lower substance quantity is a potential driver for selecting high process pressures.

3.1.2
Reaction Kinetics

As the kinetics of a chemical reaction are influenced by a multitude of different parameters such as pressure, temperature, concentrations of the reactants, molecularity and presence and type of a catalyst, the kinetics of any individually given reaction are to be evaluated empirically – sometimes including the development of an appropriate functional correlation. At the same time, there is a strong interest in the kinetics of a reaction, first to better understand the reaction mechanism and second to facilitate a basis for the optimization of reactor designs and process parameters.

For the purpose of understanding the effects of temperature and pressure, it is considered sufficient to start from a rather simple third-order law of reaction. For example, the kinetics of the reaction

$$a\text{A} + b\text{B} \rightarrow c\text{C} \tag{3.10}$$

shall be given with the reaction rate equation

$$\frac{d\xi}{dt} = k(T) \cdot [A]^2 \cdot [B] \tag{3.11}$$

where

$k(T)$ is the rate constant, and
$[A]$ and $[B]$ represent the concentrations of components A and B.

The reaction rate thus is of second order with respect to reactant A and of first order with respect to reactant B, the total order of the reaction being three. The so-called rate constant k, however, depends on the reaction temperature and thus is not a true constant. The reason is that at low temperatures, only very few molecules have sufficient energy to overcome the barrier of activation energy. The correlation of rate constant k and temperature is given by the *Arrhenius' law* for many reactions:

$$k = A \cdot e^{-E_a/RT} \tag{3.12}$$

where

A is the pre-exponential factor, and
E_a is the energy of activation.

According to the collision theory proposed by Trautz and Lewis in 1916 and 1918 [2], the pre-exponential factor of the empirical Arrhenius law can be interpreted as product of the theoretically predictable collision frequency and a steric factor. Furthermore, the collision frequency depends on the number of molecules of the reactants per volume, that is, the reactants' concentrations, which is the reason for the rate equation being formulated with concentrations instead of, for example, molar fractions. While the reactant concentrations within incompressible phases are mainly defined by the composition, temperature and pressure play important roles for gas-phase reactions. As it can be seen from the ideal gas law,

$$\frac{n}{V} = \frac{p}{RT} \tag{3.13}$$

the number of molecules per volume increases with pressure and decreases with temperature. Besides the results of Section 3.1.1, this is a second potential driver for choosing high pressures for gas-phase reactions.

It is more a theoretical point that following the transition state theory and in analogy to Eq. (3.8), a dependence of the rate constant k from the pressure can be formulated. For a few reactions, however, it is possible to achieve a significant improvement in the reaction rate on this basis, though requiring a considerable increase in pressure [3].

The above-mentioned Arrhenius law also indicates a third option to accelerate a chemical reaction – catalysis. The presence of a catalytically active substance lowers the activation energy barrier and thus allows a higher reaction rate at constant temperature as well as the same reaction rate at a reduced temperature. It is to be noted that the action of a catalyst simultaneously affects the forward and the backward reaction. It does not influence the Gibbs energy of the reaction and thus cannot shift the chemical equilibrium given by the law of mass action. In a system of parallel reactions, it is however possible to improve the selectivity.

3.1.3
Phase Equilibria and Transport Phenomena

Particularly in the field of heterogeneous reactions, the above-mentioned basics of reaction equilibrium and kinetics need to be supplemented by some considerations on phase equilibria and transport phenomena. In a gas-phase reaction involving a solid catalyst, for example, the convective and diffusive transport of the reactants to and from the catalyst surface as well as the adsorption and desorption on and from the catalyst surface affect the progress of the reaction, thus the reaction itself is only one step of a complex sequence.

Pressure affects all of the above-mentioned steps. According to the *Fick's law*, for example, the diffusive flux is correlated to the gradient in concentration and in case of gaseous systems also to pressure. Regarding adsorption and desorption, a reaction can be accelerated by a pressure-induced higher coverage of the catalyst surface with reactants; it can, however, also be slowed down due to reduced desorption of the product. As described by, for example, the *Langmuir isotherm* or the *Freundlich isotherm*, these steps also depend on pressure.

A similar situation can be found in a gas–liquid system where one reactant has to be transferred from the gas phase to the liquid phase, where the reaction finally takes place. The chemical sorption of CO_2 from a gas phase using an amine solution (e.g., MEA and MDEA) is a typical example for such a system. Again, the convective and diffusive transport has to be considered. This time, however, a gas–liquid equilibrium, which is characterized by the identity of the CO_2's chemical potential in gas and liquid phases, is involved.

Furthermore, in some cases, pressure may force two separate phases to collapse into one, causing a completely changed reaction system with respect to mass transfer resistance and composition of the reactive phase. For example, supercritical fluids are known to show distinctively increased dissolving powers.

3.2
Ammonia Synthesis Process

In this chapter, the *Haber–Bosch process* for the direct synthesis of ammonia from nitrogen and hydrogen shall be introduced as an example of a catalytic chemical high-pressure synthesis process. As we will see in the following sections, the Haber–Bosch process can be considered the first high-pressure chemical process developed according to the findings of the emerging discipline of thermodynamics, thus being a forerunner of modern high-pressure chemistry.

3.2.1
Basics and Principles

The name *ammonia* is deducted from *sal ammoniacum*, that is, ammonium chloride (NH_4Cl), which in earlier times was quarried in the West Egyptian oasis Ammon, today known as Siwa. Ammonia (NH_3) is caustic as well as toxic and characterized by a typical pungent odor. It is well soluble in water, forming ammonium hydroxide in an exothermic alkaline reaction, however, amphoteric in general. Ammonia may also be referred to as refrigerant R717 or via CAS number 7664-41-7.

Today ammonia is the main source of fixed nitrogen. The main portion of 80–85% of the world ammonia production is used for fertilization – to a large extent after processing to urea, ammonium nitrate, phosphate, and sulfate, the direct use of liquid ammonia for fertilization is less common. Chemical and miscellaneous use, for example, for pharmaceutics, polymers, and remediation of gaseous emissions such as nitrous oxides, makes up for the remaining 15–20%. This does also include the common use of fuel oil-penetrated ammonium nitrate as a blasting explosive. It is especially the increase in world population and the corresponding needs in nutrition that caused the impressive growth of world ammonia production during the last decades, as it is shown in Figure 3.1.

Ammonia is synthesized from the elements as given in Eq. (3.14) [5].

$$0.5\, N_2 + 1.5\, H_2 \rightarrow NH_3, \quad \Delta_f H^0_{(gas)} = -45.94 \pm 0.35\, kJ/mol \qquad (3.14)$$

As the reaction proceeds with reduction in molar quantity, the application of a high-pressure shifts the equilibrium to the product side. Furthermore the reaction is exothermic, thus the equilibrium can be pushed to the product by imposing a low reaction temperature. However, this is restricted by the requirements of the reaction

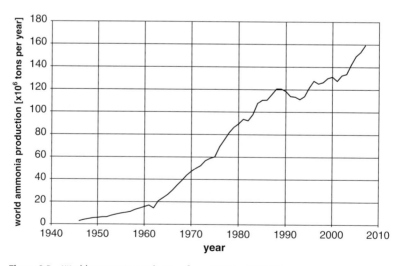

Figure 3.1 World ammonia production from 1946 to 2007 [4].

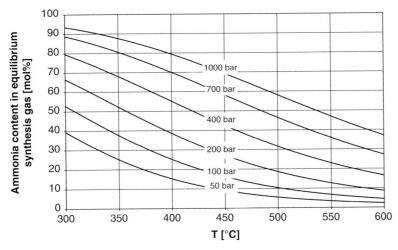

Figure 3.2 Ammonia content in equilibrium synthesis gas for various temperatures and pressures [6].

kinetics, which demand a minimum temperature to overcome the relatively high-activation energy barrier. The resulting equilibrium ammonia content in stoichiometric synthesis gas for various temperatures and pressures is given in Figure 3.2.

In favor of a reasonably low temperature and thus high-equilibrium concentration, in technical ammonia synthesis the rate of the reaction is supported by the use of a solid catalyst. A complete description of the reaction kinetics therefore has to include the kinetics of the reactants' adsorption and desorption on the catalyst surface. A first successful, integrated kinetic model of the ammonia synthesis on iron catalysts has been developed by Temkin and Pyzhev. An improved model is available from Brunauer, Love, and Keenan. A comprehensive survey of the catalysis of ammonia synthesis is given in Ref. [7].

3.2.2
History of the Ammonia Process

It was at the end of the nineteenth century that the growing demand for fixed nitrogen caused severe concerns about the possibility of future famines due to the limited availability of nitrogenous fertilizers. To overcome this perspective, the *Frank–Caro calcium cyanamide process* was developed as a first source of synthetic nitrogen compounds in 1898. Consecutively, the availability of cheap hydroelectric energy in Norway and the United States triggered the development of the *electric arc process* for the synthesis of nitrous oxides. A first plant delivered 7000 ton of fixed nitrogen in 1908. However, the energy consumption of the electric arc process corresponds to a tremendous fossil fuel consumption of about 600 GJ per ton of fixed nitrogen [6].

Though the composition of ammonia had already been ascertained by C.L. Berthollet in 1785, the direct synthesis of ammonia from the elements was considered infeasible up to the beginning of the twentieth century. It was the merit of the German chemist F. Haber to apply the new evidences from the emerging discipline of thermodynamics to the ammonia synthesis reaction. Consequently, he found that for a commercially feasible ammonia synthesis, a high-pressure recycle process would be required. Within this conclusion, the idea of a recycling process deserves some special attention, as it changed the conversion-oriented equilibrium approach of those days into a more modern space–time yield conception, considering both equilibrium and kinetics. In 1908, Haber approached BASF (Badische Anilin & Soda Fabrik), where consecutively Bosch led a team of engineers and scientists who developed a commercial process in less than 5 years. A. Mittasch contributed an economically feasible, iron-based catalyst that replaced the former osmium and uranium materials. The first plant started production in Oppau in 1913 with a daily capacity of 30 ton of ammonia. The German patents DE235421 [8] and DE238450 [9] were granted for the key inventions of this new technology.

Recognizing the work on the development of the Haber–Bosch process, the Nobel Prize for chemistry was awarded to Haber "for the synthesis of ammonia from the elements" in 1918 and also to Bosch (together with F. Bergius) for "their contributions to the invention and development of chemical high-pressure methods" in 1931.

3.2.3
Development of Process and Pressure

Following a general trend in the petrochemical industry, the early ammonia production facilities were based on coal gasification as a source of hydrogen, while later on the use of naphtha and finally natural gas became more popular. Today natural gas is the almost exclusive feedstock for the production of ammonia; correspondingly, steam reforming is the process technology of choice. Figure 3.3 gives an overview of a typical steam-reforming front end including main reactants and products.

The key elements of the front end are the primary reformer, which can be described as a set of catalyst-filled tubes being arranged in a fired box, and the secondary (autothermal) reformer, where air is fed to the process for further heating and simultaneous introduction of nitrogen. Maximum temperatures of more than 1000 °C are reached. It has to be noted, however, that there is a variety of different front end processes available from different licensors. These include processes with deviating duty allocation for primary and secondary reformer[1] as well as systems with process gas heated reformers[2] and different solutions for carbon dioxide removal[3] and synthesis gas purification.[4]

---

1) For example, Braun purifier process and AMV process (ICI) – both with increased duty of secondary reformer.
2) For example, advanced Gas Heated Reformer (Johnson Matthey), Kellogg Reforming Exchanger System and Combined Autothermal Reformer (Uhde).
3) For example, Benfield process (UOP) and aMDEA process (BASF).
4) For example, Methanation and Braun Purifier.

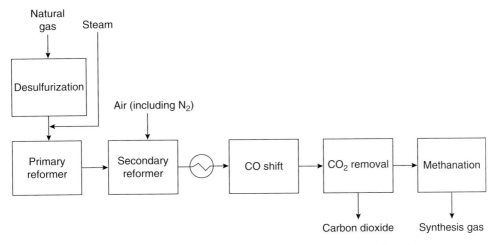

Figure 3.3 Typical ammonia plant front-end structure, including main reactants and products.

The front-end product is a mixture of hydrogen and nitrogen, typically contaminated with some inert gases like methane and argon, but virtually free of catalyst poisons. In the next step, the so-called synthesis gas is routed to the synthesis gas compressor. The compressor is the bridging element in between the medium pressure front end and the high-pressure synthesis loop. It is a key equipment of the ammonia process. Further details and steps in development will be summarized in Section 3.2.4.

From the compressor, the gas is then forwarded to the synthesis loop, which from a generalized point of view is still quite similar to the very first concepts. All synthesis loops, old or new, consist of the following elements:

- Synthesis gas preheater
- Reactor system
- Heat recovery and cooling system (most often integrating the a.m. synthesis gas preheating)
- Knockout vessel for liquefied ammonia
- Connections for addition of fresh makeup gas and as far as required extraction of purge gas
- Recirculator for unreacted synthesis gas

Figure 3.4 gives an overview of an Uhde two-converter loop, showing some typical features of a modern large-scale ammonia synthesis loop.

The synthesis gas makeup enters the loop upstream of the chilling system, is cooled down, and then mixed with the loop gas upstream of the ammonia separator. Residual traces of water are therefore washed out by the ammonia product and do not reach the sensitive synthesis catalyst. The energy of the cold gas stream is recovered in a cold exchanger. The gas is then recompressed and fed to the reactor system via the synthesis gas preheat section that includes a gas/gas heat exchanger and an interbed heat exchanger. A two-reactor system comprising three catalyst beds can often be

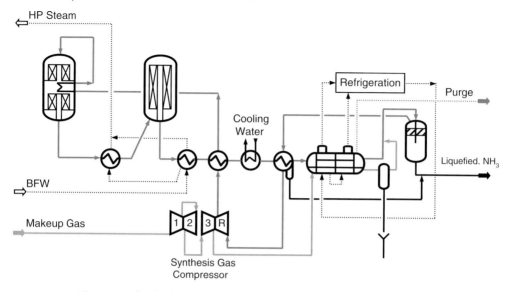

Figure 3.4 Sketch of typical ammonia synthesis loop (Uhde: two-converter loop, ~200 bar).

found in modern large-scale plants. The waste heat from the exothermic synthesis reaction is recovered in two HP steam boilers, one downstream each reactor. The synthesis gas, now rich in ammonia, is then further cooled down in the above-mentioned gas/gas heat exchanger, a water cooler, the cold exchanger, and two ammonia refrigerated chillers. In this process step, the product ammonia is liquefied and separated from the loop. A purge gas stream is extracted from the loop to maintain a reasonably low concentration of inert gases such as argon and methane. Purge gas and product are replaced with fresh synthesis gas.

Depending on the licensor and process type, there can be deviations, for example, in the number of converters and catalyst beds, heat recovery, cooling systems, and location of terminal points for makeup gas and purge. Compressors are known with three- and four-nozzle design, that is, with mixing of makeup and recycle gas in the last stage (three nozzle) or with separate sections for makeup gas compression and loop recirculation (four nozzle). Some of these deviations, in particular those being exemplary for the development of the process, will be summarized in the following paragraphs, though the main focus shall remain on the development of catalysts and loop pressure.

While for the early demonstration units, osmium and uranium had been used, it was the promoted iron (magnetite) catalyst developed by Mittasch that opened the door to commercialization of the Haber–Bosch process. Osmium had to be ruled out for cost and availability reasons, uranium is impracticable due to its sensitivity for permanent oxygen compound poisoning. Emphasizing the outstanding work of Haber, Bosch, and Mittasch, magnetite-based catalysts are still state of the art today.

As an exception that proves the rule, the KAAP (*KBR Advanced Ammonia Process*) ruthenium-on-carbon catalyst was introduced to commercial application with a Canadian plant retrofit in 1992. The manufacturer claims that the activity of the catalyst exceeds that of magnetite-based materials by about an order of magnitude. The precious metal catalyst was developed to be the technological heart of the KAAP process, operating at a loop pressure as low as 90 bar. The high activity of the catalyst lowers the barrier of activation energy and thus allows reducing the synthesis temperature. The lower temperature, however, is in favor of a higher equilibrium concentration of ammonia and thus compensates the effect of the low pressure. Even though there were two large-scale plants commissioned in 1998, the KAAP catalyst was never able to take the magnetite's place in the market.

The ICI AMV process operates at a similar pressure as the KAAP process and thus is a second example for a low-pressure ammonia synthesis, using however a promoted, still magnetite-based catalyst. A prototype plant was commissioned in Canada in 1985, further plants followed in China. Both KAAP and AMV processes derive some investment savings from the use of a single-casing turbo compressor for synthesis gas service, which however is compensated by a larger refrigeration system and higher catalyst cost.

In contrast to the above-mentioned low-pressure designs, other processes were realized with much higher pressures. Synthesis loop pressures from 300 to 1000 bar have been proposed and realized. At such high pressures, it is possible to liquefy the ammonia product with cooling water instead of using a dedicated refrigeration system. From an energetic point of view, however, the optimum synthesis loop pressure is expected to be in the range of 200 ± 50 bar [10]. Maybe for this reason, today there seems to be a settlement at about 140–210 bar loop pressure for commercially available processes using magnetite-based catalysts.

The optimum synthesis loop pressure turns out to be a trade-off in between of makeup gas compression on the one side and recirculation as well as chilling duty for product separation on the other. The energy requirements of both the recirculator and the refrigeration system, however, depend on the conversion per pass. Low conversion means high gas flow rates and low product separation temperatures, both resulting in high energy cost. High conversion reduces the gas flow rate and allows condensing significant parts of the production at cooling water temperatures and thus without loading the refrigeration system. The potential savings in operating cost, however, are counterbalanced by relatively high catalyst volumes and corresponding investment costs. It is therefore a primary objective to get the maximum conversion out of a given catalyst volume. Consequently, the reaction rate has to be kept as high as possible. In this context, low temperatures stop down the reaction kinetics, while high temperatures reduce the distance to equilibrium and thus lower the driving force. The most common way of achieving close to optimum reaction temperatures is the stepwise conversion in multibed converters with interbed gas cooling.

Figure 3.5 shows both the early technology of quench cooling (b), where the gas is cooled by the addition of cold and unreacted gas, and the preferable indirect heat transfer (c), where the hot gas is indirectly cooled in interbed heat exchangers. As the

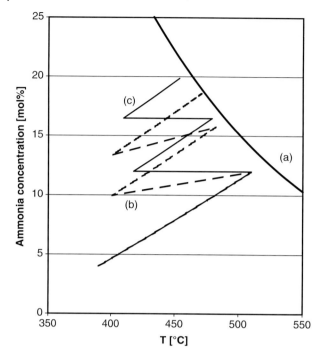

Figure 3.5 Progression of ammonia concentration and temperature in a multibed converter: (a) equilibrium curve, (b) quench cooling, and (c) indirect heat transfer.

diagram shows, for a given catalyst volume, quench cooling results in a lower conversion, since parts of the fast reacting, fresh gas are diluted with gas of a higher ammonia content.

To further reduce the power requirements of the recirculator, the former axial flow catalyst beds have been replaced with radial flow beds of a hollow cylinder geometry. Here, reduced gas velocities as well as the shortened flow path through the catalyst bed contribute to a significant reduction in pressure drop and related energy losses. Furthermore, the inner space gives a welcome opportunity to efficiently integrate the interbed heat exchangers needed for indirect cooling.

As a consequence of the high amount of waste energy from front-end and synthesis reactions, however, the overall energy consumption as far as exceeding the minimum energy for chemical conversion is predominantly determined by the quality of the waste heat system. Modern plants, therefore, involve well-integrated high-pressure steam cycles for waste heat recovery; thus, the energy consumption of a modern steam reforming ammonia plant can be as low as 28 GJ/ton$_{NH_3}$. This means that almost 66% of the energetic input to the process can still be found in the product [6].

Coming back to the development of synthesis pressure, it is an interesting, recent innovation to merge a medium-pressure synthesis section and a proven high-pressure synthesis loop within a single process concept. Following this basic idea, the Uhde dual-pressure ammonia process integrates a once-through synthesis converter in between the two casings of the synthesis gas compressor, that is, at about 100 bar, complemented by a conventional two-converter loop further downstream. The concept is considered to be the key invention of recent megascale, single-train plants, such as 3300 ton/day plant operated by the Saudi Arabian Fertilizer Company (SAFCO) in Al Jubail, Saudi Arabia. It allows a significant 50% step-up in capacity, still on the sound basis of proven equipment.

3.2.4
Special Aspects

To complement the process-related aspects as summarized in Section 3.2.3, this section shall add some mechanical points, such as development and impact of compressor technology and some basic material selection criteria as far as being specific to the synthesis of ammonia.

As already pointed out above, the synthesis gas compressor is a key equipment of the Haber–Bosch process. In the early days of direct ammonia synthesis, piston-type compressors were used, being rather flexible in outlet pressure but demanding from a maintenance point of view. Besides high maintenance cost, the low reliability and availability as well as the limitation in capacity have to be addressed as the major drawbacks of piston compressors. Consequently, the single-train capacity of an ammonia synthesis plant was limited to about 200 ton/day up to the early 1960s. It is one of the most important developments in ammonia synthesis technology that M.W. Kellogg introduced centrifugal compressors in 1963. Consecutively, the single-train capacity of new plants jumped to about 1000 ton/day within the next years. Somewhat slower, the trend to larger scale plants is still persisting today, with the above-mentioned 3300 ton/day plant in operation and plant concepts for 4000 ton/day and above being proposed. Furthermore, the changeover to turbo compressors significantly contributed to the high availability of modern plants. The failure rate of modern turbo machinery is low enough to rely on a single compressor without any redundancy. Since dry gas seals are applied to the synthesis gas compressor, a second weak point of earlier compressors, the possible contamination of synthesis loop, catalyst, and product with lube oil, can be avoided.

A potential disadvantage of turbo compressors may be found in the limited pressure ratio. Based on a typical suction pressure, the turbine tip speed and the rotor dynamics and related vibration issues limit the discharge pressure of a two-casing synthesis gas compressor at slightly above 200 bar, which, however, has proved to be absolutely sufficient. For higher pressures, a third casing would be required, which would cause a step change in investment cost and thus most often is avoided.

With respect to material selection, it is important to know about the typical process-related corrosion mechanisms. In the field of ammonia production, one should keep in mind the following:

> **Metal Dusting**: The term metal dusting describes a phenomenon being typical for synthesis gas-related processes. In reducing atmospheres containing CO at some 500–800 °C, the carburization of typical construction materials is followed

by the rapid destruction of the metallic matrix. Metal dusting, however, does not occur in the ammonia synthesis section, but in critical parts of the steam-reforming front end.

Nitriding: The nitriding (also known as nitridization) of steels is a well-known and widely used surface hardening process. It does, however, also occur in ammonia synthesis loops. Nitriding requires the simultaneous presence of a nitrogen donor such as ammonia and a sufficiently high temperature, that is, about 400 °C and above. In contrast to the controlled surface hardening process, within ammonia synthesis loops the process is continued over a long period of time. Even the thick-walled parts, particularly made from ferritic steels, can therefore show severe in-depth embrittlement, as the nitriding depth is almost proportional to the square root of the exposure time for ferritic materials. Finally, the fatal failure of equipments can follow. Therefore, for critical components, austenitic materials are to be used that are not penetrated to significant depths. Furthermore, design and operation procedures should limit alternating stresses to avoid cracks in the brittle surface, which can also accelerate the progress of nitridization.

Chemical and Physical Hydrogen Attack: Similar to the nitridization process, the chemical and physical hydrogen attacks also start from the adsorption and dissimilation of molecules from the gas phase – hydrogen in this case. The atomic hydrogen is then small enough to diffuse through the material structure, ready to recombine at any time (physical hydrogen attack). The hydrogen may also react with carbon and unstable carbides to form methane (chemical hydrogen attack). Especially in the case of methane, the resulting molecules are entrapped within the material structure, since they are too large for effective diffusion. The resulting cavities grow along the grain boundaries, leading to fatal failure in the end. The limits of use with respect to hydrogen attack can be read from Nelson curves for various steels [11].

Stress Corrosion Cracking: Another corrosion mechanism that can lead to catastrophic cracking is the so-called stress corrosion cracking. It is well known from austenitic materials already at very limited exposure to chlorides. Therefore, the stress corrosion cracking needs to be accounted for when designing, for example, the austenitic waste heat boilers. Stress corrosion cracking, however, is not limited to chloride-exposed austenitic steels. It is also known from pressurized and atmospheric vessels for ammonia storage. As stress corrosion cracking is based on chemical attack that is focused on high-stress regions, increased temperatures lead to accelerated crack propagation. The preventive measures should include but not be limited to the careful relaxation of residual stresses.

3.3
Urea Process

The urea synthesis is another example of a large-scale industrial process. The urea production is realized under high pressure between 145 and 204 bar. In contrast to the ammonia process, described in Section 3.2, no catalyst is used so far.

3.3.1
Basics and Principles

Urea was first discovered in urine in 1773 by the French chemist Hilaire M. Rouelle [12]. Pure urea ($CO(NH_2)_2$) is a white, at room temperature, crystalline compound, which represents a bulk chemical with an annual production of about 129 million ton. Urea plays a significant role in soil and leaf fertilization (more than 90% of total use) [13]. Because of its high nitrogen content (>46 wt%) and its nonhazardous character, it is today the dominant nitrogen source in agriculture. Further applications of urea are the production of urea/formaldehyde resins and melamine as well as its usage as reactant for the NO_x reduction [13].

The British chemist John Davy synthesized urea first in 1812. As Davy was unaware of his success [14], the first urea synthesis is usually credited to the German chemist *Friedrich Wöhler* [13]. In 1828, Wöhler discovered that urea can be produced from ammonia and cyanic acid, which proved that organic components can be prepared from inorganic substances. Thus, Wöhler's discovery represents one of the most important steps in the history of organic chemistry.

The industrial synthesis from ammonia (NH_3) and carbon dioxide (CO_2), which had already been suggested by the Russian chemist *Alexander Basaroff* in 1868 [12] and is exclusively used nowadays, comprises of two consecutive reactions in a liquid phase via the intermediate ammonium carbamate (NH_2COONH_4). Under industrial process conditions, the mixture of reactants, intermediate, and products form a two-phase mixture (Figure 3.6). For this reason, strong interaction between vapor–liquid and reaction equilibria can be observed [13].

In the first reaction step (Eq. (3.15)), two molecules of ammonia react with one molecule of carbon dioxide forming one molecule of ammonium carbamate [13].

$$2NH_3\,(l) + CO_2(l) \Leftrightarrow NH_2COONH_4, \quad \Delta_r H = -117\,kJ/mol \quad (3.15)$$

In the consecutive reaction (Eq. (3.16)), one molecule of urea is produced by the dehydration of one molecule of ammonium carbamate [13].

$$NH_2COONH_4 \Leftrightarrow CO(NH_2)_2 + H_2O, \quad \Delta_r H = +15\,kJ/mol \quad (3.16)$$

Both Eqs. (3.15) and (3.16) represent equilibrium reactions. The carbamate formation step (Eq. (3.15)) is exothermic and hence, according to the le Chatelier–Braun principle (see Section 3.1.1), favored by lower temperatures. At temperatures favorable

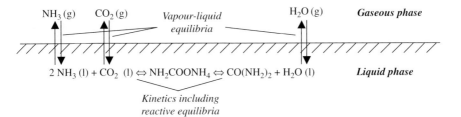

Figure 3.6 Interaction of vapor–liquid and reaction equilibria.

to reaction kinetics, the equilibrium is shifted somewhat to the reactants side. This can be compensated via higher pressure as the latter increases the solubility of the gaseous reactants in the aqueous reaction phase [13]. In the industrially relevant pressure range of 150–200 bar for the first reaction (Eq. (3.15)), the equilibrium is practically on the ammonium carbamate side [13].

The urea formation step (Eq. (3.16)) is endothermic and hence favored by higher temperature. For this reason, the heat released by the carbamate formation is partially reused. Contrary to the carbamate formation, the dehydration of carbamate represents a slow reaction as it is kinetically controlled. From this point of view, high temperatures are advantageous in increasing reaction speed, thus reducing required residence time within the reactor (industrial scale: 0.3–1 h) [15].

The overall yield of urea is maximized at reaction temperatures between 180 and 210 °C [13]. The optimal ratio of the reactants $NH_3 : CO_2$ to maximize the urea yield is somewhat above the stoichiometric ratio of 2 mol NH_3/mol CO_2 [13]. However, in industrial processes, this ratio is often increased in order to maximize the conversion of carbon dioxide [13].

In addition, the excess of ammonia reduces the required synthesis pressure, which is related to the strongly nonideal behavior of the vapor–liquid equilibrium. At a $NH_3 : CO_2$ ratio around 3, a pressure minimum azeotrope can be observed for the reactive system [13]. Starting from the composition at the pressure minimum, an increase of the $NH_3 : CO_2$ ratio results in a lower pressure rise than a decrease. This contributes to the fact that ammonia is better soluble in the liquid water/urea phase than carbon dioxide [13]. For this reason, in most of the industrial urea processes, the molar $NH_3 : CO_2$ ratio in the reactor is adjusted to be around 3 : 1 or higher.

The amount of water in the reaction mixture should be kept to a minimum as according to the law of mass action (see Section 3.1.1), water has a negative effect on the urea yield.

Due to the incomplete second reaction, the reactor outlet mixture contains significant amounts of ammonium carbamate in addition to urea and water. The ammonium carbamate is usually removed by decomposing into its constituents ammonia and carbon dioxide (reverse reaction of Eq. (3.15)) via increasing temperature and decreasing pressure [12]. Stripping using ammonia or carbon dioxide also supports ammonia carbamate decomposition [12] (see also process description in Section 3.3.2) and, in addition, removes the formed ammonia and carbon dioxide from the liquid phase.

An undesired consecutive reaction is the formation of biuret ($NH(CONH_2)_2$) by affiliation of two urea molecules (Eq. (3.17)). As biuret exerts harmful influence on vegetation, its content in fertilizers is strictly limited (in most cases <0.9 wt%) [16].

$$2CO(NH_2)_2 \Leftrightarrow NH(CONH_2)_2 + NH_3 \quad (3.17)$$

Biuret formation is also favored by low ammonia content (law of mass action) [13]. Since the partial pressure of ammonia also depends on the overall pressure, biuret formation is negligible in the high-pressure synthesis but has to be taken into consideration in the medium and low-pressure sections of the plant. In these sections, biuret is formed in significant amounts only at high temperatures, which indicates a kinetic hindrance.

Thus, the residence time at high temperatures should be reduced to minimize urea product losses through both reverse reactions of urea synthesis (Eqs. (3.15) and (3.16)) and biuret formation [13].

More information about the complex thermodynamics and kinetics of the urea synthesis is given by Chao [12] as well as Meessen and Petersen [13].

3.3.2
History of Urea Process

In the beginning of the twentieth century, urea was produced on an industrial scale by hydration of cyanamide, which can be obtained from calcium cyanamide [13]. After the development of the Haber–Bosch process (see Section 3.2.2), the synthesis from ammonia and carbon dioxide became more attractive [13]. The industrial process based on these reactants has been developed in 1922 and is called *Bosch–Meiser urea process* after its discoverers [17], which represents the basis for all industrial scale processes nowadays [13].

The very first urea synthesis processes were so-called *once-through processes* (Figure 3.7) [13, 15, 16]. To realize the high pressure in the synthesis part, the liquid ammonia is pressurized by a pump, whereas a compressor is required for the gaseous carbon dioxide. Downstream of the synthesis reactor, the pressure is released stepwise (flashed). This also supports the ammonium carbamate decomposition. Because of the limited conversion (about 35% of NH_3 and 75% of CO_2), a significant amount of nonconverted ammonia is released during flashing. The ammonia is usually neutralized by acids (e.g., nitric acid) producing corresponding ammonia salts. Both final products (urea and ammonia salt) can be applied as aqueous or solid fertilizers.

To obtain solid urea from the aqueous solution, two further process steps, namely, evaporation and shaping by prilling or granulation, are required (Figure 3.7). These process units are most independent on the previous synthesis step and, thus, can be applied in a similar way to all urea synthesis processes. A more detailed description of the shaping procedure is given at the end of this section.

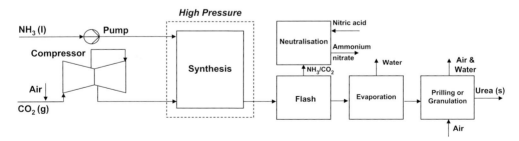

Figure 3.7 Block flow diagram of once-through process for urea synthesis with optional production of ammonia nitrate (AN).

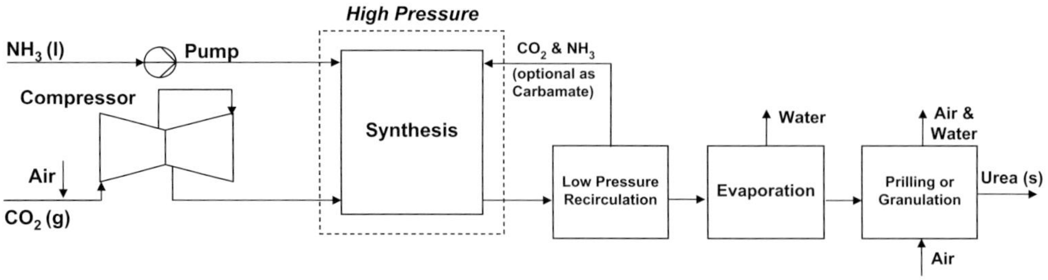

Figure 3.8 Block flow diagram of conventional total recycle process for urea synthesis with consecutive production of solid urea.

The limited carbon dioxide conversion and the huge amounts of the by-product ammonia salt represent major disadvantages with respect to production of pure urea. Thus, once-through urea synthesis processes do not have any significant industrial relevance nowadays. Modified processes have been developed, which allow recycling of the nonconverted reactants, thus increasing overall conversion and yield.

In a first developmental step, the ammonia released in the flash vapor was separated, condensed, and recycled as liquid to the reactor. This *partial recycle process* [13, 16] was replaced by the *total recycle processes* [13, 16], in which nonreacted carbon dioxide is also recycled (Figure 3.8). The recycling can be carried out via gaseous or liquid streams. Gaseous streams are disadvantageous with respect to energy efficiency (necessity of unfavorable compression work) and capital cost since compressors are required. Alternatively, the ammonia and carbon dioxide containing flash gas can be absorbed in water at the pressure level of the stripping section. The absorption of carbon dioxide is enhanced as it reacts with ammonia within the liquid phase forming ammonium carbamate (see Eq. (3.15)) (the so-called *carbamate condensation*). Compared to the recycling of gas streams with compressors, the liquid ammonium carbamate solution can be pressurized more easily by using less expensive pumps.

However, the handling of the very corrosive ammonium carbamate represents a major challenge [16]. In addition, the aqueous recycle increases the water concentration in the synthesis part, thus impeding the urea formation (see Section 3.3.1). For this reason, the total recycle process is often operated at high molar $NH_3 : CO_2$ ratio up-to 4–5 to increase carbon dioxide conversion and minimize the aqueous recycle [13].

The so-called *stripping processes* (Figure 3.9) developed by Stamicarbon [18] and Snamprogetti (nowadays Saipem) [19] in the 1960s represent the state-of-the-art urea production technology. In the stripping processes, the amount of recycle flow from outside the high-pressure section is reduced. This significantly lowers the energy consumption of the recycle pump. The stripping processes make use of the effect that ammonium carbamate can also be decomposed by stripping using carbon dioxide (applied by Stamicarbon) or ammonia (applied by Snamprogetti). This stripping can be performed under synthesis pressure within a vertical tube heat exchanger to avoid

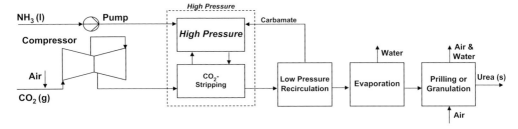

Figure 3.9 Block flow diagram of stripping process for urea synthesis with consecutive production of solid urea.

temperature decrease. The stripping processes also enable more efficient heat integration. In this way, the specific steam consumption of approximately 1.4 ton of steam per ton of urea in conventional processes is significantly reduced to 0.8–1.0 ton$_{steam}$/ton$_{urea}$ [13].

State-of-the-art urea plants based on the stripping technology are designed up to 3850 ton of urea per day. The daily feedstock requirements of such a plant are 2200 ton of ammonia and 2800 ton of carbon dioxide.

An overview of some commercially used processes with related synthesis conditions is given in Table 3.1. The listed NH_3/CO_2 ratios refer to the stream composition at reactor inlet.

To obtain solid urea, in the first step, the water content of the urea solution is reduced by evaporation. The evaporator is operated with vacuum as fairly low temperatures reduce biuret formation (see Section 3.3.1). In addition, attention should be given to short residence times. If very low water concentration is required (e.g., for prilling), it can be necessary to realize two successive evaporation steps with decreasing operating pressure [13]. This is related to the complex crystallization behavior of the water/urea melt, which can lead to undesirable crystallization within a single-step evaporation at low water concentration.

Solidification of dehydrated urea melt is realized in the last process step by prilling or granulation [13, 15]. Prilling requires a water content as low as 0.25 wt%, whereas for granulation, boiling down to a water content of 1–5 wt% is sufficient [13]. In both processes, air is essential to remove the heat of crystallization.

In the prilling process, the urea melt is fed to the prilling tower top, where small droplets are formed by the use of showerheads or rotating, perforated buckets [13]. In addition, a countercurrent airflow is passed through the tower. The evaporation of water and cooling of the melt during the free fall of the droplets lead to urea crystallization. At the tower bottom, the solid prills are removed.

Nowadays, granulation processes are mainly realized by using fluidized beds. The bed is fed with recycled urea granules, which are screened out of the final urea product due to undersize or oversize (the latter is crushed before recycling). At the bottom of the fluidized bed, the urea melt is sprayed into the bed. The liquid gets in contact with the granules and solidifies on their surface. In this way, the granules

Table 3.1 Selected overview of historic and presently used urea processes with approximate synthesis conditions [12, 13, 15, 20].

Process type	Process name	Company/companies	Pressure (bar)	Temperature (°C)	Characterization
Once-through processes		Chemico, CPI-Vulcan, Inventa, Stamicarbon, Weatherly, and so on	200–240	180–190	Two-stage flashing for carbamate decomposition
Partial recycle processes		Chemico, CPI-Allied, Inventa, Montecatini, and so on			Recycling of ammonia after liquefaction
					Two-stage flash for carbamate decomposition
Total recycle processes	Gaseous recycle	CPI-Allied, Chemico	200–210	About 200	High $NH_3 : CO_2$ ratio (4–5 mol/mol) to minimize CO_2 recycle Two-four flash stages
	Liquid recycle	Chemico, Inventa, Lonza-Lummus, Snamprogetti Stamicarbon, and so on	200–210	About 200	In addition to CPI-Allied, condensation of NH_3 and CO_2 and recycling as ammonium carbamate solution
Stripping processes	CO_2 stripping process	Stamicarbon	140–180	170–185	Stripping with CO_2 feed $NH_3 : CO_2$ ratio: 2.8 mol/mol
	Ammonia and self-stripping process	Snamprogetti/Saipem	150–156	170–200	Stripping with evaporating NH_3 (self-stripping) Higher $NH_3 : CO_2$ ratio of 3.3–3.8 to provide NH_3 for stripping
	Advanced process for cost and energy saving (ACES)	Mitsui Toatsu Chemicals	175	185–190	Stripping with CO_2 $NH_3 : CO_2$ ratio: 4 mol/mol
	Isobaric double-recycle process (IDR)	Montedison	200	190–200	Successive stripping with NH_3 and CO_2 at different pressure levels $NH_3 : CO_2$ ratio: 4–5.1 mol/mol

gradually increase in size. Granulation processes usually require additives like formaldehyde [13] to reduce dust generation during granulation and enhance the storage stability.

Prilling was popular for a long time (low investment and operating costs), but it is limited to small particles (<2.1 mm) and generates a lot of dust, which has to be removed. In addition, prilled urea is less robust concerning crushing strength compared to granulated urea [13]. For these reasons, granulated urea is predominantly used for agriculture application, whereas prilled urea is mostly used by the chemical industry.

3.3.3
Integration of Ammonia and Urea Processes

In practically all urea production plants operated today, the ammonia and urea process plants are fairly independent of each other. The stand-alone operation of each plant during shutdown periods of the other plant is usually possible, provided that the feedstocks are available and the products can be stored.

In principle, the ammonia and urea processes offer some potential for integration, associated with considerable reductions in energy consumption as well as capital investment. Several different process schemes have been suggested (e.g., by Bonetti [21] and Pagani and Umberto [22]). The main step in these concepts is to replace the traditional CO_2 removal section in the ammonia synthesis gas generation by an ammonia or ammonia/water scrubber. Thus, the carbamate forming step would essentially become a part of the ammonia process. The proposed process schemes vary in the design of this carbamate forming section as well as in the treatment of both the carbamate solution and the ammonia synthesis section.

So far, none of these alternative process schemes has been realized [13]. Fundamental reasons for this have not become public. However, the envisaged integration of the process is likely to reduce operational flexibility of the plants as a whole and can be expected to make the plants more difficult to operate [13].

3.3.4
Special Construction Materials

During material selection, special attention has to be given to the intermediate product ammonium carbamate. This component can cause active corrosion rates up to 50 mm per year [13]. Therefore, equipment, which comes into contact with ammonium carbamate solution, has to be made of proper material. In addition to good corrosion resistance, the materials applied have to meet further requirements like sufficient mechanical properties, weldability, price, and so on. Often stainless steels like AISI 316L and 317L have been used [13].

The sensitivity to corrosion of stainless steels can be reduced by adding a certain amount of the so-called passivation air to the carbon dioxide or ammonia stream to maintain a minimum oxygen concentration in the synthesis system. The oxygen enables the formation of a protective oxide layer on the metal surface [13]. To reduce the risk of corrosion, the fluents containing significant amounts of ammonia carbamate are usually forwarded by gravity or by means of ejectors [16]. However, high temperatures can lead to spontaneous activation of passivated steel [13].

The most critical corrosion conditions are present within the high-pressure strippers. Therefore, even higher grade materials like 25Cr22Ni2Mo (Stamicarbon) or duplex alloys (Mitsui Toatsu Chemicals) are required [13].

Recently, completed research revealed new materials like Safurex® from Sandvik Materials Technology (applied in Stamicarbon synthesis) and DP28W™ from Sumitomo Metals (applied in ACES synthesis) or composite materials like Omegabond™ from ATI Wah Chang (applied in Snamprogetti synthesis). The application of such materials can have additional positive effects on the urea synthesis. Due to higher corrosion resistance, the amount of passivation air can be minimized or even be eliminated. This reduces the amount of process off-gases and increases the process capacity. In addition, the wall thickness can be reduced due to both higher strength and lower corrosion rates lowering construction weight and overall material cost.

The application of the new (composite) materials can be combined with the concept of multilayer vessels (see Section 3.4.1). In the urea synthesis, the idea of multilayer vessels is often applied to the urea reactor due to its large dimensions. In doing so, the choice of the inner wall material can be focused more on corrosion resistance than on material strength.

3.4
General Aspects of HP Equipment

In general, for all processes dealing with high pressure, the wall thickness of equipment results in increased weights and costs. In the design phase of high-pressure equipment, consideration has to be given to shape and material. High-strength materials may reduce wall thickness but may not be resistant to process media with respect to corrosion or other deterioration. High-pressure processes often go together with high temperatures that further increase the wall thickness. Therefore, consideration must also be given to the thermal insulation of the process media from the pressure-bearing parts by internal insulation or internal baskets (e.g., ammonia converters) designed for differential pressure only.

For corrosive service, the use of cladding or loose liners may also be applied. In some cases, leak detection devices may be required for ensuring safe operation of the vessel during its lifetime. In urea plants, corrosion rates may reach meters per year for ferritic steels in contact with process media. Therefore, the selection of corrosion-resistant materials in contact with such media is obligatory. Use of cladding or lining allows selection of high-strength materials for the pressure-bearing wall that would fail due to corrosion in case of direct contact with the process medium.

3.4.1
Multilayered Vessels

Multilayered or multiwall vessels were specially developed to cope with high design pressures. Multilayer means a multiple of sheets (~8 mm thick) that are arranged in concentric layers around a core pipe that is made from material resistant to process conditions. The outer layers prestress the layers below due to weld shrinkage of the longitudinal welds of each layer. The weld positions are staggered to each other and the welds must be ground smooth properly before applying the next layer with a temporary bandage to ensure proper fit before welding. All layers apart from the core pipe have vent holes to avoid cavities that may have adverse influences during operation, for example, by hydrogen diffusion through the layers. Also, spirally wound multilayer design and strip wound vessels are known.

A multiwall vessel means a multitude of several shells of ~45–50 mm thickness. Each shell is fabricated separately after finishing the inner layer and grinding the longitudinal welds flush with the base metal. The next outer layer is then manufactured to fit to the existing inner shell. After grinding the longitudinal welds, the outer shell is put in a furnace for heating up to light red color near to annealing temperature and then the inner shell is inserted into the heated shell. Shrink fit is achieved just by cooling down the outer shell. Also, in this case, vent holes are provided for each outer shell. Both designs have in common that the inner shells are subject to compressive stresses under pressureless conditions. During pressurization, the compressive stresses are reduced and in general the tensile stresses at the inner surface are lower than for a solid wall vessel.

The multilayer design has the further advantage that a postweld heat treatment (PWHT) for stress relieving is not required due to the low thickness of each layer. For multiwall design, PWHT is typically required. The multilayer design has therefore some advantages for high-pressure vessels that may have to be delivered in several parts as no PWHT is required at site.

3.4.2
Recommendations to Vessel Design

As a general rule, one can say that long vessels with small diameter are more economic than short ones of equal volume with large diameter because the thickness of shells and heads increases with diameter. This is also influenced by the number of necessary circumferential welds between single-shell barrels. The use of hemispherical heads is advantageous as the thickness is only half of the shell thickness or the thickness of dished heads. However, for smaller vessel diameters and lesser wall thickness of heads, a single-wall dished head in "Korbboden" or elliptical head formed from a plate without additional weld may be more economical. Hemispherical heads are typically fabricated in crown and petal design that require welding. Contrary to other dished heads, also formed from crown and petals, the welding of petal weld seams can be performed with automated processes.

In case of heat exchangers where both shell and tube sides are connected by uninterrupted pipelines and other equipment where isolation of one side by valves or blind is not possible, the tube sheets may be designed for differential pressure only. Also, internals provided with insulation to prevent contact of hot process media with the pressure wall of the equipment can be designed for a differential pressure only and save money for the pressure vessel shell as it can be designed for a lower temperature, whereas the internals must be designed for the high temperatures but for lower pressures.

3.4.3
Gaskets and Bolting

For all high-pressure processes, gaskets are normally arranged in grooves or with male–female flange connections. Also, solid metal gaskets as ring joint or lens gaskets are used. All these gaskets have in common that they need a prestress for getting tight. This can be avoided by using welded gaskets. These gaskets need no prestress for ensuring tightness, but need more effort when the flange connection needs to be opened. So this is the gasket of choice for applications where opening is required only every few years.

For very high pressures, "Bretschneider" closing or double-cone gaskets were used. For these designs, the internal operation pressure additionally stresses the metallic gasket ring to ensure tightness. For a smooth fit with the flange surface, the gasket rings are equipped with aluminum foils that get extruded into surface unevenesses during pressurization of the HP vessel. Typically, these closings had some leakages during raising the pressure during start-up. Therefore, care has to be taken by tightening the bolts with the help of hydraulic bolt-tensioning devices. In general, bolt loads increase with the pressure and large vessel diameters increase the size of bolting. Typically, bolting larger than 2″ or 3″ is tightened by hydraulic tensioning device. Torque-controlled tightening becomes more unexact as the tightening torque converts into bolt-tensioning forces involving friction between nuts and flange. Depending on surface smoothness and greasing, the achieved values can be influenced. For the hydraulic bolt tightening, settling of the flange and bolt contact surfaces under load may also have an influence on the exact bolt force.

In order to avoid too large bolt sizes, several companies attempted designs reducing both bolt size and bolt forces. ABB Lummus Heat Transfer invented the so-called Breech Lock closing that is manufactured under license by companies, for example, Koch Heat Transfer Inc. Superbolt™ developed a bolt design that also reduced bolt forces for tightening of flanged connections. However, the last one is mainly used for machinery, for example, turbine casings rather than pressure vessels. This is due to design codes for vessels that do not know such bolt design.

References

1 Atkins, P.W. (1990) *Physikalische Chemie*, Wiley-VCH Verlag GmbH, Weinheim (translation from Atkins, P.W. (1986) *Physical Chemistry*, 3rd edn, Oxford University Press).
2 Trautz, M. (1916) Das Gesetz der Reaktionsgeschwindigkeit und der Gleichgewichte in Gasen. Bestätigung der Additivität von Cv-3/2R. Neue Bestimmung der Integrationskonstanten und der Moleküldurchmesser. *Zeitschrift für anorganische und allgemeine Chemie*, **96** (1), 1–28.
3 Isaacs, N.S. (1995) *Physical Organic Chemistry*, Prentice Hall.
4 http://minerals.usgs.gov/ds/2005/140/nitrogen.xls (accessed July 24, 2009).
5 Cox, J.D., Wagman, D.D., and Medvedev, V.A. (1984) *CODATA Key Values for Thermodynamics*, Hemisphere Publishing Corp., New York.
6 Appl, M. (1999) *Ammonia, Principles and Industrial Practice*, Wiley-VCH Verlag GmbH, Weinheim.
7 Nielsen, A. (1995) *Ammonia, Catalysis and Manufacture*, Springer, Berlin.
8 Badische Anilin- & Soda-Fabrik (1908) Verfahren zur synthetischen Darstellung von Ammoniak aus den Elementen, German Patent No. DE235421, filed October 13,1908 and issued June 8,1911.
9 Haber, F. (1909) Verfahren zur Darstellung von Ammoniak aus den Elementen durch Katalyse unter Druck bei erhöhter Temperatur, German Patent No. DE238450, filed September 14, 1909 and issued September 28, 1911.
10 Bakemeier, H., Huberich, T., Krabetz, R., Liebe, W., Schunck, M., and Mayer, D. (1985) Ammonia, in *Ullmann's Encyclopedia of Industrial Chemistry*, 5th edn, vol. **A2**, Wiley-VCH Verlag GmbH, Weinheim, pp. 143–242.
11 API Recommended Practice 941 (2004) *Steels for Hydrogen Service at Elevated Temperatures and Pressures in Petroleum Refineries and Petrochemical Plants*, American Petroleum Institute, Washington.
12 Chao, G.T.-Y. (1967) *Urea: Its Properties and Manufacture*, Chao's Institute, West Covina.
13 Meessen, J.H. and Petersen, H. (1996) Urea, in *Ullmann's Encyclopedia of Industrial Chemistry*, 5th edn, Wiley-VCH Verlag GmbH, Weinheim.
14 McKie, D. (1944) Wöhler's 'synthetic' urea and the rejection of vitalism: a chemical legend. *Nature*, **153**, 608–610.
15 Bosch, C. and Meiser, W. (1922) Process of manufacturing urea. US Patent 1,429,483, filed July 9, 1920 and issued September 19, 1922.
16 Chauvel, A. and Lefebvre, G. (1998) *Petrochemical Processes*, Gulf Publishing Company Book Divisions, Houston.
17 UN Industrial Development Organization and International Fertilizer Development Center (eds.) (1998) *Fertilizer Manual*, Kluwer Academic Publishers.
18 Inc. Stamicarbon N.V. (1964) Preparation of urea. GB Patent 952,764, filed April 8, 1960 and issued March 18, 1964.
19 Inc. Snam S.p.A (1966) Method for producing urea starting from carbon dioxide, GB Patent 1,031,528 (A), filed March 22, 1962 and issued June 2, 1966.
20 Reference Document on Best Available Techniques for the Manufacture of Large Volume Inorganic Chemicals – Ammonia, Acids and Fertilisers; European Commission, 2007.
21 Bonetti, A. (1977) Integrated urea–ammonia process. US Patent 4,012,443, filed June 12, 1975 and issued March 15, 1977.
22 Pagani, G. and Umberto, Z. (2001) Process for combined production of ammonia and urea. EP Patent 0 905 127, filed September 20, 1997 and issued November 28, 2001.

4
Low-Density Polyethylene High-Pressure Process

Dieter Littmann, Giulia Mei, Diego Mauricio Castaneda-Zuniga, Christian-Ulrich Schmidt, and Gerd Mannebach

4.1
Introduction

4.1.1
Historical Background

Low-density polyethylene (LDPE) is the oldest commercially available polyethylene resin that was first discovered by ICI in 1933 during screening experiments on various chemicals that apply high-pressure conditions. By compressing ethylene to 1400 bars in the presence of benzaldehyde and traces of oxygen, a white solid was formed that proved to be LDPE. In 1936, ICI obtained a patent (Gibson *et al.*, GB 471590) for the chemical polymerization process whose nature is of a free radical type.

LDPE attained rapid popularity due to its electrical insulating properties, which were exploited for the development of radar technology during the Second World War. After the discovery of LDPE's excellent film forming properties in the late 1940s, a sudden increase in demand took place, which triggered the worldwide commercialization of the resin and related technologies.

4.1.2
Properties and Markets

With a consumption of about 20 million ton [1] per year in 2008, LDPE is one of the world's most important resin types. The majority of LDPE is used for film applications. This is due to the specific rheological properties of LDPE that are determined by its branching structure and molecular weight distribution. LDPE has a statistical distribution of short- and long-chain branches, formed intrinsically through intra- and intermolecular hydrogen transfer reactions. Polymer properties are correlated with the temperature and pressure conditions during polymerization, and can be fine-tuned by controlling these values.

4.1.3
Polyethylene High-Pressure Processes

Two different high-pressure processes using autoclave or tubular reactors are applied for LDPE production. The autoclave process was developed by ICI, whereas the tubular reactor process was developed by BASF Aktiengesellschaft (predecessor of LyondellBasell's *Lupotech* T process). Monomer conversion rates of the adiabatic autoclave process can reach 25% compared to values up to 40% for tubular reactors, where the heat of polymerization can be partly removed through the jacketed reactor tubes via circulating cooling water.

4.1.4
Latest Developments

With the invention of linear low-density polyethylene (LLDPE) by coordinative catalysis using ethylene and alpha-olefins as comonomers in the 1970s, many consultants predicted a significant replacement of the entire LDPE product portfolio by the novel LLDPE resins. As a consequence, only limited activities for further development of the LDPE processes have been undertaken by the market players.

This picture did change when it became clear about two decades later that the negative market expectations for LDPE did not materialize. The favorable LDPE resin properties for film applications could not be matched by LLDPE. As a consequence, a new wave of investments in LDPE plant capacity has been initiated by the petrochemical industry.

Parallel to these new investments, various process and equipment improvements have been introduced to increase the overall process competitiveness. From this new wave of investments in LDPE capacity, the tubular reactor technology could mainly benefit due to the more efficient reactor scale-up. Today, tubular reactor capacities up to 450 kiloton/annum (Company: SEPC, KSA; Technology: *Lupotech* T from LyondellBasell) are in operation, whereas the largest capacity of an autoclave reactor is about 200 kiloton/annum.

4.2
Reaction Kinetics and Thermodynamics

The free radical polymerization process for LDPE manufacture can be described by a detailed kinetic scheme. This scheme allows the calculation of structural properties such as molecular weight distribution and branching frequencies. Therefore, it distinguishes several reaction steps, for example, initiator decomposition, radical chain propagation, chain transfer to monomer and to modifier, intra- and intermolecular chain transfer, β-scission of secondary radicals, and chain termination.

The overall reaction takes place in the following steps:

$$\text{Initiation} \rightarrow \text{propagation} + \text{chain transfer} \rightarrow \text{termination}$$

4.2.1
Initiation

The polymerization of ethylene under high pressures can be initiated by the following:

- Compounds that decompose into free radicals (e.g., peroxides).
- Oxygen.

The simplest type of initiation takes place by decomposition of an organic peroxide into two radicals:

$$\text{R-O-O-R'} \xrightarrow{k_d} \text{R-O}^* + {}^*\text{O-R'}$$

For initiation and molecular weight control in free radical polymerization of ethylene, a combination of several substances can be applied. Typically used free radical initiators belong to the classes of di-alkyl peroxides, peroxy alkyl esters, peroxy-carbonates, or di-acyl peroxides. The choice of the initiator mainly depends on its half lifetime at application temperature. To generate a more or less constant radical concentration level over a wide range of temperatures (e.g., 150–300 °C), a combination of different initiators is commonly applied. A typical mixture consists of a low and high temperature decomposing peroxide dissolved in hydrocarbons.

4.2.2
Propagation

The polymer is formed by multiple addition of the monomer to the free radical end of a growing polymer chain:

$$\text{-R}^* + \text{CH}_2 = \text{CH}_2 \xrightarrow{k_p} \text{-R-CH}_2\text{-CH}_2{}^*$$

$$\text{-R-CH}_2\text{-CH}_2{}^* \xrightarrow{+n(\text{CH}_2=\text{CH}_2)} \text{-R-}(\text{CH}_2\text{-CH}_2)_n\text{-CH}_2\text{-CH}_2{}^*$$

The heat generated during the polymerization reaction is about 3600 kJ/kg polymer.

When the reaction mixture contains not only ethylene but also the other so-called comonomers such as propylene, vinyl acetate, or acrylates, the reaction is more complex:

$$\text{-CH}_2\text{-CH}_2^* + \text{CH}_2 = \text{CH}_2 \xrightarrow{k_{11}} \text{-CH}_2\text{-CH}_2^*$$

$$\text{-CH}_2\text{-CH}_2^* + \text{CH}_2 = \text{CRH} \xrightarrow{k_{12}} \text{-CH}_2\text{-CRH}^*$$

$$\text{-CH}_2\text{-CRH}^* + \text{CH}_2 = \text{CRH} \xrightarrow{k_{22}} \text{-CH}_2\text{-CRH}^*$$

$$\text{-CH}_2\text{-CRH}^* + \text{CH}_2 = \text{CH}_2 \xrightarrow{k_{21}} \text{-CH}_2\text{-CH}_2^*$$

Each of these reactions is characterized by its specific rate constant k_{ij}. The composition of the polymer is determined by the copolymerization parameters $r_1 = k_{11}/k_{12}$ and $r_2 = k_{22}/k_{21}$. If $r_1 = r_2 = 1$, the comonomer is randomly distributed in the polymer chain. This is largely the case when vinyl acetate is copolymerized with ethylene.

4.2.3
Chain Transfer

Besides reacting with ethylene or a comonomer, a growing polymer chain may also react with transfer agents (modifiers). Modifiers are chemical substances, which easily transfer an H-atom to the free radical end of a growing polymer chain. By this reaction the modifier becomes a radical itself. This radical can start a new polymer chain, while the growth of the polymer chain to which the H-atom is transferred is stopped.

$$-CH_2-CH_2^* + HX \xrightarrow{k_{tr}} -CH_2-CH_3 + X^*$$

$$X^* + CH_2 = CH_2 \xrightarrow{k_p} X-CH_2-CH_2^*$$

The free radical on the growing polymer chain is not eliminated from the reaction but just transferred to a new molecule.

The effectiveness of a modifier depends on its chemical structure, concentration, temperature, and pressure. A concentration-independent measure for its effectiveness is the chain transfer constant, defined as the ratio of kinetic coefficients for the transfer reaction to this substance and radical chain propagation reaction. Usually the effectiveness of chain transfer agents is increased with rising temperature and reduced pressure. The chain transfer constant of modifiers falls from aldehydes, which are more effective than ketones or esters, to hydrocarbons. Unsaturated hydrocarbons typically have higher transfer constants than saturated hydrocarbons and a strong effect on polymer density must be considered because of the ability to copolymerize that give a higher frequency of short-chain branches in the polymer.

Even the polymer itself can react as a chain transfer agent. In the latter case, one has to distinguish between intramolecular and intermolecular chain transfer.

Intramolecular chain transfer leads to short-chain branching with mainly butyl groups at the branches:

Intermolecular chain transfer forms long-chain branches:

$$\text{-CH}_2\text{-CH}_2^* + \text{H-CR}'\text{R}''\text{H} \xrightarrow{k_{LCB}} \text{CH}_2\text{-CH}_3 + {}^*\text{CR}'\text{R}''\text{H}$$

4.2.4
Termination

The growth of polymer chain radicals can be stopped by various reactions:

i) Two chain radicals may combine with saturation of the free valences:

$$\text{R}'\text{-CH}_2\text{-CH}_2^* + {}^*\text{CH}_2\text{-CH}_2\text{-R}'' \xrightarrow{k_t} \text{R}'\text{-CH}_2\text{-CH}_2\text{-CH}_2\text{-CH}_2\text{-R}''$$

This reaction forms very large molecules.

ii) Two chain radicals may form a terminal double bond (disproportionation):

$$\text{R}'\text{-CH}_2\text{-CH}_2^* + {}^*\text{CH}_2\text{-CH}_2\text{-R}'' \xrightarrow{k'_t} \text{R}'\text{-CH} = \text{CH}_2 + \text{CH}_3\text{-CH}_2\text{-R}''$$

The chain length is unaffected by this reaction.

iii) A chain radical reacts with an initiator radical:

$$\text{R}'\text{-CH}_2\text{-CH}_2^* + {}^*\text{I} \xrightarrow{k''_t} \text{R}'\text{-CH}_2\text{-CH}_2\text{-I}$$

In all three cases, two free radicals disappear from the reaction mixture. In order to keep the concentration of radicals in the reaction mixture constant, a new initiator molecule must decompose into free radicals.

4.2.5
Reaction Kinetics

All the above-described reactions take place simultaneously in a polymerization reactor. However, the rate of each reaction depends on the concentration of the reactants and on the individual reaction rate constants k_i. These rate constants largely depend on pressure and temperature.

$$k_i = A_i \exp\left[-{}^A E_i \frac{+\Delta V_i(P-P_0)}{RT}\right]$$

Each rate constant k_i is characterized by a particular value of the factor A_i, the activation energy ${}^A E_i$, and the activation volume ΔV_i.

For polymerization at constant pressures, we can write

$$k_i = A'_i \exp[-{}^A E_i/RT]$$

4.3
Process

4.3.1
General Process Description

The overall process description is the same for both technologies and can be divided into the following steps/process units:

- Compression
- Polymerization
- Polymer/gas separation and unreacted gas recycle
- Extrusion
- Degassing

Refer to Figure 4.1. Fresh ethylene and a chain transfer agent (modifier) are mixed with the low-pressure recycle stream previously compressed in a booster compressor, from almost atmospheric conditions up to 20–50 bar (depending on fresh ethylene conditions). The combined stream is then compressed to an intermediate pressure of 200–300 bar approximately (supercritical conditions) in a multistage primary compressor. Optionally, comonomers such as vinyl acetate, acrylic acid, and methyl acrylate can be used for the production of ethylene copolymers, and in this case are typically fed at the discharge of the primary compressor.

The booster and primary compressors are alternative multistage compressors, equipped with interstage coolers. The number of stages of both compression steps depends on ethylene supply pressure at battery limits as well as on the high-pressure recycle conditions.

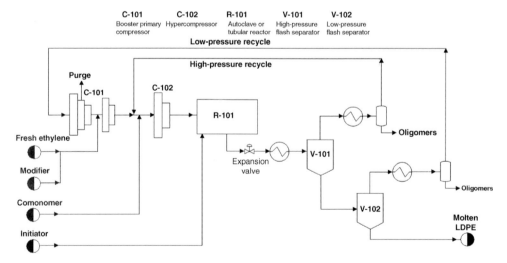

Figure 4.1 Flow sheet LDPE industrial production.

The outlet stream from the primary compressor is mixed with the high-pressure recycle gas stream and compressed in the hypercompressor up to a final pressure of approximately 1200–2500 bar for autoclaves and 2000–3300 bar for tubular reactors. The hypercompressor is a two-stage reciprocating compressor with interstage cooling.

Downstream of the hypercompressor, the gas is heated or cooled depending on the selected technology. Independent of the type of reactor, initiators (e.g., oxygen, peroxides, mixture of peroxides, or mixture of oxygen and peroxides) are injected in one or more injection points to start the polymerization. Starting temperature requirements are 170–190 °C for oxygen-initiated reactions and 150–170 °C for peroxides.

Modifiers control the molecular weight and act as chain transfer agents. The most common are aldehydes (e.g., propionic aldehyde), ketones, esters, and hydrocarbons such as propane, propylene, and 1-butene. The required quantity of modifiers can vary according to the type and depending on the melt flow rate (MFR). Typical aldehyde or ketone consumption figures are in the range of 1–4 kg per ton of product, while the consumption of α-olefins is about 10 times higher.

One pass monomer conversion is between 15% and 25% in an autoclave reactor and between 25% and 40% in a tubular reactor; therefore, unreacted monomer and modifier have to be separated from the product. Normally two separation steps operating at different pressures are provided.

The homogeneous mixture of LDPE + ethylene that leaves the reactor is expanded through a pressure control valve, from reaction conditions down to 200–400 bar, where the molten polymer precipitates from the supercritical ethylene. Due to a reverse Joule–Thomson effect, the expansion of ethylene at these conditions leads to a temperature increase, which requires a proper cooling system before the first separation is carried out. Most of the available manufacturing processes provide a product cooler to reduce the mixture temperature down to 250–300 °C. Alternatively, a direct gas ethylene side stream coming from the hypercompressor is used to cool down the mixture.

Downstream of the cooling, a high-pressure separation step is provided to separate the polymer from the unreacted gas. High-pressure separation is typically operated between 230 and 290 barg and temperatures between 220 and 280 °C. The recovered monomers are cooled down, separated from oligomers, and fed back to the process as a high-pressure recycle gas stream at the inlet of the hypercompressor.

The mixture of LDPE and residual ethylene from the bottom of the high-pressure separation vessel is then flashed almost to atmospheric conditions. Low-pressure separation takes place at pressures in the range of 0.2–2.5 barg and temperatures in the range of 200–260 °C.

The gaseous phase is cooled through different cooling steps during which comonomer, if any, and by-products (e.g., oligomers, waxes, and oils) are separated from the gas. Comonomers can be recovered eventually through a purification unit, while by-products are disposed. The gas stream is compressed up to the fresh ethylene supply pressure through a multistage reciprocating compressor (booster compressor).

To avoid accumulation of inerts/poisons, a small amount of gas is normally purged from the booster compressor and can be recovered into the cracker if this is available.

The polyethylene melt, which still contains residual ethylene and comonomer, is directly discharged to the extrusion section. Final removal of ethylene before bagging is ensured by degassing in the extruder or/and in downstream dedicated degassing silos.

Additives can be added to the polymer in the extruder depending on the final application.

4.3.2
Autoclave Reactor

The continuous stirred autoclave was originally the first viable manufacturing process developed by Imperial Chemical Industries Ltd – ICI – (UK) in 1939 for the production of LDPE.

The autoclave reactor is a single or a multiple stage continuous stirred tank reactor (CSTR), as shown in Figures 4.2 and 4.3, with its characteristic residence time distribution. The different reaction zones can be isolated by means of proper baffles at the agitator itself. The number of zones in a multistage autoclave varies between two and six.

The process is adiabatic and the reaction heat ($\Delta H_R = 100 \, \text{kJ/mol}$) is mainly removed by the feed gas, since the heat transfer through the vessel wall is negligible. For this reason, the gas from the discharge of the hypercompressor is precooled before entering the reactor.

The total conversion is given by the temperature difference between the reactor inlet and outlet. Pressure influence on the conversion can be neglected.

Multizone autoclaves give the advantage, with respect to the single stage, of operations with different temperatures in each zone (profile increasing from the top to the bottom). This results in the manufacture of products with a broader molecular weight distribution and increases the overall conversion working at higher

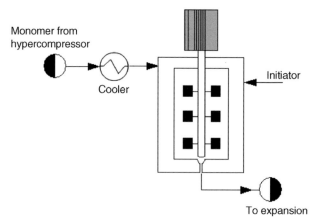

Figure 4.2 Autoclave reactor: single zone reaction.

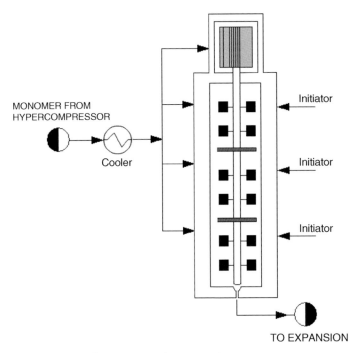

Figure 4.3 Autoclave reactor: multizone reaction.

temperatures. The temperature in the different zones is controlled by acting on the amount of initiator and cold gas injected. Different peroxides are injected in the different zones to reproduce the temperature profile required by the product. The maximum operating temperature is about 300 °C.

The reactor can be operated in a single-phase or in a two-phase region, which can be beneficial for special product applications.

The agitator motor can be installed outside or inside the reactor. In the second case, it is cooled down by part of the ethylene feed stream.

The current maximum size of autoclave reactors is about 2500 l for a capacity of approximately 200 kiloton/year of LDPE. The length to diameter ratio L/D can be as low as 2 for a single zone and up to 20 for multizone reactors. Residence time ranges between 8 and 60 s.

4.3.3
Tubular Reactor

In 1942, the company Badische Anilin & Soda Fabrik Aktiengesellschaft – now BASF SE (Germany) – started the production of LDPE using a tubular reactor process.

The tubular reactor is a plug flow reactor (PFR) provided with an external jacket cooling water system. The polymerization takes place in a single homogeneous phase.

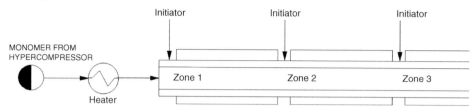

Figure 4.4 Tubular reactor: single ethylene feed or "S-reactor."

Two different possible configurations are available:

- Single ethylene feed or "S-Reactor" (Figure 4.4), where the total ethylene flow from the discharge of the hypercompressor is fed to the inlet of the reactor after being preheated.
- Multiple feed reactor (Figure 4.5), where part of the total ethylene flow is fed to the inlet of the reactor after being preheated and the rest is fed in one or more points along the reactor after being cooled. In this case, the reactor can be designed with a stepwise increase in diameter in order to compensate for the different flow rates in the different sections.

In both configurations, the gas fed to the inlet of the reactor is preheated at 150–190 °C, depending on the type of the initiator, and the first initiator injection is right at the inlet of the reactor creating the first reaction zone. As the mixture cools after the first reaction peak, additional initiator injections along the reactor are provided that create a sequence of different reaction and cooling zones (up to five).

In the multiple feed configuration, part of the reaction heat is also removed by means of the cold ethylene gas injections along the reactor. The split between the hot gas fed at the inlet and the cold gas can be from 40/60 up to 70/30.

The temperature profile as well as the peak temperature in each reaction zone depends on the type of initiator as well as the properties of the desired product. The temperature profile in the multiple feed reactor is characterized by a steep decrease at

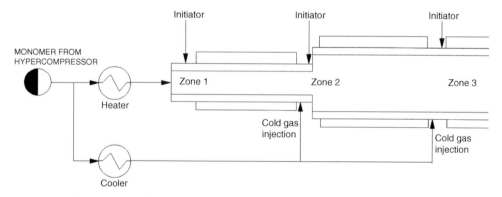

Figure 4.5 Tubular reactor: multiple cold ethylene feed points.

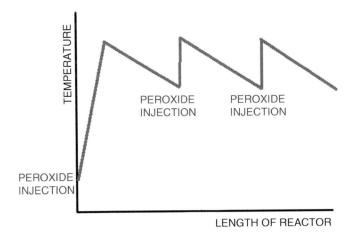

Figure 4.6 Typical temperature profile for a single tubular reactor (single tube length 10 m).

the injection of cold gas. Typical temperature profiles for single and multiple feed reactors are shown in Figures 4.6 and 4.7.

The velocity inside the reactor for both configurations varies from a minimum of 6 m/s up to a maximum of about 20 m/s.

The maximum conversion depends on starting and peak temperatures in the different reaction zones, the jacket cooling water temperature, and the overall heat transfer coefficient. Maximum peak temperatures are about 310 °C, usually higher with oxygen than with peroxides. Depending on the product, conversion can vary in a range of 25–40%.

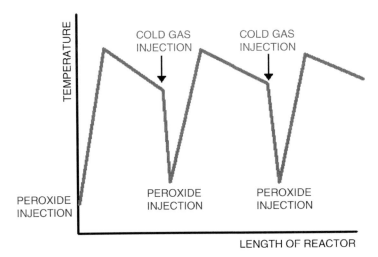

Figure 4.7 Typical temperature profile for a multiple feed tubular reactor.

Product properties are mainly affected by peak temperature, modifier concentration, and pressure:

- Density decreases with increasing peak temperature.
- Density increases with increasing pressure at constant peak temperature.
- MFI increases with the concentration of modifier in the reactor.

Current plants in operation have installed capacities for a single line up to 450 kiloton/annum, and extension of the design up to 500 kiloton/annum is already considered an achievable target. The tubular reactor consisting of a large number of jacketed tubes, arranged in a serpentine-like structure, can reach total lengths higher than 2 km, with inner diameters between 10 and 100 mm and length to diameter ratios from 10 000 to 40 000.

4.3.4
Safety

In the high-pressure LDPE reactor, runaway reactions can occur with very fast increases in temperature (>500 °C) and pressure (maximum achievable pressure limited by safety devices) and this can lead to ethylene decomposition. Thermal decomposition of ethylene is a radical reaction leading to the formation of carbon, hydrogen, and methane. The decomposition reaction may occur at temperatures above 330 °C when the operating pressure is above 2500 barg. At lower pressures, higher temperatures are required to have spontaneous decomposition. Other factors such as impurities or the presence of iron oxides or other metal oxides can facilitate the initiation of this reaction.

Ethylene decomposition is strongly exothermic ($\Delta H_{decomp\ ethylene} = 125$ kJ/mol) and its associated reaction heat causes simultaneous temperature and pressure increases inside the reactor within a few seconds, further accelerating the decomposition reaction itself.

To prevent such a scenario, an automatic interlock system actuated by redundant temperature and pressure measurements is in most cases provided in order to totally or partially depressurize the reactor and/or the surrounding process systems as soon as an abnormal situation is detected.

Reactor systems are equipped with pressure relief devices (safety valves and bursting disks) as an ultimate protection, which enable the release of all the reactor contents into the atmosphere. A hot gas mixture is dispersed into the atmosphere into one or more stacks with a nitrogen blanketing and/or water quenching system to separate the polymer and cool down the gas.

Controlled release of gases from the reactor to the flare is also possible under certain failures not implying decomposition risks. In this case, the pressure release can be done in such a way that makes the peak flow rate to stay within the acceptable limit of the flare. Benefits include reduced shutdown times and environmental impact.

Other countermeasures that increase the overall safety of the existing high-pressure technologies include the implementation of hydrocarbon detection systems in all the process areas to detect any eventual leakage as soon as possible. Also, proper material

4.4
Products and Properties

Theoretically, an almost infinite number of different grades can be produced in a high-pressure polyethylene plant, due to the high flexibility in polymerization conditions and the relative short transition times. In contrast to the low-pressure technology, where the polymerization is carried out by Ziegler/Natta, chromium, or metallocene catalysts, the high-pressure process can also use polar comonomers such as vinylacetate, acrylic acid, or N-butylacrylate to modify product properties. With all these opportunities, the resin producer can adjust the product portfolio to cover the market needs at the current economic environment. LDPE homopolymer resins with densities 0.917–0.934 g/cm^3 at MFRs 0.15–100 g/10 min are available in the market.

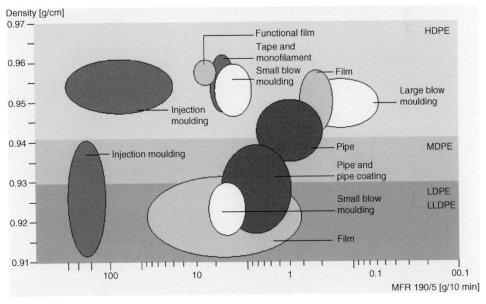

Due to the unique polymer structure of LDPE with significant long-chain branching, they can easily be processed to show good thermal stability in the melt phase. LDPE can be used in a large number of applications; with more than 60% of applications in blown film, this has become the most important segment for LDPE, followed by extrusion coating, injection molding, wire and cable, and blow molding.

4.4.1
Blown Film

LDPE resins cover a broad application range based on the type of density and melt flow rate combinations used; for example, agricultural films, shrink films, and heavy duty bags have densities of 0.918–0.924 g/cm^3 and melt flow rates of 0.15–0.30 g /10 min. A special feature of LDPE resins is that with increasing density, the molecular mass distribution gets more narrow, resulting in better optical properties and lowering the already very low gel level, which makes the grades suitable for the production of label films and surface protection films. There are two major trends in the film market. One trend is to use higher density LDPE that offer converters the potential for film thickness reduction; the other is the substitution of pure LDPE films with blends of LDPE, LLDPE, and/or mLLDPE and the use of multilayer films.

4.4.2
Extrusion Coating

LDPE resins from autoclaves are used for extrusion coating due to their better processing behavior related to the different kinds and higher amounts of long-chain branching. This different molecular structure is a consequence of the specific polymerization conditions of autoclave reactors. Autoclave grades have lower neck-in during processing and better web stability during extrusion.

4.4.3
Injection Molding

LDPE homopolymers and copolymers for injection molding are characterized by their softness and flexibility. Due to the polymer structure, these resins show excellent processability and flowability at low melt temperatures, resulting in world-class organoleptic properties.

Typically, injection molding grades are used for masterbatches. One of the main masterbatches used in the polyolefins industry is LDPE-based silica masterbatches, since SiO_2 is used in film applications as an antiblocking additive.

4.4.4
Wire and Cable

LDPE homopolymers include everything from power cables, insulation for coaxial cables, and telephone cable cores to sheathing material. Especially, for medium- and high-voltage cable insulation, a contamination-free resin with a very low gel level and excellent dielectric properties is required.

4.4.5
Blow Molding

An additional advantage of LDPE resins is that no heavy metal catalysts are used for the polymerization, and therefore stabilization with antioxidants is not required. This feature makes LDPE a candidate for pharmaceutical applications, such as infusion bottles made by the BFS (blown-fill seal) process. The worldwide market trend in this application is the use of higher density LDPEs, since in most BFS production lines the sterilization step is the bottleneck and a stiffer LDPE with a higher melting point offers higher sterilization temperatures and therefore shorter cycle times.

4.4.6
Copolymers

An increasing amount of incorporated polar comonomers such as vinylacetate leads to reduced crystallinity combined with improved low-temperature properties, improved toughness and flexibility, improved impact strength, and reduced flexural modulus (stiffness). Typical applications for EVA grades include footwear, films, hot melt adhesives, and wire and cable. The improved low temperature properties are utilized in, for example, deep-freeze bags.

4.5
Simulation Tools and Advanced Process Control

4.5.1
Introduction

Compared to other polymer production processes and even among polyolefin processes, the high-pressure LDPE production technology is highly suitable for model applications for two reasons.

First, the radical polymerization kinetics is well known and has been extensively studied in the past and continuously updated. The kinetics of oxygen and peroxide initiation are relatively easy to be determined experimentally.

In addition, the reaction conditions are such that complicated phase transitions and mass transfer driven by concentration gradients can be ignored. High-pressure tubular reactors offer the challenge and attractiveness of handling a system with two independent variables, namely, the time and space, being modeled as plug flow reactor, whereas all other reaction processes in the polyolefins industry using loops, stirred tank reactors, or fluidized beds as reactors are modelwise, mostly treated as the ideal stirred tank reactor (CSTR).

In this chapter, the focus is on practical model applications in an industry aiming to increase production rates, narrow specification ranges, accelerate grade transitions, reduce specific operating costs, and improve scale-up.

Various commercial software packages are used industrially that ease the computational burden.

4.5.2
Off-Line Applications

4.5.2.1 Flow Sheet Simulations
Almost all licensors and contractors use commercial flow sheet simulation software either for primary design or for changes to an existing plant. Various software

vendors offer tools for this purpose. Common to all of them is the application of mass and heat balances using multicomponent flow streams. Essential to this is the proper choice of thermodynamic models that are offered by these generic tools. The proprietary know-how of a licensor is necessary to customize the commercially available models for high-pressure technology requirements.

4.5.2.2 Steady-State Simulation of the Tubular Reactor

Steady-state simulations of the high-pressure tubular reactor for the production of LDPE with distributed parameters are used for reactor design of a new plant as well as for optimization of the operating parameters in an existing plant – for product development and reduction of production costs.

Such models based on detailed kinetic parameters and heat and mass balances provide a reactor temperature profile, overall heat transfer coefficient, one-pass conversion rates, product molecular structure, and related properties such as MFR and density. Figure 4.8 shows the ethylene conversion profile along the reactor for a tubular reactor with three peroxide initiation points provided by a commercial design package.

4.5.2.3 Dynamic Simulation of the Process

Dynamic simulations are used for design and even more for operator training. Operators feel like running the real plant since the human machine interface is a copy or an emulation of the DCS image, and the model reflects the dynamic behavior of the plant. Operator training is most effective when rarely encountered situations such as reactor start-up, grade transitions, and disturbance rejection can be repeated as often as desired. Training on the simulator prior to start-up of a new plant is highly beneficial. Reactor behavior can be demonstrated but the controller can also be tuned

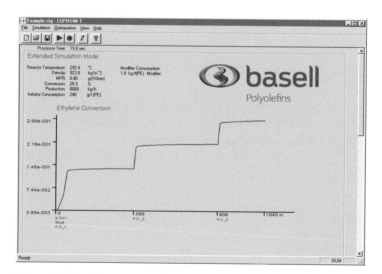

Figure 4.8 Sketch from process design tool: ethylene conversion example.

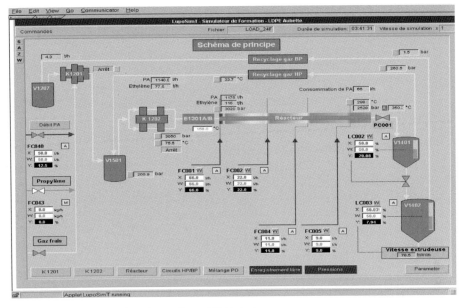

Figure 4.9 Training simulator image for an LDPE plant.

before the first product leaves the plant. Such operator training simulators (OTS) are often applied in new world-scale plants with inexperienced personnel.

Figure 4.9 shows an image of a generic simulator with animated numbers and additional information such as online product properties.

4.5.3
Online Application

4.5.3.1 Soft Sensors

Online applications of models describing reactor or plant behavior are used for operator information. The most interesting applications include those parameters that are difficult to measure or can only be measured with time delay such as melt flow index and polymer density. Soft sensors eliminate the time delay that is extremely critical during plant start-up and grade transitions. Even the so-called online analysis with automatic sampling has a certain time delay, as shown in Figure 4.10.

Online calculations of MFR and polymer density are developed for several processes.

Density soft sensors are first-principle models comprising kinetics, thermodynamics, and mass balances. They use online information from the plant, such as peroxide flow rates, modifier, and cooling water, as well as the measured temperature profile. Model execution can be organized on a separate computer connected to the DSC and the results can be trended in the DCS for operator use. Figure 4.10 shows the density soft sensor results for a grade transition toward a lower polymer density.

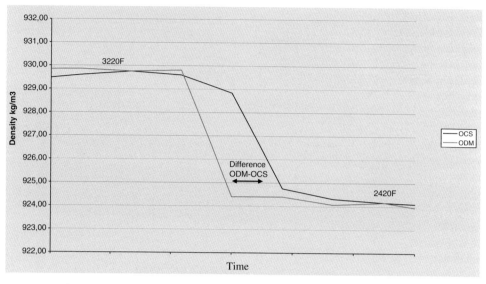

Figure 4.10 Time gain in accessing transition data for product density, comparing online model application (soft sensor)) versus online measurements (OCS).

MFR soft sensors are often data-driven models that use information from the reactor and from the extruder. Figure 4.11 shows the comparison with an online analytic system for a grade transition.

4.5.3.2 Advanced Process Control

Advanced process control (APC) is a vague term often with different interpretations. In this chapter, it is meant to be a controller that is linear or nonlinear in nature,

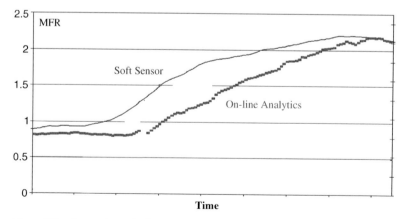

Figure 4.11 Comparison of soft sensor results for MFR with online analytics for a grade transition.

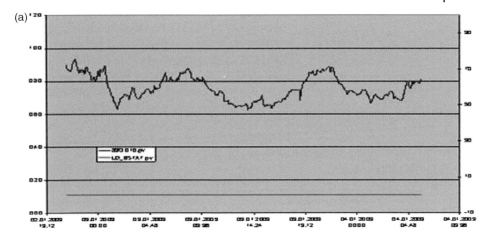

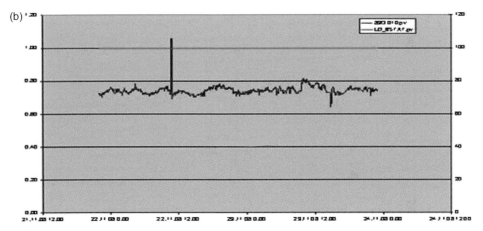

Figure 4.12 (a) MFI trend for a certain grade without APC. (b) MFR trend for a certain grade with APC.

multivariate, constrained, and model based. The models are mainly empirical, but it is possible to integrate reliable, fast, and easy to solve first-principle models such as mass balances.

An APC system can control melt flow index and polymer density within specification limits, with maximum production rates under certain constraints. In order to control the polymer qualities, the above-mentioned soft sensors can be applied. The APC system can increase production rates, speed up grade transitions, and allow the plant to run more consistently within a grade with less variance. APC installations can be realized with the software of a commercial APC vendor.

Figure 4.12 shows the reduction of MFR variance when running with APC for a certain grade. This allows the plant to run at either the upper or the lower limit of the specification with economic benefits.

References

1 Mähling, F.-O., Klimesch, R., Schwibach, M., Buback, M., and Busch, M. (1999) *Chemie-Ingenieur-Technik*, **71**, 1391.
2 Kiparissides, C., Verros, G., and McGregor, J.F. (1993) *Journal of Macromolecular Science: Reviews in Macromolecular Chemistry & Physics*, **C33**, 437.
3 Schmidt, C.-U., Busch, M., Lilge, D., and Wulkow, M. (2005) *Macromolecular Materials and Engineering*, **290**, 404.

Further Reading

1 Buchelli, A., Call, M., Brown, A., Bokis, C., Ramanathan, S., and Franjione, J. (2002) Physical properties, reactor modeling and polymerization kinetics in the low-density polyethylene tubular reactor process. *Industrial & Engineering Chemistry Research*, **41** (5), 1017–1030.
2 Castañeda, D.M., Finette, A.A., and Wolf, C. (2008) Evaluation of the high pressure separation of LDPE/supercritical ethylene based on PC-SAFT equation of state in combination with the simulation of a production plant of LDPE. Basell Polyolefine GmbH, Wesseling.
3 Crossland, B., Bett, K.E., Ford, H., and Gardner, A.K. (1986) Review of some of the major engineering developments in the high pressure polyethylene process 1933–1983. *Proceedings of the Institution of Mechanical Engineers Part A: Power and Process Engineering*, **200** (4), 237–253.
4 Finette, A.A. and ten Berge, G. (2004) Basell Lupotech T technology for LDPE and EVA-copolymer production, in *Handbook of Petrochemical Production Processes* (ed. R.A. Meyers), McGraw-Hill Professional Publishing, Chapter 14.8.
5 Gropper, H. (1980) Polyolefine: LD-Polyäthylen, in *Ullmanns Encyclopädie der Technischen Chemie. Band 19: Polyacryl Verbindungen bis Quecksilber*, 4th edn, vol. 19, Verlag Chemi, Weinheim, p. 167.
6 Klimesch, R., Littmann, D., and Mähling, F.O. (2004) Polyethylene: high pressure, in *Encyclopedia of Materials: Science and Technology* (eds K.H.J. Buschow, R.W. Cahn, M.C. Flemings, B. Ilschner, E.J. Kramer, and S. Mahajan), Elsevier Science, New York, pp. 7181–7184.
7 McHugh, M. and Krukonis, V. (1994) *Supercritical Fluid Extraction: Principles and Practice*, 2nd edn., Butterworth-Heinemann, Boston.
8 Mirra, M. (2004) POLIMERI Europa polyethylene high-pressure technologies, in *Handbook of Petrochemical Production Processes* (ed. R.A. Meyers), McGraw-Hill Professional Publishing, Chapter 14.5.
9 Pertsinidis, A., Papadopoulos, E., and Kiparissides, C. (1996) Computer aided design of polymer reactors. *Computers Chemical Engineering*, **20** (Suppl.), S449–S454.
10 Schuster, C.E. (2004) EXXONMOBIL high-pressure process technology for LDPE, in *Handbook of Petrochemical Production Processes* (ed. R.A. Meyers), McGraw-Hill Professional Publishing, Chapter 14.4.

5
High-Pressure Homogenization for the Production of Emulsions

Heike P. Schuchmann, Née Karbstein, Lena L. Hecht, Marion Gedrat, and Karsten Köhler

5.1
Motivation: Why High-Pressure Homogenization for Emulsification Processes?

High-pressure homogenizers (HPHs) are broadly used in technical emulsification, especially in the pharmaceutical, cosmetic, chemical, and food industry. Milk, dairy products, fruit juices, and concentrates are examples of high-pressure homogenized commodities. Functional products or formulations containing encapsulated, dissolved, or emulsified bioactives are high-pressure homogenized specialties. Submicrometer-sized emulsion droplets as templates in hybrid particle formation or as nanoreactors for well-defined chemical reaction conditions present future applications already in research. However, high-pressure homogenization processes require high energy input, investment, and maintenance costs. Furthermore, products and active ingredients are subjected to intense mechanical and thermal stresses. Both disadvantages are accepted for several reasons, making high-pressure homogenizers unique in diverse applications:

- Compared to chemical–physical methods as the phase inversion temperature (PIT) method as used, for example, in the cosmetic industry, the high-pressure homogenization can be applied for a wide range of ingredients. It is extremely flexible in terms of phase viscosities and interfacial tension.
- Compared to other mechanical emulsification apparatus as stirred vessels, colloid mills, or gear rim devices, the mechanical stresses are significantly increased. This is important for effective droplet disruption, especially if the submicrometer- and nanometer-sized range is targeted for low viscous products.
- Compared to ultrasound emulsification, the high-pressure homogenizers are able to realize similar local stresses, abrasion is reduced, heat control is easier, and continuous-process solutions at volume streams of up to 50 000 l/h exist, which is the basis for industrial applications.

5.2
Equipment: High-Pressure Homogenizers

5.2.1
Principal Design

High-pressure homogenizers were developed approximately 100 years ago [1] during the industrialization. The basic idea of combining a high-pressure pump and a disruption system, like a valve, was presented on the universal exposition in Paris and endures to date. However, we see ongoing developments of the pumps and valves motivated either by daily application problems or new product challenges.

Current technique permits volume streams of up to 50 000 l/h and pressures up to 10 000 bar; however, homogenization pressures in industrial applications today are in the range of 50–2000 bar. Mainly piston pumps serve as high-pressure pumps. In bench scale, commonly a single piston pump delivers the volume stream or rather the pressure, but in production plants up to eight piston pumps are found. The single piston pump has the disadvantage that the pressure and the volume stream vary extremely over timescale, which results in a pulsation of the stresses on the product. Inhomogeneous stresses act on the droplets and thus product properties are difficult to control. To reduce the pulsation, several pistons are combined phase-shifted. Valves are used to control the different pistons usually not influencing the quality of the emulsion.

5.2.2
Disruption Systems for High-Pressure Homogenization

The main part of a high-pressure homogenizer is the disruption system. Here, the pressure built up by the pump is expanded resulting in specific flow conditions used for droplet disruption. The disruption systems that are available on the market can be divided into valves and nozzles.

5.2.2.1 Valves
Valves, also known as radial diffusers, are common high-pressure disruption systems. The fundamental idea is to reduce the flow diameter with a valve plunger that is pushed to a valve seat forming a small gap. The fluid enters the valve via a central hole in the valve seat pumped by the high-pressure pump. It is then deflected by 90° and flows through a radial gap between the valve plunger and the valve seat. Often flat valves are fitted with an impingement ring to deflect the fluid a second time before it leaves through a drain hole. The impingement ring forms a defined outlet cross section and protects the valve housing against damaging fluid mechanics, like cavitation.

Figure 5.1a depicts an old-generation flat valve. In this valve type, the parallel area between the plunger and the valve seat, called land, is larger compared to the land in the newer flat valve system (see Figure 5.1b). The reduction of the land size reduces

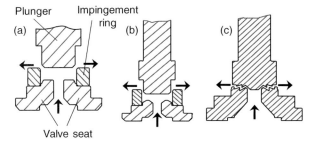

Figure 5.1 (a) Old-generation flat valve. (b) New-generation flat valve. (c) Liquid whirl valve (APV, 2010, Types of SPX APV valves, personal communication).

the flow resistance, which results in decreased pressure before the valve as well as increased turbulence and cavitation in its outlet. Furthermore, the valve seat is cone-shaped being responsible for elongational stresses in the inlet improving droplet disruption [2]. The liquid whirl depicted in Figure 5.1c deflects the fluid stream between the plunger and the seat several times, which enhances the shear stresses on the droplets.

The advantage of using valve systems is that the pressure can be easily adjusted independent of the volume stream. This can be realized by varying the plunger's position and thus the gap size between the valve seat and plunger. Furthermore, the pressure can be adjusted automatically to the volume stream by using a spring to pressure the plunger. Flow conditions within flat valves were intensely investigated and published in Refs [2–6].

5.2.2.2 Orifices and Nozzles

An orifice presents the technically simplest possibility to increase the pressure and transfer this energy into high-flow velocities (Figure 5.2a) [7]. In contrast to valves, simple orifices are constructed without any movable part. This has the advantage that orifices are easy to manufacture. At constant viscosity of the emulsion, the

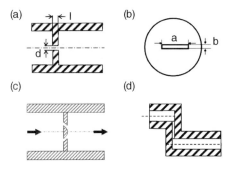

Figure 5.2 Schematic drawing of a simple orifice (a) and modified types (b–d) as published in Refs [2, 10, 11].

homogenizing pressure is adjusted by the volume stream or the orifice hole diameter, respectively, and the cross-sectional area. Increasing the volume stream at target pressure loss requires a numbering up of the orifices as realized, for example, in Refs [8, 9]. Numbering up is only limited by a minimum distance between the holes being in the order of 10 [10]. To ensure a constant homogenizing pressure even for fluctuating volume flow rates, the number of orifices also has to be automatically adapted.

Bayer (now BTS) was among the firsts to patent and commercialize orifices for high-pressure homogenization applications [12]. Basic research on the flow conditions in orifices of circular cross section and their effect on droplet disruption were first published in Ref. [11]. This was followed by intense research in several groups also resulting in several patents [2, 10, 11, 13–22]. Typical dimensions of the orifice are hole diameters of $d = 0.1–1$ mm and a thickness of $l = 0.4$ mm to several millimeters. By modification of the orifice, the so-called nozzles were developed. In Figure 5.2b, a trench is used instead of a hole. The trench has the advantage that just the smaller edge has an impact on the homogenization result and, thus, the larger one can be used for an increase of the cross-sectional area and thus the volume flow rate [10]. This is limited only by the production accuracy of the smaller edge. Impinging the free jet that develops in the orifice's outlet section on a plate or a second liquid jet (Figure 5.2c) improves the droplet breakup by inducing turbulent disturbances [10]. Similar effects are found for orifices with internal steps deflecting the flow (Figure 5.2d) [10, 13, 23, 24].

5.2.3
Flow Conditions

5.2.3.1 **Flow Conditions in the Disruption System**
Generally, a laminar flow is found after the high-pressure pump. Due to the reduction of the cross-sectional area in front of a disruption system, the stream is accelerated and elongated, which results in elongational and shear stresses. From a critical homogenization pressure, the stream detaches on the inlet edge and thus produces the first depression area. In this depression area, cavitation may occur. Furthermore, the detaching of the flow depicts the instability in the stream and may also induce a turbulent transition or a back flow area. Inside the hole, the core of the stream stays laminar, but on the boundaries first eddies can rise.

Within a short orifice-type valve, elongational and later shear stresses dominate the flow. On the outlet edge of the hole, the flow detaches. This detaching results again in a depression area and cavitation. Depending on the outlet geometry, a free jet develops. At the boundaries of the free jet, transitional or turbulent flow regions may develop and induce shear, elongational, and rotational stresses. Boundary effects in smaller outlet channels also influence the turbulence and cavitation behavior [25].

Geometric parameters influence local flow conditions. In flat valves, as designed to date, the inlet edge has no 90° but rather around 45°. This extends the time of the elongational flow in the inlet and reduces the effects of the detaching on the hole's inlet. Thus, a reduction of the abrasion can be achieved and elongation acts for a longer period on droplets. An additional deflection of the fluid inside the hole (Figure 5.2d) results in an additional detaching point and therefore higher turbulences and cavitation around the valve [6, 26, 27].

5.2.3.2 Effect of Flow Conditions in Homogenization Valves on Emulsion Droplets

Droplets are deformed and disrupted by tensions, which result from different flow conditions and act on their interfaces. The high-pressure disruption system creates the required local flow conditions. The resulting flow conditions also depend on the emulsion's material parameters like the viscosity of the phases or the viscosity ratio between droplets and continuous phase, respectively [19, 21]. Laminar shear flow, elongational flow, as well as the turbulent flow and cavitation-induced microturbulences are usually found in industrial homogenization valves.

Droplet disruption due to laminar shear flow has been widely investigated [19, 20]. However, it is restricted to a narrow range of viscosity ratios (between disperse and continuous phase η_d/η_c for single droplet disruption or between the disperse phase and the emulsion η_d/η_e for emulsions, respectively [21]. Laminar elongational flow is advantageous if high-viscous disperse phases have to be disrupted [22]. It is usually found in the inlet of homogenization valves. Specific valves are designed for increased elongational inlet flow.

In the turbulent flow, as found within the flat valves or in the free jet after orifice-type valves, droplets are deformed and disrupted mostly by inertial forces that are generated by energy-dissipating small eddies. Besides, droplets may be stretched and broken up also by shear and elongational forces within the regions of bigger eddies (turbulent viscous regime) [28]. Due to internal viscous forces, droplets try to regain their initial form and size [29].

Cavitation is a huge challenge for the service life of homogenization system due to the abrasion induced, but is also effective in disrupting emulsion droplets. By collapsing of the vapor or gas bubbles and induced microjets pressure, fluctuations and microturbulences are produced that result in the necessary stresses for the droplet disruption. The cavitation can be influenced significantly by a back pressure that can be produced by a second homogenization stage. For this, a simple counterpressure valve or a second disruption system can be applied.

5.2.4
Simultaneous Emulsification and Mixing (SEM) Systems

A simultaneous emulsification and mixing (SEM) system combines two process steps within one disruption system: emulsification and mixing. In a SEM system, one stream (Figure 5.3, (a) in pictogram 1) passes valve, nozzle, or orifice at high pressures and is therefore responsible to produce flow patterns, which are necessary for droplet disruption (emulsification). The second or mixing stream (Figure 5.3, (b) in pictogram 1) is added directly or some millimeters behind the disruption system into the emulsification stream. This means that the mixing stream enters directly into the zone of droplet disruption. In case of simple nozzles used in the emulsification stream, the mixing stream enters directly into the free jet after the nozzle. With this, the mixing stream enters in the zone, where new droplets are formed and thus influence either their formation or their stabilization, or both, depending on its location. Local turbulences in the free jet are responsible for high local micromixing efficiency [30]. Construction details are published in Refs [30–32]. The location and angle at which the mixing stream enters not only decide about the mixing time and efficiency but also influence the droplet disruption.

SEM systems offer the possibility to use the disperse phase fraction in the emulsification and mixing stream as an additional process parameter. In general, the disperse phase fraction can be in the range of 0–100% in both streams (see Figure 5.3). When both streams have the same phase content of 0% or 100%, no emulsion will be produced. If the three concentrations of $\varphi = 0\%$ (pure continuous phase), $\varphi = 100\%$ (pure disperse phase), and $0\% < \varphi < 100\%$ (premix) are distinguished, seven operational modes are possible and described below for o/w emulsions.

Operational modes 1 and 2, depicted in Figure 5.3, use a continuous-phase stream to produce the disruptive stresses and mix a premix or the pure disperse phase into it. In the operational modes 3–5, an emulsion premix is the emulsification stream and a continuous phase, premix, or disperse phase, is mixed into it. The pure disperse phase produces the disruptive stresses in operational modes 6 and 7 and a continuous phase or premix is mixed into it. The operational modes 2 and 6 have the advantage that no premix needs to be prepared prior to the process.

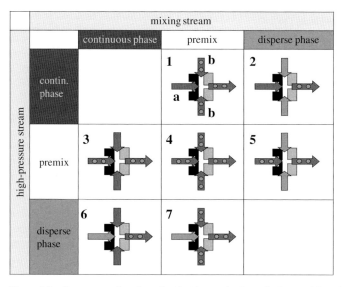

Figure 5.3 Seven operational modes for the production of o/w emulsions in SEM valves.

5.3
Processes: Emulsification and Process Functions

5.3.1
Droplet Disruption in High-Pressure Valves

Droplets are disrupted, if they are deformed over a period of time t_{def} that is longer than a critical deformation time $t_{def,cr}$ and if the deformation, described by the Weber number, We, exceeds a critical value [19]:

$$We > We_{cr} \quad \text{and} \quad t_{def} > t_{def,cr} \tag{5.1}$$

with

$$We = \frac{\sigma}{p_c} \tag{5.2}$$

$$p_c = \frac{4\gamma}{x} \tag{5.3}$$

$$t_{def,cr} = \frac{\eta_d}{\sigma - p_c} \tag{5.4}$$

Herein, σ is the external tension, responsible for droplet deformation, p_c is the capillary pressure, γ is the interfacial tension, x is the droplet diameter, and η_d is its viscosity.

As the external tensions and the deformation time cannot be calculated for commercial valves used in industrial processes to date, measurable process parameters have to be found to characterize the droplet disruption, as the volumetric energy density or specific disruption energy E_V proposed by Refs [33, 34]. In the general definition as given in Eq. (5.5), E and P are the energy and power, respectively, being supplied by the emulsification machine, V is the emulsion volume, $\dot{V}$ is the emulsion volume throughput, and t_{res} is the residence time within the zone of disruptive stresses. In the specific disruption energy E_V, the applied power is related to the volume stream, both parameters being measurable in industrial processes.

In high-pressure systems, the disruption power is the homogenization pressure multiplied by the volume stream $\dot{V}$. Therefore, the specific disruption energy is simply measurable by using the pressure difference Δp over the disruption system.

$$E_V = P_V \cdot t_{res} = \frac{E}{V} = P\dot{V} = \frac{\Delta p \cdot \dot{V}}{\dot{V}} = \Delta p \tag{5.5}$$

When passing the emulsion several (n) times through the valve, orifice, or nozzle, this has to be multiplied by the number of passes n. With the specific disruption energy E_V, the Sauter mean droplet diameter $x_{3,2}$ can be calculated in case of well-defined flow conditions by the following process functions:

Laminar shear flow [22]:

$$x_{3,2} \propto E_V^{-1} \cdot f(\eta_d/\eta_e) \quad \text{or} \quad x_{3,2} \propto \Delta p^{-1} \cdot f(\eta_d/\eta_e) \qquad (5.6)$$

Laminar elongational flow [19, 35, 36]:

$$x_{3,2} \propto E_V^{-1} \quad \text{or} \quad x_{3,2} \propto \Delta p^{-1} \qquad (5.7)$$

Isentropic turbulent flow [29] and microturbulences in cavitational flow [37–40]:

$$x_{1,2} \propto E_V^{-0.25 \cdots -0.4} \cdot \eta_d^{0 \cdots 0.75} \quad \text{or} \quad x_{3,2}\infty \propto \Delta p^{-0.25 \cdots -0.4} \cdot \eta_d^{0 \cdots 0.75} \qquad (5.8)$$

Independent of local flow conditions within a high-pressure valve, the achievable mean droplet diameter gets smaller with increasing energy density or pressure difference applied. In defined flow conditions, the exponent of the energy density or pressure difference is either −1 or between −0.25 and −0.4. For industrial homogenization valves, exponents in the range of −0.6 are published, for example, in Ref. [19], depicting the mix of flow conditions found in homogenization valves (see Section 5.2.3).

In commercial homogenization valves, turbulence and cavitation dominate droplet disruption. Here, droplets of a low viscosity are more easily deformed and disrupted than droplets of higher viscosities. Therefore, the overall equation describing the process functions for high-pressure homogenization processes is

$$x_{3,2}\infty E_V^{-b} \cdot \eta_d^c \quad \text{or} \quad x_{3,2}\infty \Delta p^{-b} \cdot \eta_d^c \qquad (5.9)$$

with b being in the range of 0.25–1 and $c < 0.75$.

Using Eq. (5.9), different homogenization valves can be compared. In Figure 5.4, disruption systems presented in Figures 5.1 and 5.2 are compared by producing corn oil droplets in water.

In all cases, the droplet diameter decreases with increasing specific energy input (corresponding to the pressure difference applied). The new generation flat valve geometry results in decreased Sauter mean diameters due to the longer elongation in the inlet as well as higher turbulence and cavitation in the outlet. This in turn decreases the slope of the curves in the double logarithmic diagrams. Results are similar to the simple orifice geometry. The Microfluidizer® geometry, equipped with the double-stage disruption system and applying the back pressure, allows improving homogenization results further. Comparable slopes of the curves depict comparable flow conditions being responsible for droplet breakup.

5.3.2
Droplet Coalescence in Homogenization Valves

In breaking up droplets, a large surface area is created that has to be stabilized against agglomeration, sticking (agglomeration and flocculation), and coalescence (mergers of droplets). To estimate the influence of coalescence and agglomeration on the

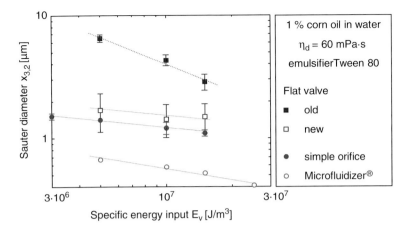

Figure 5.4 Influence of the specific energy input and the geometry of the disruptive system on the Sauter mean diameter. Flat valves of old and new generation types are depicted in Figure 5.1, the simple orifice nozzle in Figure 5.2. The Microfluidizer was equipped with a double-stage interaction chamber, applying additional back pressure. This was not done for the other valves or orifices in this figure.

process, agglomeration or coalescence rates can be applied. The coalescence rate Ω expresses the quantity of droplet coalescences per volume and time and can be calculated by the collision frequency C multiplied by the coalescence probability p_{coal} (Eq. (5.10)) [36].

$$\Omega = p_{coal} \cdot C \tag{5.10}$$

The collision frequency C can be calculated in case of isentropic turbulent flow with Eq. (5.11):

$$C = k \cdot u \cdot x_{3,2}^2 \cdot n^2 \tag{5.11}$$

where k is the velocity constant and u is the droplet velocity, the latter depending on the specific energy dissipation ε. The velocity constant k depends on the flow type, for example, viscous simple shear, fine-scale turbulence, and inertial subrange turbulence and can be calculated as described in Ref. [36]. In Eq. (5.11), note that the Sauter mean diameter $x_{3,2}$ and the number of droplets n have a strong influence on the collision frequency. Detailed information regarding droplet coalescence is found in Ref. [41]. As both the collision rate and the coalescence probability parameters depend on the specific energy input and increase with it, intense droplet coalescence is usually found for high-pressure homogenization processes working at elevated pressures.

Once emulsion droplets are disrupted in a high-pressure nozzle, they will be stabilized by adsorbing emulsifier molecules. This, however, is a process taking some time. This time depends on the emulsifier molecule structure and chemical nature of the phases [42]. The kinetics of the surfactant(s) also influences the probability of coalescence and thus the coalescence rate [28].

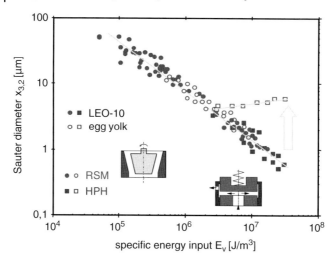

Figure 5.5 Influence of the specific energy input and the emulsifier adsorption kinetics on the emulsification result when using either a rotor–stator machine (RSM) or a high-pressure homogenizer (HPH) [33].

For droplet coalescence, a certain time t_{coal} is necessary to allow drainage of the liquid film between two colliding droplets [41]. In orifice-type high-pressure homogenization, coalescence is found in the orifice outlet only when the free jet relaminarizes and resultantly the contact time between droplets increases [43]. The coalescence rate in high-pressure homogenizer valves thus depends on several material and valve design parameters such as the droplet concentration (disperse phase content), the droplet size after the breakup (breakup conditions), local energy release (droplet velocity), and emulsifier stabilization kinetics and the length of the zones of different energy release (film drainage time). Controlling it is a challenge to date. Thus, it is a challenge to control coalescence.

In Figure 5.5, the achievable droplet diameter in a high-pressure homogenizer is depicted over the specific energy input for two different emulsifiers in comparison to results achieved by producing the same formula in a rotor–stator machine (RSM).

With the fast stabilizing emulsifier LEO-10 (laurylethyleneoxide 10), the homogenization result is dominated by the droplet disruption process and therefore only limited by the specific energy input. At the same specific energy input, the RSM and the HPH achieve comparable homogenization results, with HPH being able to work at higher specific energy inputs. By using slow stabilizing emulsifier molecules as found, for example, in egg yolk – a food grade emulsifier, however, the RSM homogenization result is furthermore dominated by the specific energy input, while it is dominated by coalescence processes in high-pressure homogenizer. This effect can be explained by long residence times (range of seconds) at decreased disruptive stresses in the RSM, which gives the emulsifier enough time to stabilize the new droplet interfaces. In contrast to this, the residence time in high-pressure valves is in the range of milliseconds, while disruptive stresses are higher by several powers.

This results in high coalescence rates and thus in large droplets. This effect increases with increasing specific energy input as more and smaller droplets are created. Thus, the Sauter mean diameter increases with increasing specific energy input. This depicts that high-pressure homogenization is limited to emulsions of either low disperse-phase content or those containing emulsifiers of fast stabilization kinetics or both.

5.3.3
Droplet Agglomeration in Homogenization Valves

The agglomeration rate Ω_{aggl} is described very similar to the coalescence rate (Eq. (5.10)), as found in Refs [44, 45]:

$$\Omega_{aggl} = \beta_{lam/turb} \cdot p_{aggl} \tag{5.12}$$

The collision frequency in isentropic turbulent flow β_{turb} is given by Eq. (5.13) depending on the number of particles n and the shear rate σ [46, 47]:

$$\beta_{turb} = \begin{cases} n_i \cdot n_j \sqrt{8\Pi} \cdot \sigma \cdot R^3 & \text{for } R_i, R_j < \lambda \\ K \cdot n_i \cdot n_j \cdot \Pi \cdot \varepsilon^{1/3} \cdot (R_i + R_j)^{1/3} & \text{for } R_i, R_j > \lambda \end{cases} \tag{5.13}$$

and the agglomeration probability p_{aggl} by Eq. (5.14):

$$p_{aggl} = \left[\frac{18 \cdot \Pi \cdot \eta \cdot R^3}{4A} \right] \tag{5.14}$$

with $K = 1.37$, η is the viscosity, ε is the dissipation rate, and A is the Hamaker constant [36].

Agglomeration of droplets is well known in dairy homogenization processes. Here, fat globules tend to agglomerate, as the emulsifiers available (being mainly caseins) tend to adsorb at several fat globules in the same time and thus form casein bridges. This effect increases with increasing fat content and dominates the homogenization result for butter fat contents above 13% [27].

5.4
Homogenization Processes Using SEM-Type Valves

5.4.1
Dairy Processes

Dairy processes are one of the oldest industrial high-pressure homogenization processes and up to now the ones with the biggest volume streams. In conventional processing, raw milk is separated prior to homogenization into a low-fat phase (0.03–0.3 vol% fat, "skim milk") and a fat-enriched phase (13–42 vol% fat, "cream") using a separator [48]. In the conventional "full stream" homogenization process, milk is first standardized to the final product fat content by mixing these two phases, the product of this is then homogenized by typical homogenization pressures around 100 bar. Also, conventionally applied are "partial stream" homogenization processes, in which the cream is diluted with skim milk to a fat content of 13–17 vol%, homogenized and afterward standardized again to the target fat concentration of, for example, 3.5 vol% in full cream milk. This reduces the required energy as less continuous phase has to be compressed to nearly the same homogenization pressure.

This two-step remixing process interrupted by high-pressure homogenization is required as aggregation of fat globules (casein bridging) is found in homogenized cream with more than 13 vol% of fat, as described in Section 5.3.3. The negative impact of aggregation can be compensated by an increased homogenizing pressure up to fat contents of 17 vol%. At fat contents higher than that, the process is controlled by coalescence and mainly aggregation of the fat globules leading to dissatisfying homogenization results (see Figure 5.6, Δ). Coalescence of newly formed fat droplets is found until adsorbing dairy proteins have stabilized the droplets. In stabilization of milk fat globules, a secondary droplet membrane is built up by adsorbing casein micelles and submicelles as well as lactoalbumins and lactoglobulins [49, 50]. As adsorbed casein micelles tend to adsorb at more than one fat globule at the same time and also strongly interact, bridges between fat globules are formed at increased fat content resulting in high aggregation rates. These fat globule aggregates can be partially destroyed in a second homogenizing stage [51], as it is realized in conventional technical processes. With increasing fat globule concentration, however, coalescence and aggregation rates also increase [30, 52, 53] limiting partial

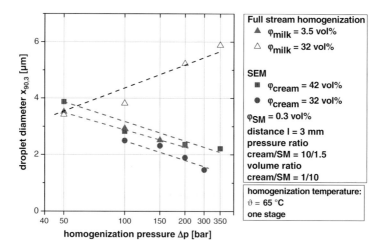

Figure 5.6 Influence of the fat content of homogenized cream and the homogenization pressure on the characteristic maximum droplet diameter $x_{90,3}$ for full stream and partial homogenization processing. When SEM valves are used, cream of 32 and 42 vol% fat were mixed with skim milk SM ($\varphi = 0.3$ vol%) within the valve. In conventional full stream homogenization, milk (volume fat content $\varphi = 3.5$ vol%) and cream ($\varphi = 32$ vol%) were homogenized as "full stream" [32, 54].

homogenization to 17 vol% of fat. Thus, the energy reduction potential cannot be fully exploited.

In the common dairy process, the standardization (which in fact is done by mixing) is located several meters behind the homogenizer. SEM valves enable to (i) combine the homogenization and standardization steps into one process unit, (ii) dilute the fat globules directly after their production, and (iii) add additionally emulsifier molecules (here, dairy proteins) in the moment of their need (droplet breakup) with high mixing intensity. For this, SEM valves are run in operational mode 3 (see Figure 5.3), with a premix, here cream at a fat content of 32–42 vol%, being homogenized as emulsifying stream, and the skim milk runs as mixing stream at pressures being 0.1–20% of the homogenizing pressure.

Comparing the SEM partial homogenization results at 32 and 42 vol% fat, respectively, with those of conventional full stream homogenization at 3.5 vol% fat, product quality is fully maintained by the same homogenization pressure (with a slight, but negligible improvement at 32 vol% and a slight loss in 42 vol% fat) (Figure 5.6).

However, the new SEM process requires only 20% of the energy input compared to the full stream process, and only 60% of the energy is applied in conventional partial homogenization processing at comparable process parameters. This results in considerable energy and cost savings in dairy processing without any loss in product quality. In addition, two mixing units can be eliminated from the process line resulting in less investment, cleaning, and maintenance costs (process intensification).

5.4.2
Pickering Emulsions

In 1903, Ramsden [54, 55] and in 1907 Pickering [55] reported that solid colloidal particles can be used to obtain the stability of emulsions. Particle-stabilized emulsions (PSE), also called Pickering emulsions, have recently regained interest in scientific literature [56] what is partially due to the recent advances in nanotechnology and the associated increased availability of suitable particles. Nanoparticles used for PSE stabilization are mostly inorganic and range in size of 10 nm to some hundred nanometers. These particles are highly abrasive and can damage the plant, especially the pump and valve system, within minutes of processing [57]. Thus, adding particles to emulsion premixes is not trivial, especially if continuous processing is targeted as in high-pressure homogenization.

To prevent the abrasion on the high-pressure pump and valve, Pickering emulsions stabilized by highly abrasive silica nanoparticles can be produced in SEM valves by running pure disperse or continuous phase or an emulsion premix as emulsifying stream and the nanoparticle suspension as mixing stream (operational modes 1, 3 and 6, in Figure 5.3).

In general, Pickering emulsions can be produced using all three operational modes (Figure 5.7). In all of them, high standard deviations were found at low pressures. In operational mode 3, this can be explained by the missing stabilization of the emulsion premix droplets. Segregation and coalescence in the premix on its

5 High-Pressure Homogenization for the Production of Emulsions

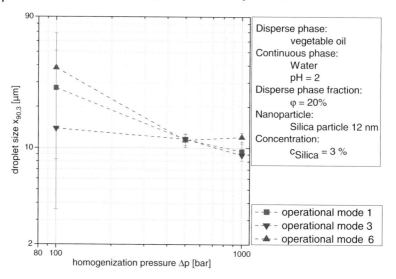

Figure 5.7 Production of Pickering emulsions stabilized by silica nanoparticles of 12 nm diameter using the SEM valve. Comparison of the homogenization results (characteristic value $x_{90,3}$ of droplet size distributions) obtained with the operational modes 1, 3, and 6.

way from the vessel to the orifice result in a high time-dependent fluctuation of the disperse-phase fraction and thus an unstable process. This consequently results in some large droplets, which tend to cream significantly. In order to decrease the fluctuation of the disperse-phase fraction, the two phases can be fed directly in front of the valve, for example, with a static mixer, instead of preparing the Pickering emulsion premix in a separate vessel, like it was done in these experiments. At higher pressures and therefore higher volume throughputs, this problem is also reduced. The samples obtained at higher pressures are smaller in droplet size ($x_{90,3} \approx 10\,\mu m$ at pressures of 500–800 bar) and characterized by low standard deviations.

Operational mode 1 also shows a high standard deviation at low pressures, most probably due to the high viscosity of the Pickering emulsion premix added to the side stream. The mixing stream tended to stick resulting in an instable mixing process. An additional pump or a geometric modification of the SEM valve could improve this operational mode at low pressures. At higher homogenization pressures, this operation also ran at a constant flow and became a stable process.

Operational mode 6 led to nearly the same results compared to mode 1, but ran more stable due to the lower viscosity of the mixing stream (no sticking of the mixing stream) and the pure disperse phase in the high-pressure stream (no phase separation within the inlet).

All three operational modes may be applied and do not significantly differ in the resulting particle-stabilized droplet sizes. Operational mode 6 ran most stable, especially at low homogenizing pressures. Because no premix has to be produced, it seems to be an attractive possibility for future applications.

Nanoparticles for stabilizing Pickering emulsions should not always be inorganic, especially for applications in Life Sciences. Here, organic nanoparticles are highly preferred, especially those that do not need any stabilization by emulsifiers or stabilizers. Using SEM valves such organic, emulsifier-free nanoparticles can be produced by realizing a rapid local temperature drop directly in the homogenization valve.

5.4.3
Melt Homogenization

Organic nanoparticles for PSE applications should be spherical in shape and free of emulsifiers for dense packing at the droplet interfaces they have to stabilize. Thus, a process is proposed where nanoparticles are produced in organic solids of considerable melting temperatures. After melting them, they are emulsified by high-pressure homogenization and stabilized by rapid cooling. In Figure 5.8, results are given for different waxes with melting temperature ranges between 50 and 62 °C, homogenized at pressures between 100 and 1000 bar in the SEM valve. Cold water of 20 °C is used as side stream in order to realize an instant cooling of the droplets after their disruption. The cold water is sucked into the mixing zone by the depressurization directly behind the valve outlet. Resulting product temperatures are below the melting points (40–60 °C depending on process conditions). No emulsifier is required for stabilization.

Figure 5.8 represents cumulative particle size distributions of different waxes after SEM melt homogenization at 130 °C and pressures of 300 bar. Monomodal particle

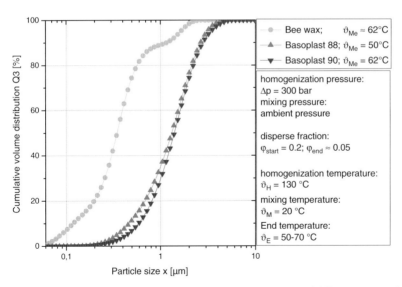

Figure 5.8 Homogenization results for SEM melt homogenization of different waxes with increasing melting temperatures at a homogenization temperature of 130 °C. Stabilization was realized by instant mixing with cold water (20 °C) in the SEM valve [58]. No emulsifier was added.

size distributions with mean values $x_{3,50}$ in the order of 1 μm can be reached with waxes of higher melting points, for example, Basoplast 88 and 90 (melting points: 50 and 62 °C, respectively). Bee wax (a natural wax used, for example, in food and pharmaceutical products for coating microencapsulated actives, melting at approximately 62 °C) even results in smaller particles (mean particle size $x_{3,50} \approx 500$ nm). Most probably this is due to the different natural compounds of the bee wax, which may be interfacial active, even if their concentration is quite low. However, the small difference between the final product temperature after SEM homogenization and the melting point results in some larger particles ($\approx$10 vol% found in particles of 1–2 μm), which could be eliminated by working at different temperatures.

Today, small particles of waxes are commonly produced by wet milling [59, 60]. Stirred ball milling, however, is difficult for these cases as their stickiness limits the application. This limitation can be solved by melt emulsification in general. Melt homogenization, in turn, is usually limited as emulsifiers, required for droplet stabilization, and have to be thermally stable at elevated emulsification temperatures. Using the SEM technology as represented here, droplets can be stabilized by instant cooling only, and thus the emulsifier problem does not arise.

A disadvantage of the SEM technology results from the fact that a second cold stream is required for instant droplet cooling, which dilutes the homogenized emulsion and thus reduces the product yield. For this, the resulting particle suspension has to be concentrated, with the possibility to recycle the continuous phase [58, 61]. However, instant diluting reduces droplet collision rates. Thus, melt homogenization can be run at elevated dispersed phase fractions (as for cream). Additional energy required for melting the products has to be taken into account considering processing costs. The reduction in processing time and energy compared to stirred ball milling, however, is significant and compensates this by far.

5.4.4
Emulsion Droplets as Templates for Hybrid (Core–Shell) Nanoparticle Production

Core–shell nanoparticles (CSNs) can be produced by miniemulsion polymerization [62, 63]. So far the research on the preparation of core–shell nanoparticles via miniemulsion polymerization in laboratory scale has made immense progress in the last years. There are several different applications for polymer-coated particles, for example, for medical products, paints, and catalysts. The polymer coating prevents the particles from agglomerating that results in a better color intensity or in an improvement of catalytic performance [64]. In medical applications, organic coating is required to depress toxic reactions [65].

The complete miniemulsion polymerization process, as can be seen in Figure 5.9, requires first a homogeneous dispersion of the nanoparticles in the monomer. This nanoparticle suspension is then emulsified in water and homogenized to a target droplet size $d < 500$ nm. In the following process step, the resulting droplets are polymerized, keeping their size, shape, and inner structure (especially the particle load). However, the breakup of nanoparticle-loaded droplets is a nontrivial task. The nanoparticle load is leading to an increase of viscosity and a complex rheological

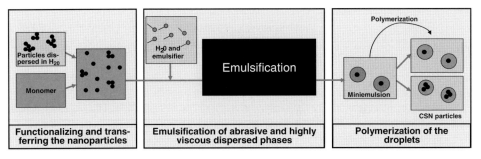

Figure 5.9 Flow chart for the production process of core–shell nanoparticles (CSN) via miniemulsion polymerization.

behavior [66], which is strongly influencing deformation and breakup of the droplets [10, 67]. In addition, inorganic nanoparticles, especially metallic oxides as investigated here, are likely to cause strong abrasion leading to short dispersing unit (mainly valve) lifetime and high maintenance costs [57].

The established emulsifying process for preparing a submicrometer-sized monomer-in-water emulsion in lab scale is ultrasonication in a batch mode [68–70]. Ultrasound, however, deals with several disadvantages in process scale-up, for example, locally poorly controllable conditions in droplet breakup resulting in broad and often bimodal droplet size distributions. High abrasion due to cavitation leads to short process unit lifetimes and product contamination. Furthermore, controlling product temperature in scale-up is quite difficult as locally heating rates are rather high [40, 71]. High-pressure homogenization may overcome these problems and result in miniemulsions of comparable size range.

Figure 5.10 gives an example. Monomers (methylmethacrylate) at a dispersed phase content of 20% were high-pressure homogenized in water. Adding the corresponding poly(methylmethacrylate) (PMMA) of different molecular weight to the monomer in concentrations ranging from 0% to 34% simulates the effect of an increased viscosity due to a nanoparticle load with only very little change in chemical composition and resulting physical characteristics (interfacial tension and droplet density). With an increasing viscosity, a significant increase in droplet and thus particle size after polymerization can be observed. But even for high-viscous droplets, a mean Sauter diameter below 500 nm can be reached at reasonable homogenization pressures (<1000 bar). Furthermore, it can be seen that the SEM geometry that allows operating the process with heavily decreased abrasion (see Section 5.2.4) also results in droplet size distributions comparable to those achieved by standard valve geometries, like a simple orifice valve.

The particle size distribution after polymerizing a 5% SiO_2 particle in monomer suspension in water is depicted in Figure 5.11. Particles range in the size of 100 nm and distributions are rather narrow. As can be seen in the transmission electron microscopy (TEM) picture, the silica nanoparticles are encapsulated in PMMA. As the particle content of the suspension was low, some polymer particles remain nonfilled.

114 | 5 High-Pressure Homogenization for the Production of Emulsions

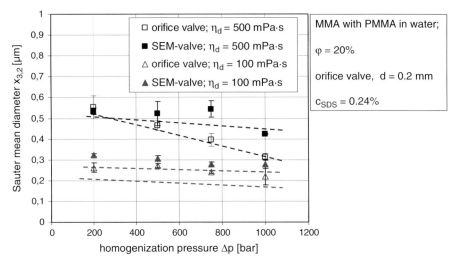

Figure 5.10 Sauter mean diameter for different viscosities of the disperse phase and different valve geometries. SEM system was applied in operating mode 1. Droplet size distributions were measured by dynamic light scattering.

5.4.5
Submicron Emulsion Droplets as Nanoreactors

Nanoparticles enjoy great popularity in chemical, pharmaceutical, cosmetic, and textile industry due to their unique properties related to the small particle size ($x < 100$ nm) and high specific surface area. Nevertheless, the control of nucleation

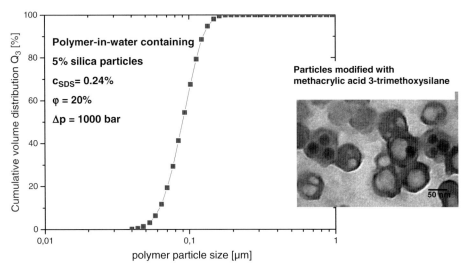

Figure 5.11 Particle size distribution measured with static laser light diffraction of PMMA particles filled with SiO_2 nanoparticles and TEM picture of the core–shell nanoparticles.

and growth of nanoparticles is still a challenge with so far applied processes. Ideal reactors take control of both particle nucleation as well as particle growth and furthermore allow a simple scale-up range in the submicrometer size. These are found in miniemulsions (Sauter mean diameter $d_{3,2} = 100–1000$ nm) of the water-in-oil type where each submicrometer-sized droplet serves as a nanoreactor.

There are two process paths to precipitate nanoparticles with the miniemulsion technique. The first requires the production of two water-in-oil emulsions where the dispersed phases are loaded with one precipitation precursor, respectively. Controlled coalescence of droplets containing the precipitation reactants results in a product of low solubility and – in case of supersaturation – homogeneous nucleation is induced. As high-pressure homogenization results in high coalescence rates (see Section 5.3.2), this is the process of choice.

The second process path involves the preparation of one emulsion. A water-soluble reactant is provided within the emulsion droplets, whereas an oil as well as a water-soluble reactant is mixed into the emulsion and the precipitation reaction is induced once the reactant diffuses via the droplet interface into the emulsion droplet. This process is considered to be controlled by mass transfer [72]. Mixing efficiency may influence the particle size and structure of the final products in precipitation reactions [73]. Influencing mixing efficiency and time is realized by SEM nozzles.

In Figure 5.12 an example is given for zinc oxide precipitated with the mass transfer-controlled process path under different mixing conditions. Both reactants were provided in a molar concentration yielding complete precursor consumption. The continuous phase, 57 w/w% of the emulsion, consisted of n-decane, whereas the watery precursor suspension served as dispersed phase (40 w/w%). The miniemulsions were stabilized by the emulsifier Glissopal® EM-23.

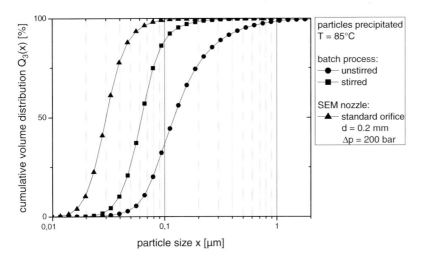

Figure 5.12 Influence of mixing efficiency on resulting nanoparticle size using miniemulsion droplets as nanoreactors for the precipitation. The precipitation agent is introduced to the emulsion without stirring or under stirring. These results are compared to a particle size distribution obtained within the SEM nozzle allowing an integrated processing.

In case of precipitating particles with a batch-like procedure, the emulsion was prepared with a high-pressure device (M – 110Y Microfluidizer, Microfluidics, United States). Following the high-pressure homogenization, the oil-soluble reactant was added to the emulsion (stirred/unstirred) to induce the precipitation reaction and the particle formation. For the precipitation of the particles with the SEM nozzle, the ratio of premix and n-decane/reactant suspension was kept at a mixing ratio of 1 : 1. The premix was pumped through the disruption system, whereas the n-decane/oil-soluble reactant suspension was added after the disruption unit. The material composition and the experimental procedure were adjusted to obtain identical emulsion composition, droplet size distribution, and reactant concentration in the resulting emulsion compared to processing in stirred/unstirred vessels.

Results presented in Figure 5.12 clearly depict the importance of mixing time for mass-controlled nanoparticle precipitation in miniemulsion droplets. High-pressure homogenization using a SEM-type nozzle allows creating the droplets (nanoreactors) and – in the same process unit – homogeneous adding and distributing the reactant in the emulsion. This leads to the formation of nanoparticles being smaller and having a narrower particle size distribution compared to particle sizes obtained with a conventional two-step process using batch-like mixing procedures.

5.4.6
Nanoparticle Deagglomeration and Formulation of Nanoporous Carriers for Bioactives

Dispersing particles homogeneously in a solution is a key requirement for a great variety of industrial products and processes, including pigments in coating applications or fillers in polymers. Especially for nanoparticles, this often is a challenge as they often are highly aggregated due to van der Waals interactions and sintering

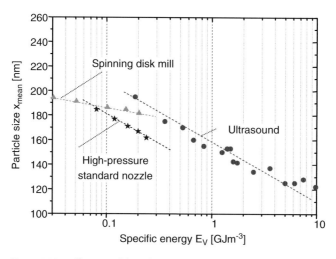

Figure 5.13 Efficiency of deagglomeration in nanoparticle dispersion using different dispersing systems.

bonds, requiring the exertion of high stresses for deagglomeration and deaggregation. High-pressure dispersing [74] is a promising new technology being effective in deagglomeration and reducing process time and energy (see Figure 5.13 taken from Ref. [75]) compared to conventional processing as stirred ball milling, roller milling, using dissolvers, kneaders [76], or spinning disk mills [75] or applying ultrasound [77]. Homogenizing pressure and nozzle geometry are found to be key process parameters in high-pressure dispersing [74]. Typical pressures required for dispersing nanoparticles are in the range of 500–3000 bar. Nozzles of simple hole geometry can be applied with increased throughput and deagglomeration efficiency at increased nozzle diameter. However, significant abrasion is typically found when inorganic nanoparticles are dispersed in high-pressure nozzles. In order to overcome this drawback, a HPPF (high-pressure postfeeding) nozzle configuration similar to the SEM disruption system was developed for dispersing inorganic nanoparticles such as silica, alumina, or titanic oxides in aqueous phases enhancing nozzle service life and thus enabling industrial applications [57].

As it is found for homogenized droplets, deagglomerated nanoparticles usually have to be stabilized in order to prevent their reagglomeration. Agglomeration is enhanced as nanoparticles are characterized by an extremely high surface area. This results in an increase of van der Waals forces in relation to competing interactions as gravitational forces. For this reason, in commercial applications, additives are used, which adsorb at the surface and by this induce steric and/or electrostatic interaction forces. Regarding alumina oxide nanoparticles, for example, stability can also be induced by electrostatic interparticle forces only. For this material matter, reagglomeration is found in resting solutions around the isoelectric point IP and up to zetapotentials of ± 30 mV. In case of dynamic situations, as found in dispersion nozzles, high collision rates and speeds result in even worse conditions in terms of particle stability. Reagglomeration of dispersed aluminum oxide nanoparticles was found even for zeta-potentials up to IP ± 80 mV [78]. This could be effectively used for controlled reagglomeration within an HPPF-type nozzle resulting in nanostructured agglomerates of different sizes. For this, a modified HPPF nozzle was designed allowing to feed pH-controlling solutions at different locations after the nozzle outlet [77]. With this, pH values could be changed at different locations within the free jet after the nozzle outlet, starting the reagglomeration process. Depending on the location of pH change, agglomerate sizes can be varied ranging from 100 nm to 10 µm. For details, see Ref. [75]. Active ingredients or carrier molecules (e.g., cyclodextrin-based ones) being dispersed or dissolved in the continuous phase in the moment of reagglomeration could be successfully embedded in the nanostructured agglomerates [77].

5.5
Summary and Outlook

High-pressure homogenizers are commonly used today in the chemical, pharmaceutical, and cosmetic as well as the food industry. They are equipped typically with one–eight piston pumps to produce the pressure required and with a disruption system to expand the fluid and disrupt emulsion droplets. The most common disruption systems are valves. New-generation flat valves have been optimized for efficient droplet breakup at high throughput. The volume stream can be adjusted independent of the homogenization pressure and the risk of blocking is rather low. Orifices and nozzles often achieve better results in terms of smaller droplet diameters, especially for low-viscous emulsions, but the scale-up to date is limited. Depending on the geometry of the disruption systems, different flow patterns dominate the droplet breakup and thus the parameters available for process control.

Coalescence and agglomeration are a challenge to be met in high-pressure homogenization processes. Droplets are disrupted within milliseconds and emulsifiers available often do not meet the target in adsorption and stabilization kinetics. When stabilization can be ensured, as often found when short-chain low-molecular weight emulsifiers are used, the homogenization result is dominated by the energy input or the homogenizing pressure, respectively. Otherwise, agglomeration and coalescence strongly control high-pressure homogenization processes. If this is the case, the disperse phase content and the emulsifier molecular structure are the parameters controlling homogenization results. The high-pressure homogenization can then not be applied for high concentrated emulsions. In foods, this is found when natural emulsifiers as dairy proteins or egg yolk are used as emulsifiers.

With simultaneous emulsification and mixing valves, these problems can be eliminated. Furthermore, energy savings of up to 90%, for example, in dairy homogenization, can be realized. SEM valves also allow the production of hybrid particles as Pickering emulsions or core–shell nanoparticles via high-pressure homogenization processes. The high-pressure homogenizers may then be used even for the deagglomeration of nanoparticles produced by flame pyrolysis (as inorganic metal oxides) and they can also be used as carrier systems for actives with retard effect. These examples depict the broad range of applications for high-pressure homogenization processes in future.

References

1 Gaulin, A. (1899) Appareil et Procédé pour la Stabilisation du Lait, Patent Brecet No. 295596.

2 Freudig, B. (2004) Herstellen von Emulsionen und Homogenisieren von Milch in modifizierten Lochblenden. Dissertation. Universität Karlsruhe (TH).

3 Phipps, L.W. (1974) Cavitation and separated flow in a simple homogenizing valve and their influence on the break-up of fat globules in milk, *Journal of Dairy Research*, **41** (1), 1–8.

4 Kiefer, P. and Treiber, A. (1975) Prall und Stoß als Zerkleinerungsmechanismen bei der Hochdruck-Homogenisation von O/W-Emulsionen. *CIT*, **47** (13), 573.

5 Treiber, A. and Kiefer, P. (1976) Kavitation und Turbulenz als Zerkleinerungsmechanismen bei der Homogenisation von O/W-Emulsionen. *Chemie Ingenieur Technik*, **48** (3), 259.

6 Innings, F. and Tragardh, C. (2005) Visualization of the drop deformation and break-up process in a high pressure homogenizer. *Chemical Engineering & Technology*, **28** (8), 882–891.

7 EN ISO 5167-1: *Durchflussmessung von Fluiden mit Drosselgeräten in voll*

durchströmten Leitungen mit Kreisquerschnitt Teil 1: Allgemeine Grundlagen und Anforderungen (ISO 5167-1:2003); Deutsche Fassung EN ISO 5167-1:2003, 2003.

8 Bayer, A.G. (1997) Verfahren und Vorrichtung zur Herstellung einer parenteralen Arzneistoffzubereitung, Patent No. WO9717946.

9 Bayer, A.G. (2001) Dispersion Nozzle with variable throughput. Patent. No. WO 01/05517 A1

10 Aguilar, F.A., Köhler, K., Schubert, H., and Schuchmann, H.P. (2008) Herstellen von Emulsionen in einfachen und modifizierten Lochblenden: Einfluss der Geometrie auf die Effizienz der Zerkleinerung und Folgen für die Maßstabsvergrößerung. *Chemie Ingenieur Technik*, **80** (5), 607–613.

11 Stang, M. (1998) Zerkleinern und Stabilisieren von Tropfen beim mechanischen Emulgieren. Dissertation. Universität Karlsruhe (TH).

12 Bayer A.G. (1991) Preparation of pharmaceutical or cosmetic dispersions, Patent No. US 4,996,004, Feb. 26.

13 Penth, B. (2000) Method and device for carrying out chemical and physical processes, PCT/DE2000/001061, Patent No. WO/2000/061275.

14 Tesch, S. (2002) Charakterisieren mechanischer Emulgierverfahren: Herstellen und Stabilisieren von Tropfen als Teilschritte beim Formulieren von Emulsionen. Dissertation. Universität Karlsruhe (TH).

15 Kolb, G.E. (2001) Zur Emulsionsherstellung in Blendensystemen. Dissertation. Universität Bremen.

16 Muschiolik, G., Roeder, R.-T., and Lengfeld, K. (1995) Druckhomogenisator, Patent No. DE 1,953,0247, Aug. 17.

17 Cook, E.J. and Lagace, A.P. (1985) Apparatus for forming emulsions, biotechnology development, Patent No. 4,533,254.

18 Floury, J., Bellettre, J., Legrand, J., and Desrumaux, A. (2004) Analysis of a new type of high pressure homogeniser: a study of the flow pattern. *Chemical Engineering Science*, **59** (4), 843–853.

19 Walstra, P. (1983) Formation of emulsions, in *Encyclopedia of Emulsion Technology*, vol. 1 (ed. P. Becher), Marcel Dekker Inc., pp. 57–128.

20 Stone, H.A., Bentley, B.J., and Leal, L.G. (1986) An experimental-study of transient effects in the breakup of viscous drops. *Journal of Fluid Mechanics*, **173**, 131–158.

21 Armbruster, H. (1990) Untersuchungen zum kontinuierlichen Emulgierprozeß in Kolloidmühlen unter Berücksichtigung spezifischer Emulgatoreigenschaften und der Strömungsverhältnisse im Dispergierspalt. Dissertation. Universität Karlsruhe (TH).

22 Grace, H.P. (1982) Dispersion phenomena in high-viscosity immiscible fluid systems and application of static mixers as dispersion devices in such systems. *Chemical Engineering Communications*, **14** (3–6), 225–277.

23 Cook, E.J. (1985) Microfluidizer (Teil I), Patent No. 4,533,254.

24 Cook, E.J. (1990) Microfluidizer (Teil II), Patent No. 4,908,154.

25 Schlichting, H. and Gersten, K. (2006) in *Grenzschicht-Theorie* (eds H. Schlichting and K. Gersten), Springer, Berlin.

26 Casoli, P., Vacca, A., and Berta, G.L. (2010) A numerical procedure for predicting the performance of high pressure homogenizing valves. *Simulation Modelling Practice and Theory*, **18** (2), 125–138.

27 Köhler, K., Karasch, S., Schuchmann, H.P., and Kulozik, U. (2008) Energiesparende Homogenisierung von Milch mit etablierten sowie neuartigen Verfahren. *Chemie Ingenieur Technik*, **80** (8), 1107–1116.

28 Vankova, N., Tcholakova, S., Denkov, N.D., Ivanov, I.B., Vulchev, V.D., and Danner, T. (2007) Emulsification in turbulent flow, 1. Mean and maximum drop diameters in inertial and viscous regimes. *Journal of Colloid and Interface Science*, **312**, 363–380.

29 Arai, K., Konno, M., Matinaga, Y., and Saito, S.J. (1977) Effect of dispersed-phase viscosity on the maximum stable drop size for breakup in turbulent flow. *Chemical Engineering of Japan*, **10** (4), 325–330.

30 Köhler, K., Aguilar, F.A., Hensel, A., Schubert, K., Schubert, H., and Schuchmann, H.P. (2007) Design of a microstructured system for homogenization of dairy products with high fat content. *Chemical Engineering & Technology*, **30** (11), 1590–1595.

31 Köhler, K., Aguilar, F.A., Hensel, A., Schubert, H., and Schuchmann, H.P. (2009) Design of a micro-structured system for the homogenization of dairy products at high fat content. Part III: Influence of geometric parameters. *Chemie Ingenieur Technik*, **32** (7), 1120–1126.

32 Köhler, K., Aguilar, F.A., Hensel, A., Schubert, K., Schubert, H., and Schuchmann, H.P. (2008) Design of a microstructured system for the homogenization of dairy products at high fat content. Part II: Influence of process parameters. *Chemical Engineering & Technology*, **31** (12), 1863–1868.

33 Karbstein, H. (1994) Untersuchungen zum Herstellen und Stabilisieren von Öl-in-Wasser-Emulsionen. Dissertation. Universität Karlsruhe (TH).

34 Schuchmann, H.P. (2005) Tropfenaufbruch und Energiedichtekonzept beim mechanischen Emulgieren, in *Emulgiertechnik* (ed. H. Schubert), Behr's Verlag, pp. 171–205.

35 Bentley, B.J. and Leal, L.G. (1986) An experimental investigation of drop deformation and breakup in steady two-dimensional linear flows. *Journal of Fluid Mechanics*, **167**, 241–283.

36 Chesters, A.K. (1991) The modelling of coalescence processes in fluid–liquid dispersions: a review of current understanding. *Chemical Engineering Research & Design*, **69** (4), 259–270.

37 Bechtel, S., Gilbert, N., and Wagner, H.G. (1999) Grundlagenuntersuchungen zur herstellung von emulsionen im ultraschallfeld. *Chemie Ingenieur Technik*, **71** (8), 810–817.

38 Bechtel, S., Gilbert, N., and Wagner, H.G. (2000) Grundlagenuntersuchungen zur herstellung von emulsionen im ultraschallfeld teil 2. *Chemie Ingenieur Technik*, **72** (5), 450–459.

39 Behrend, O., Ax, K., and Schubert, H. (2000) Influence of continuous phase viscosity on emulsification by ultrasound. *Ultrasonics Sonochemistry*, **7** (2), 77–85.

40 Behrend, O. (2002) Mechanisches Emulgieren mit Ultraschall. Dissertation. Universität Karlsruhe (TH).

41 Danner, T. (2001) Tropfenkoaleszenz in Emulsionen. Dissertation. Universität Karlsruhe (TH).

42 Miller, R. (1990) Adsorption kinetics of surfactants at fluid interfaces: experimental conditions and practice of application of theoretical models. *Colloids and Surfaces*, **46**, 75–83.

43 Kempa, L., Schuchmann, H.P., and Schubert, H. (2006) Drip-reducing and drip coalescence in mechanical emulsifying with high pressure homogenisers. *Chemie Ingenieur Technik*, **78** (6), 765–768.

44 Schubert, H. (2003) Agglomerationsprozesse, in *Handbuch der Mechanischen Verfahrenstechnik* (ed. H. Schubert), Wiley-VCH Verlag GmbH, pp. 433–498.

45 Pietsch, W. (2001) *Agglomeration Processes: Phenomena, Technologies, Equipment*, Wiley-VCH Verlag GmbH.

46 Saffman, P.G. and Turner, J.S. (1956) On the collision of drops in turbulent clouds. *Journal of Fluid Mechanics*, **1** (1), 16–30.

47 Dodin, Z. and Elperin, T. (2002) On the collision rate of particles in turbulent flow with gravity. *Physics of Fluids*, **14** (8), 2921–2924.

48 Kessler, H.G. (2002) *Food and Bio Process Engineering: Dairy Technology*, Verlag A. Kessler, München.

49 Walstra, P. and Oortwijn, H. (1982) The membranes of recombined fat globules. 3. Mode of formation. *Netherlands Milk and Dairy Journal*, **36** (2), 103–113.

50 Dalgleish, D.G., Tosh, S.M., and West, S. (1996) Beyond homogenization: the formation of very small emulsion droplets during the processing of milk by a microfluidizer. *Netherlands Milk and Dairy Journal*, **50** (2), 135–148.

51 Darling, D.F. and Butcher, D.W. (1978) Milk-fat globule membrane in homogenized cream. *Journal of Dairy Research*, **45** (2), 197–208.

52 Walstra, P. (1999) Casein sub-micelles: do they exist? *International Dairy Journal*, **9** (3–6), 189–192.

53 Ogden, L.V., Walstra, P., and Morris, H.A. (1976) Homogenization-induced clustering of fat globules in cream and model systems. *Journal of Dairy Science*, **59** (10), 1727–1737.

54 Ramsden, W. (1903) Separation of solids in the surface-layers of solutions and 'suspensions' (observations on surface-membranes, bubbles, emulsions, and mechanical coagulation). *Proceedings of the Royal Society of London (1854–1905)*, **72**, 156–164.

55 Pickering, S.U. (1907) Emulsions. *Journal of the Chemical Society, Faraday Transactions*, **91**, 2001–2021.

56 Aveyard, R., Binks, B.P., and Clint, J.H. (2003) Emulsions stabilised solely by colloidal particles. *Advances in Colloid and Interface Science*, **100**, 503–546.

57 Sauter, C. and Schuchmann, H.P. (2008) Materialschonendes Hochdruckdispergieren mit dem high pressure post feeding (HPPF)-system. *Chemie Ingenieur Technik*, **80** (3), 365–372.

58 Schuchmann, H.P. and Köhler, K. (2010) Verfahren zur Herstellung einer Dispersion und Vorrichtung hierzu, Patent Nos DE 102,009,009,060, DE 102,009,009,060.

59 Mende, S., Stenger, F., Peukert, W., and Schwedes, J. (2003) Mechanical production and stabilization of submicron particles in stirred media mills. *Powder Technology*, **132** (1), 64–73.

60 Schilde, C., Gothsch, T., Quarch, K., Kind, M., and Kwade, A. (2009) Effect of important precipitation process parameters on the redispersion process and the micromechanical properties of precipitated silica. *Chemical Engineering & Technology*, **32** (7), 1078–1087.

61 Lortz, W., Batz-Sohn, C., and Penth, B.(31-1- 2006) Process for producing dispersions, US Patent No. 6,991,190.

62 Antonietti, M. and Landfester, K. (2002) Polyreactions in miniemulsions. *Progress in Polymer Science*, **27** (4), 689–757.

63 Landfester, K. and Musyanovych, A. (2007) Core–shell particles, in *Macromolecular Engineering: Precise Synthesis, Materials, Properties, Applications*, vol. **2** (eds Y. Matyjaszewski, Y. Gnanou, and L. Leibler), Wiley-VCH Verlag GmbH, pp. 1209–1243.

64 Erdem, B., Sudol, E.D., Dimonie, V.I.., and El-Aasser, M.S. (2000) Encapsulation of inorganic particles via miniemulsion polymerization. I. Dispersion of titanium dioxide particles in organic media using OLOA 370 as stabilizer. *Journal of Polymer Science Part A: Polymer Chemistry*, **38** (24), 4419–4430.

65 Jordan, A., Wust, P., Scholz, R., Tesche, B., Fähling, H., Mitrovics, T., Vogl, T., Cervós-Navarro, J., and Felix, R. (1996) Cellular uptake of magnetic fluid particles and their effects on human adenocarcinoma cells exposed to AC magnetic fields *in vitro*. *International Journal of Hyperthermia*, **12** (6), 705–722.

66 Pahl, M. et al. (1991) *Praktische Rheologie der Kunststoffe und Elastomere*, VDI-Verlag.

67 Desse, M., Wolf, B., Mitchell, J., and Budtova, T. (2009) Experimental study of the break-up of starch suspension droplets in step-up shear flow. *Journal of Rheology*, **53** (4), 943–955.

68 Costoyas, A., Ramos, J., and Forcada, J. (2009) Encapsulation of silica nanoparticles by miniemulsion polymerization. *Journal of Polymer Science Part A: Polymer Chemistry*, **47** (3), 935–948.

69 Mirzataheri, M., Mahdavian, A.R., and Atai, M. (2009) Nanocomposite particles with core–shell morphology IV: an efficient approach to the encapsulation of Cloisite 30B by poly (styrene-*co*-butyl acrylate) and preparation of its nanocomposite latex via miniemulsion polymerization. *Colloid and Polymer Science*, **287** (6), 725–732.

70 Theisinger, S., Schoeller, K., Osborn, B., Sarkar, M., and Landfester, K. (2009) Encapsulation of a fragrance via miniemulsion polymerization for temperature-controlled release. *Macromolecular Chemistry and Physics*, **210** (6), 411–420.

71 Kentish, S., Wooster, J., Ashokkumar, A., Balachandran, S., Mawson, R., and Simons, L. (2008) The use of ultrasonics for nanoemulsion preparation. *Innovative*

Food Science & Emerging Technologies, **9** (2), 170–175.

72 Gedrat, M. and Schuchmann, H.P. (2011) Precipitation of metal oxide nanoparticles via a miniemulsion technique. Particuology, **9** (5), 502–505.

73 Schlomach, J., Quarch, K., and Kind, M. (2006) Investigation of precipitation of calcium carbonate at high supersaturations. Chemical Engineering & Technology, **29** (2), 215–220.

74 Sauter, C. and Schuchmann, H.P. (2007) High pressure for dispersing and deagglomerating nanoparticles in aqueous solutions. Chemical Engineering & Technology, **30** (10), 1401–1405.

75 Schuchmann, H.P. and Sauter, C. (2009) Modified emulsion technologies for the production of nano-sized suspensions: potentials and draw-backs in ultrasonic, spinning disc and high pressure dispersing technologies. cfi/Ber.DKG, **86** (13), pp. 27–31.

76 Schilde, C., Breitung-Faes, S., and Kwade, A. (2007) Dispersing and grinding of alumina nano particles by different stress mechanisms. Ceramic Forum International 12–17.

77 Sauter, C., Emin, M.A., Schuchmann, H.P., and Tavman, S. (2008) Influence of hydrostatic pressure and sound amplitude on the ultrasound induced dispersion and de-agglomeration of nanoparticles. Ultrasonics Sonochemistry, **15** (4), 517–523.

78 Sauter, C., Schuchmann, H.P., and Bräse, P. (2008) Einschluss von Cyclodextrin durch kombinierte Dispergierung und der Reagglomeration nanoskaliger Partikel. Chemie Ingenieur Technik, **80** (10), 1539–1543.

6
Power Plant Processes: High-Pressure–High-Temperature Plants

Alfons Kather and Christian Mehrkens

6.1
Introduction

Approximately 42% of the global power generation is based on the coal-fired steam power plant process, in which the chemical energy of the fuel is used for evaporation and superheating of the thermodynamic working fluid water/steam at high pressures (HP) and temperatures. By expanding the steam in the steam turbine, the working fluid's enthalpy is first converted into mechanical energy at the turbine shaft and finally into electrical energy in the generator.

Compared to other fossil fuels, the use of coal for power generation leads to high specific CO_2 emissions between 750 and 1100 g/kWh. As in the near term power generation from fossil fuels is mandatory, significant efforts of the industry were made since the beginning of the 1990s to reduce the CO_2 emissions. The two main approaches considered are the increase of the power plant efficiency and the sequestration and underground storage of carbon dioxide (CCS: carbon capture and storage). However, CCS leads to a significant reduction in the overall power plant efficiency up to approximately 10% points. In the context of fossil-fired power plants, increasing the power plant efficiency is therefore also known as "no-regrets" strategy.

In Figure 6.1, the historical development of the net efficiency of coal-fired power plants since the 1970s is shown. While from the mid-1970s until the mid-1980s net efficiencies in the range of approximately 35–40% (based on lower heating value (LHV)) were achieved, since the beginning of the 1990s a significant increase up to values above 46% is evident. In addition to other measures, almost 1/6 of this efficiency increase is attributed to the raise of the main process parameters, such as the live steam pressure and temperature, up to values of currently 30 MPa/600 °C. Figure 6.2 summarizes the evolution of the live steam parameters of coal-fired steam power plants in the past 40 years in Germany. While from the 1980s to the beginning of the 1990s in power plants usually live steam parameters of approximately 20 MPa/535–545 °C were common, since the end of the 1990s a significant increase in these steam parameters up to, 27.5 MPa and temperatures of about 580 °C (Niederaußem power station unit K, Germany, commissioned in 2003 [1]) were

---

Industrial High Pressure Applications: Processes, Equipment and Safety, First Edition. Edited by Rudolf Eggers.
© 2012 Wiley-VCH Verlag GmbH & Co. KGaA. Published 2012 by Wiley-VCH Verlag GmbH & Co. KGaA.

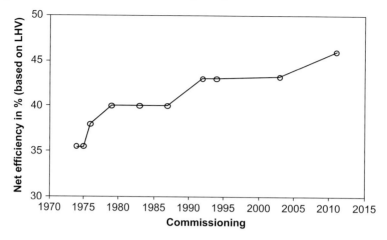

Figure 6.1 Historical development of the net efficiency of coal-fired steam power plants in Germany.

realized. More than 10 years ago, power plants with steam temperatures up to 600 °C were developed and built in Japan [2]; even in Europe and China, power plants with these steam parameters are currently under construction or are already operating. At the same time, a worldwide trend toward higher single unit power outputs of up to 1100 MW$_{el}$ (gross) can also be observed. For the next generation of coal-fired steam power plants, new concepts and materials for the application of steam parameters of more than 35 MPa/700 °C are under currently development by different research institutes and the power generation industry [3–5].

Except for coal-fired power plants, the steam power plant process is also used for the direct or indirect power generation from other fuels. In principle, this process forms the thermodynamic basis for power generation from nuclear fuels or biomass

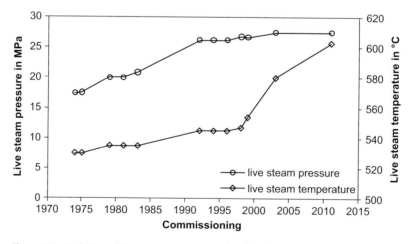

Figure 6.2 Evolution of the steam parameters of coal-fired steam power plants in Germany.

or is used as basic process for waste heat recovery from gas turbine exhaust gas in combined cycle power plants. Due to the different requirements for safety (nuclear power plants) and lifetime issues for components exposed to the hot flue gas (corrosion in incineration and biomass plants), in these power plants comparatively low steam parameters are applied. Since today the highest steam parameters are achieved in coal-fired steam power plants, the main focus here is on the current 600 °C class of steam power plants. In addition to the general basic plant layout of today's modern steam power plants, the development status of the most highly stressed components will be mainly considered. These critical components consist of the high-pressure part of the steam generator and the high-pressure steam turbine.

6.2
Coal-Fired Steam Power Plants

For power generation in coal-fired steam power plants, preferably hard coal, lignite, or low volatile coals such as anthracite are used. As interface between the flue gas side and the water/steam cycle serves the steam generator, where the chemical energy from the coal is converted into useful heat for preheating and evaporating the water, as well as superheating and reheating the steam.

To put it simply, the water/steam cycle consists of several high-pressure and intermediate-pressure (IP) heating surfaces, a steam turbine (including the condenser), and several low- and high-pressure pre-heaters, as well as a condensate pump and a feedwater pump. The superheated steam produced in the steam generator is expanded in the steam turbine powering a generator, which converts the enthalpy difference of the working fluid water/steam into electrical energy. Because the highest process pressures occur within the thermodynamic cycle, in the following the water/steam side of coal-fired power plants will be considered in more detail. For details on the arrangement and design of the flue gas side, one is referred to the literature [6–8]. The feedwater pump and the HP pre-heaters are exposed to the highest process pressures occurring in the steam power plant. Due to the pressure drops in the HP pre-heaters, piping, and the HP heating surfaces in the steam generator, the live steam pressure between steam generator and HP steam turbine is up to approximately 5 MPa lower than that directly after the feedwater pump. The high steam temperatures of approximately 600 °C at the steam generator outlet and the HP steam turbine inlet, respectively, make the components in this area require a careful design due to their low allowable material strength values.

6.2.1
Thermodynamics and Power Plant Efficiency

Simplified the net efficiency η_{net} of coal-fired steam power plants can be written as product of the partial efficiencies of the ideal thermodynamic cycle $\eta_{th,id}$, steam generator η_{SG} (boiler efficiency), steam turbine η_T, generator η_G, and the equivalent efficiency of the electrical auxiliary power consumption η_{aux} as well as other efficiencies η_{other}, which are negligible:

$$\eta_{net} = \eta_{th,id} \times \eta_{SG} \times \eta_T \times \eta_G \times \eta_{aux} \times \eta_{other}$$

While the steam generator efficiency ($\eta_{SG} \sim 0.92$), the steam turbine efficiency ($\eta_T = 0.9\text{--}0.92$), the generator efficiency ($\eta_G > 0.98$), and the corresponding efficiency for the power plant's electrical auxiliary power consumption (η_{aux} up to 0.93) all have relatively high values above 0.9, the net efficiency of a coal-fired steam power plant is mainly affected by the thermal efficiency of the cycle $\eta_{th,id}$, which today is about 0.55. Besides the minimization of losses by an optimized flue gas heat utilization, the application of optimized turbine blading, or the reduction of the auxiliary power consumption, the main potential for enhancing the overall power plant efficiency can be seen in increasing the live steam parameters.

As ideal thermodynamic cycle for steam power plants, the Rankine cycle with the working fluid water/steam is commonly used for converting thermal energy into mechanical energy (Figure 6.3). At first, the working fluid in liquid phase is brought to a higher pressure level in the feedwater pump ($0 \rightarrow 1$) and then is preheated, evaporated, and superheated under isobaric conditions in the steam generator ($1 \rightarrow 4$). The vaporized working fluid is then expanded under isentropic conditions in a steam turbine, where the enthalpy difference $h_5 - h_4$ between the inlet and outlet is converted into mechanical energy. To close the thermodynamic cycle, it is necessary to liquefy the saturated steam under isobaric conditions in a condenser ($5 \rightarrow 0$) further downstream. The mechanical energy at the turbine shaft corresponds to the difference between the heat supplied to the cycle q_{in} and the heat removed q_{out}, so that the thermal efficiency of the Rankine cycle $\eta_{th,CR}$ can be calculated from the following equation:

$$\eta_{th,CR} = \frac{q_{in} - |q_{out}|}{q_{in}} = 1 - \frac{h_5 - h_0}{h_4 - h_1}$$

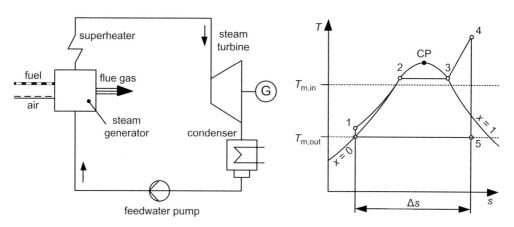

Figure 6.3 Simplified view of a steam power plant process and the corresponding T, s-diagram.

To identify optimization potentials within the thermodynamic cycle, in the following the basic definition of the thermal efficiency of the Rankine cycle is converted into an equivalent definition according to Carnot's efficiency, which depends only on the mean temperatures of the heat supplied and removed (Figure 6.3):

$$\eta_{th,CR} = 1 - \frac{T_{m,out}}{T_{m,in}}$$

To put it simply, these mean temperatures can be derived from the specific enthalpies h and specific entropies s of the Rankine cycle by the following equations:

$$T_{m,in} = \frac{q_{in}}{\Delta s_{in}} = \frac{(h_4 - h_1)}{(s_4 - s_1)} \quad \text{and} \quad T_{m,out} = \frac{q_{out}}{\Delta s_{out}} = \frac{(h_0 - h_5)}{(s_0 - s_5)}$$

In addition to applying a single reheat with an equal steam temperature as used for the live steam and a multistage preheat train, the main potential for enhancing the thermal efficiency can be seen in raising the mean temperature of the heat supplied to the process by further increasing the live steam enthalpy (pressure and temperature). The mean temperature corresponding to the heat removed from the cycle is primarily determined by the available condenser pressure, which mainly depends on the cooling water conditions prevailing at the site and hence can hardly be influenced by process design improvements.

6.2.2
Configuration of Modern Steam Power Plants

Figure 6.4 shows a simplified diagram of a modern 600 °C supercritical steam power plant. In this figure, the air/flue gas side and the water/steam side of a coal-fired steam power plant with a power output of 600 MW$_{el}$ (gross) using best available technology in 2004 are shown. This concept was developed as a result of an engineering study for the German "Reference Power Plant North Rhine Westphalia" [9], carried out in 2004 by the VGB PowerTech (European association for power generation) with contribution of leading German power plant manufacturers and operators. The main scope of this study was the development of a modern economic power plant concept with a net efficiency of more than 45%. The basic concept of this power plant has been adopted almost unchanged to currently projected or recently built 600 °C coal-fired power plants in Europe, however, with an increased power output from 800 MW$_{el}$ to about 1100 MW$_{el}$.

Although the highest system pressure occurs in the HP part of the water/steam cycle, in the following a short overview of the air/flue gas side of a coal-fired power plant will also be given. At first, the coal is ground in a coal pulverizer and then is pneumatically transferred, utilizing the preheated primary air as carrier gas, to the main burners located in the steam generator furnace. In the steam generator, the chemical energy of the pulverized coal is converted into heat by combustion and then is transferred to different heating surfaces by means of radiant and convective heat transfer to the water/steam side. Depending on the fuel type and national

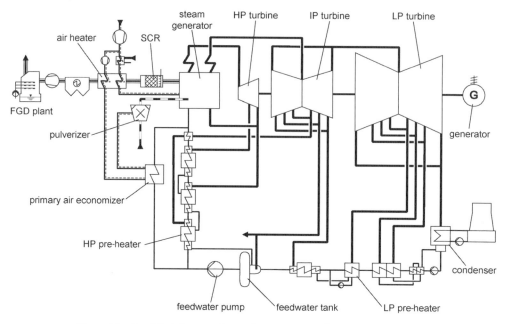

Figure 6.4 Simplified diagram of a modern 600 °C coal-fired steam power plant.

regulations, an SCR (selective catalytic reduction) reactor for the reduction of nitrogen oxides (NO_x) is placed before the air heater in the flue gas path. In the air heater, the primary air and the secondary air are heated up to temperatures of about 320–350 °C and hence the flue gas is cooled down to approximately 115–160 °C, depending on the coal's sulfur content. The fly ash from the flue gas is almost completely retained by an electrostatic precipitator, before the flue gas is transferred by the induced draft fan into the flue gas desulfurization (FGD) plant. In the latter, most of the sulfur dioxide (SO_2) is removed by chemical absorption, before the treated flue gas is released to the environment either through the stack or through the cooling tower.

The water/steam side of a steam power plant normally consists of the following key components:

- LP (low pressure) and HP (high pressure) preheat train
- Feedwater pump
- Flue gas heated HP and IP parts of the steam generator
- Steam turbine system (HP, IP, and LP turbines)
- Condenser for heat removal

At first, the condensate is extracted from the condenser at a pressure of about 2–6 kPa, before being transferred by the condensate pump to the feedwater tank through the four-stage LP preheat train. The LP preheat train represents the first part of the preheat system, comprising a total of eight stages. For preheating purposes, steam from the IP and LP turbines is extracted at different pressure levels. The extracted steam is first desuperheated and then condensed and subcooled in different vertically or horizontally arranged shell-and-tube heat exchangers, in which the condensate, flowing through the tubes, is preheated simultaneously. After this, the condensate is directly supplied to the feedwater tank. The feedwater tank is a steam-heated direct contact heater in which the water is heated up to saturation at a corresponding vessel pressure of approximately 1.0–1.2 MPa. Besides preheating of the condensate, the feedwater tank also serves the following additional purposes: on the one hand, the majority of the noncondensable gases dissolved in the water are released and, on the other hand, it serves as a storage vessel for the compensation of transient operation conditions in the water/steam cycle, for which reason the tank is designed with a sufficient capacity.

From the feedwater tank, the condensate is led to one or two feedwater pumps in parallel, in which the working fluid, which is now designated as feedwater, is pressurized up to 35 MPa and then flows through the HP preheat train directly into the steam generator. Depending on the power plant's operational mode and power output, multistage centrifugal pumps are applied as feedwater pumps that are driven either electrically or by a separate steam turbine (preferably in larger steam power plants). The spray water for controlling the hot reheat steam temperature is extracted from in-between these stages.

The HP preheat train consists of three heating condensers as well as one externally located desuperheater for the first HP heating condenser, in which the steam extracted from the HP and IP turbines is first desuperheated and then condensed and subcooled for preheating the feedwater up to temperatures of about 290–320 °C. In addition to the HP preheat train shown in Figure 6.4, a primary air economizer is also integrated parallel to the water/steam cycle, in which heat from the flue gas is recovered by further preheating of a partial feedwater stream. In this concept, the primary air is brought to a temperature as high as possible, hence the flue gas temperature after the steam generator can be cooled further down by using the waste heat for preheating. The extraction of water for the HP steam attemperator and live steam temperature control is normally done after the last HP pre-heater.

Next, the feedwater is preheated further by flue gas in the economizer and afterward evaporated and superheated up to 600 °C at a pressure of approximately 27–30 MPa in the evaporator and in several SH (superheater) heating surfaces. The live steam is then passed through the control valves, in which the pressure before HP steam turbine is adjusted according to the operational mode of the power plant (pressure reserve for frequency control, etc.). In the HP steam turbine, the live steam is expanded to a pressure of typically around 5.5–6.5 MPa at temperatures of approximately 340–370 °C and then is returned to the steam generator for reheating of the steam to temperatures up to 600–620 °C. The IP steam then is led to the single or double-flow IP steam turbine, in which it is expanded to approximately 0.4–0.6 MPa. The IP exhaust steam is then passed through one or more crossover pipes to two or three double-flow LP turbines, where the steam is expanded into vacuum with a moisture content up to approximately 8–15%. Finally, the LP exhaust steam is condensed in a water-cooled condenser located directly below the LP steam turbine casing.

6.3
Steam Generator

In coal-fired steam power plants, the steam generator represents the key component for the heat transfer from the flue gas to the water/steam side of the thermodynamic cycle. For this reason, in this part of the power plant, the highest process pressures and temperatures are obtained in combination. In particular during load changes, certain components in the HP section of the steam generator are exposed to both high mechanical and thermal stresses, by varying process pressures and temperatures. Since all components used in steam generators are designed for a lifetime of more than 200 000 h, the proper selection and development of materials are of particular importance. As a consequence of the live steam parameters increasing during the last years, the main focus was placed on the steam generator membrane wall, the final SH heating surface, and the final SH outlet header.

Besides describing the overall layout of today's coal-fired steam generators for steam parameters of up to 30 MPa/600 °C, in the following sections the design and selection of materials for the just mentioned highly stressed components in steam generators will be discussed in detail.

6.3.1
Steam Generator Design

Figure 6.5 shows the cross-sectional view of a modern hard coal-fired steam generator with live steam temperatures of 600 °C. As usual in Europe and particularly in Germany, the steam generator is designed as tower-type boiler with a rectangular or square-cut furnace cross section, in which the radiant and convective heating surfaces are all arranged in a single pass. In the second pass of the steam generator, the SCR reactor for the reduction of the NO_x emissions and the air heater are located. Usually the steam generator is suspended directly from the boiler supporting grid, so that the steam generator can expand freely in longitudinal direction during operation (e.g., the downward thermal expansion of a lignite-fired steam generator with a height of 160 m is around 800 mm). In Asia and Northern America, predominantly two-pass steam generators are used, which are basically characterized by a low height and a compact design but have some disadvantages due to horizontal thermal expansions.

The steam generator's furnace wall is designed as a gas-tight, welded membrane wall (tube-fin-tube), cooled by water/steam, and serves as evaporator and SH heating surface. All Flue gas heated heat exchangers in the convection area of the steam generator are normally designed as in-line tubular bundle heat exchangers, suspended from several hundreds of hanger tubes. For the realization of supercritical steam parameters, modern steam generators are designed as once-through boilers, in which the feedwater is preheated, evaporated, and superheated in one way through, without using an additional circulation system for evaporation. For steam generators with subcritical steam parameters, commonly used in industrial applications or for waste heat recovery from gas turbine exhaust gases,

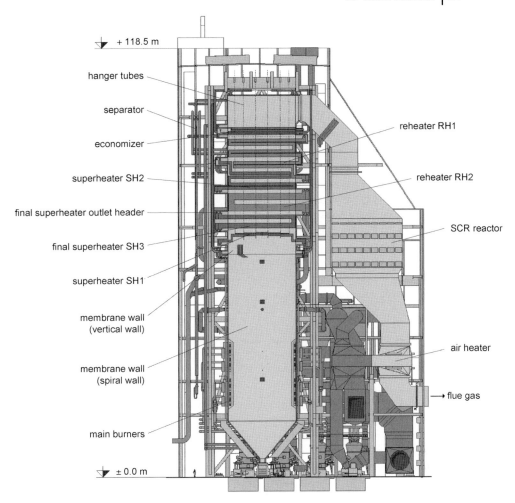

Figure 6.5 Cross-sectional view of a hard coal-fired steam generator of a modern 600 °C power plant (EnBW Kraftwerke AG: new build of unit 8 of Rheinhafen-Dampfkraftwerk power station, Karlsruhe, Germany). Copyright: Alstom 2011.

predominantly drum-type boilers are used, in which the steam is evaporated by natural or forced circulation.

The HP system of a steam generator consists of differently connected furnace wall and convective tubular bundle heating surfaces, in which the feedwater is further preheated, evaporated, and superheated, before being finally sent to the steam turbine as live steam. At first, the preheated water is fed to the upper part of the steam generator, where it is supplied to the economizer that represents the last convective heating surface downstream along the flue gas path. There the feedwater is preheated to a temperature of about 330–350 °C, before being led downward through connecting pipes to the furnace hopper. Regarding the low driving temperature difference between feedwater and flue gas in hard coal-fired boilers, usually a fin tube economizer is used, made of low-alloy steel 16Mo3.

To avoid film boiling due to high heat fluxes in the evaporator, in smooth tubes a minimum mass flux at the water/steam side of approximately 700–800 kg/(m² s) must always be maintained. Hence, the lower part of the evaporator is designed as a spiral wall consisting of several hundreds of inclined tubes, in which the water flows upwards? while being evaporated. Alternatively, particularly Japanese-built steam generators are designed with vertical internally riffled evaporator tubes for more than 10 years. This tube configuration impacts an additional rotation to the water/steam stream, which leads to higher wetting and enhanced cooling of the tube inner wall at lower part loads and thus enables lower water/steam mass fluxes, and due to that, vertical tubing can be realized in the evaporator. When using smooth tubes, at around half height of the steam generator, the spiral wall of the evaporator is converted into a vertical tube wall using special transition pieces. The related reduction of the water/steam mass flux can be done due to the lower heat fluxes in this area. At sliding pressure operation and boiler loads that correspond to evaporator pressures below the critical pressure of 22.1 MPa, the water along the evaporator membrane wall is preheated, evaporated, and finally superheated. Between the load corresponding to the critical pressure and the boiler once-through minimum load of around 40% for this evaporator system, the location of the final point of evaporation varies. At loads corresponding to pressures above the critical pressure, no phase change takes place anymore. With respect to low medium temperatures in the lower part of the membrane wall (lower part of the spiral tube wall), low-alloyed materials, like 16Mo3 and 13CrMo4–5 can be used. In the upper part of the membrane wall (vertical tube wall), outlet temperatures of approximately 450–480 °C are reached, so higher alloyed materials must be used here.

At the end of the vertical tube wall, the water/steam mixture – or above once-through minimum load only steam – is first collected and then supplied to the separators, in which the water is separated from the steam and returned to the economizer inlet. After passing through the separators, the steam is superheated in counterflow via several hanger tubes supporting the tubular bundle heat exchangers of the steam generator. At the end, the hanger tubes form the primary superheater heating surface SH1 (applied materials here are, for example, T24 or HCM12). The steam superheated to 500–520 °C is collected and then supplied to the next heating surface SH2, which is located normally between the primary and secondary reheaters RH1 and RH2 along the flue gas path. In SH2, which is made of materials like Super 304H, the steam is further superheated in counterflow to approximately 540–560 °C before being sent to the final superheater heating surface SH3. The final SH heating surface is located in an area of high flue gas temperatures and therefore is usually arranged in parallel flow. In the outlet header, the steam parameters are about 27–30 MPa and 600–605 °C. Both the final SH tubes and the thick-walled final SH outlet header are exposed to high pressures and the highest temperatures occurring in the process, thus high-alloyed austenitic materials such as HR3C or Super 304H are used for the final SH and the martensitic materials like P92 for the final SH outlet header and live steam piping.

After expanding the live steam in the HP turbine, the exhaust steam is returned to the steam generator to be reheated to a temperature of 620 °C in the intermediate pressure system that consists of the primary and secondary reheaters RH1 and RH2. To control the superheater and reheater outlet temperatures, spray coolers are located between the bundle heat exchangers injecting preheated feedwater for desuperheating of the steam.

In Figure 6.6, the historical and recent developments in materials used for the highly stressed components in steam generators, like the membrane wall, the final SH tubes and the final SH outlet header are shown (see also Refs [9–11]). In the membrane wall, only material without a postwelding heat treatment should be used. While for live steam parameters up to about 27.5 MPa/580 °C, the steel 13CrMo4–5 was the highest alloyed material in 1995, the further raised steam parameters demanded for new materials such as the European T24 or the Japanese T23 (HCM2S) steels that shall not necessitate a heat treatment, either. The austenitic alloys well known from power plants with steam parameters of 27.5 MPa/580 °C could still be used as final SH tube materials for today's 600 °C power plants. However, with respect to the maximum creep strength of the materials used, at steam parameters of 35 MPa/700 °C in the final SH and final SH outlet header, nickel alloys must be applied. For the SH outlet header and the live steam piping of today's 600 °C power plants, in the 1990s martensitic alloys like P92 were developed and qualified. Figure 6.7 shows the average creep strength values for 100 000 h, as function of the material temperature, for selected alloys used in modern 600 °C steam generators. For the design of steam generator components with a lifetime of 200 000 h, these creep strength values must be divided by a safety factor of 1.5, whereas 200 000 h values have to be divided by only 1.25.

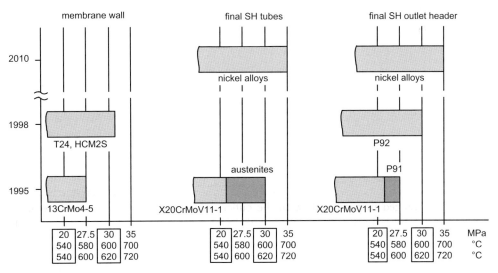

Figure 6.6 Development of different materials used in steam generators, shown against the evolution in the corresponding live steam and reheat steam parameters.

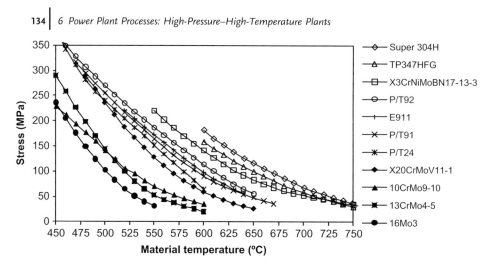

Figure 6.7 Average 100 000 h creep strength values of selected materials used in current steam generators of the 600 °C power plant class.

6.3.2
Membrane Wall

Modern steam generators are designed exclusively with welded membrane walls. In order to avoid hydrogen-induced stress corrosion cracking, according to former German TRD 201, the hardness of 350 HV 10 should not be exceeded in the heat-effected zones of the membrane wall welds. For higher-alloyed materials than 10CrMo9–10, this was only possible by means of a heat treatment, after the welding process has been completed. According to the former German regulations, this heat treatment was required also for the material 10CrMo9–10, due to the temperatures existing in the wall. Although it is possible to carry out this subsequent heat treatment in the workshop, it raises some practical problems during erection and repair. Therefore, materials are preferred for the construction of the membrane wall that do not require a subsequent heat treatment. On the basis of the material 13CrMo4–5, used before 2000 in membrane walls, and the usual temperature additions, a limit value of approximately 435 °C for a tube outside diameter of 38 mm and a wall thickness of 6.3 mm results for the design of the maximum steam temperature in the wall. If a larger wall thickness is selected, this limit value can be increased to about 450 °C [10, 12].

For steam parameters of 26 MPa/590 °C, the temperatures at the transition from spiral to vertical wall are not higher than those in the walls of plants built in the 1990s [12]. Under these steam conditions, the spiral wall can therefore still be built from the material 13CrMo4–5. In the area of the vertical tubes above the transition, however, allowable limits may be exceeded, since in this area of the steam generator, a considerable heat radiation at relatively high steam temperatures is present.

Thus, it becomes evident that the membrane wall was one of the limiting parts in case of a further increase of the steam parameters. The measures mentioned below, which all aim at reducing the medium temperature in the wall, help to counteract this effect to a more or less high degree:

- Minimization of the injection mass flow in the superheater system, in order to increase the mass flow through the membrane wall.
- Use of a flue gas recirculation.
- Reduction of the feedwater temperature.
- Reduction of the fin tube economizer with a simultaneous, clear enlargement of the reheater heating surface and maintaining the flue gas temperature downstream of the economizer.
- Increase of the furnace outlet temperature.

With increased steam parameters, the medium temperature also increases in the steam generator membrane wall, so that the limitation for the exclusive application of 13CrMo4–5 was reached and finally even exceeded when coming to the 600 °C power plant technology. For this reason, it was necessary to develop new evaporator materials with higher creep strength values, which do not require a postwelding heat treatment during construction. Results of this development were the new materials T23 and T24, which can be used with temperatures up to approximately 50 K higher than 13CrMo4–5. The material T24 (7CrMoVTiB10–10) is used in nearly all European 600 °C steam generators shortly before commissioning and has to demonstrate within the next years its qualification as furnace wall material.

6.3.3
Final Superheater Heating Surface

The 100 000 h creep strength average values of the austenitic tube materials generally used today as well as of the formerly used martensitic material X20CrMoV11–1 are shown in Figure 6.8. It can be seen that the austenitic material Super 304H has

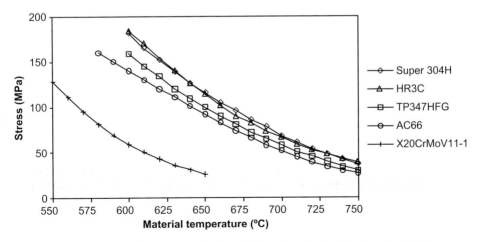

Figure 6.8 Average 100 000 h creep strength values of selected steels utilized in the past and today in the final SH heating surfaces.

the same strength values as the X20CrMoV11–1 used until the mid-1990s at 100–120 K higher design temperatures. Besides the strength, the flue gas sided high-temperature corrosion and the steam-sided oxidation of the used material also have an impact on the design of the heating surface at high steam parameters.

The resistance against high-temperature corrosion on the flue gas side strongly depends on the tube material used and thus is primarily influenced by its chrome content. Therefore, in the final SH heating surface of current steam generators with parameters of 30 MPa/600 °C, solely austenitic materials with high chromium content are used, such as Super 304H or HR3C.

Due to higher steam temperatures, the rate of oxidation on the steam side of heated tubes also increases. The interior oxide layer leads to a deterioration of the heat transmission, causing a further increase of the tube wall temperature, which in turn may produce both a higher creep strength damage and, in particular, also a reinforced high-temperature corrosion. As for high-temperature corrosion, in this case also higher chrome contents result in improvements.

However, this advantage of austenitic over martensitic materials generally applies only to base-load operation of the steam generator. In case of operation with daily start-ups and shutdowns, the austenitic steels can show worse corrosion properties than the martensitic ones. The resulting higher corrosion rates can be explained by the large difference between the thermal expansion coefficients of the austenitic tube material and of the oxide deposits, which lead to spalling of the oxide layer and thus result in an increased corrosive attack.

6.3.4
Final Superheater Outlet Header and Live Steam Piping

Due to the higher steam parameters, the design of the SH outlet header leads to larger wall thicknesses and thus considerable restrictions of the load change rate during start-up. One possibility to reduce the necessary wall thickness is to select austenitic materials. However, these have the disadvantage of showing a worse temperature change behavior than the martensites, due to the higher thermal expansion coefficient and the lower thermal conductivity, if the wall thickness is the same. These relations are shown qualitatively in Figures 6.9 and 6.10. In these diagrams, the header wall thickness on the one hand and the rate of temperature change at the beginning of the cold start-up on the other hand are presented against the SH outlet temperature, for the SH outlet pressures of 26 and 30 MPa. In this case, the materials examined are X20CrMoV11–1, P91 (X10CrMoVNb9–1), and austenite X3CrNi-MoBN17–13–3 (1.4910). For the material X20CrMoV11–1 at an outlet pressure of 30 MPa, the maximum allowable SH outlet temperature of 548 °C is already exceeded (see also Figure 6.7). With P91, the outlet temperature can be increased to about 572 °C. With the austenite 1.4910, even for 30 MPa/600 °C the diameter ratio limits according to the regulations are still not reached [12, 13].

Due to the considerably increased wall thicknesses, the martensites lose their advantage over the austenites in terms of the allowable rate of temperature change. At 600 °C, this rate amounts for P91 only half the value for the austenite. Therefore, the

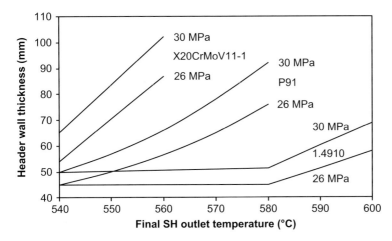

Figure 6.9 Wall thickness of SH outlet headers as function of the material, SH outlet temperature, and pressure [12].

following conclusions can be drawn from Figures 6.9 and 6.10 for the SH outlet header:

- Compared to X20CrMoV11–1, with P91 approximately 15–25 K, higher operating temperatures can be realized. However, P91 is still no substitute for austenites at, 30 MPa/600 °C.
- Despite the unfavorable physical properties of austenites (thermal expansion coefficient and thermal conductivity) compared to ferritic/martensitic materials, larger rates of temperature change can be realized during the start-up of the plant, on account of the high creep strength at high temperatures.

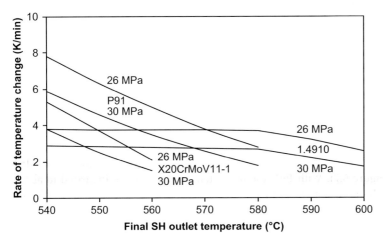

Figure 6.10 Acceptable rates of temperature change for the beginning of cold start-up for SH outlet headers as function of the material, SH outlet temperature, and pressure [12].

For economic reasons, in the development of new materials for next-generation power plants since 2000, a fundamental requirement comprised the application of nonaustenitic materials for the live steam piping. To protect at the same time the thick-walled weld between the live steam piping and the final SH outlet header against overstress due to differential thermal expansion of the materials used, the SH outlet header had to be manufactured also from martensitic materials [11]. The SH outlet header is equipped with martensitic connecting tubes so that the weld between the austenitic SH heating surfaces and the martensitic SH outlet headers is done via these tubes and therefore with comparatively small welds that show less problems due to differential thermal expansions of the applied materials.

For power plants built since 2000 with steam parameters of 27.5 MPa/580 °C and power plants currently under construction with parameters of 30 MPa/600 °C, new martensitic materials for the final SH outlet header and the live steam piping had to be developed. While the martensitic material E911 (X11CrMoWVNb9–1–1), with creep strength values comparable to P91 (Figure 6.7), has its application limit at steam parameters of approximately 27.5 MPa/580 °C, the newly qualified martensitic P92 is barely applicable to steam parameters up to 30 MPa/600 °C [10]. For this reason, in currently planned or recently built power plants in Europe, the material P92 (X10CrWMoVNb9–2) is applied predominantly, for both the final SH outlet header and the live steam piping. However, for the further increase of the steam parameters up to 35 MPa/700 °C, nickel-based materials must be used.

6.4
High-Pressure Steam Turbines

In coal-fired steam power plants, the HP turbine and the high-pressure system of the steam generator are exposed to the highest mechanical and thermal stresses occurring in the process. Compared to the steam generator, the HP turbine is exposed at 100% load to steam parameters slightly less heavily (due to temperature imbalances of approximately 20 K in the steam generator and losses in the steam piping); however, in recently built or currently projected power plants, steam temperatures of approximately 600 °C and pressures up to 28–29 MPa are reached.

6.4.1
Configuration of Modern Steam Turbines

Basically modern steam turbines consist of a HP, IP, and LP turbine, in which the steam's enthalpy is first converted into mechanical energy at different pressure levels and subsequently into electrical energy in the generator. For large steam power plants, two different concepts of shaft arrangement are to be distinguished, in principle. Whereas in Europe large power plants are generally built in single-shaft configuration (e.g., Niederaußem power station unit K, Germany, gross power output 1012 MW$_{el}$), in Japan and Northern America the double-shaft arrangement is widely used [14].

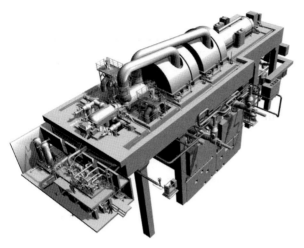

Figure 6.11 Three-dimensional view of a modern steam turbine of Alstom with a gross power output of approximately 1100 MW$_{el}$. Copyright: Alstom 2011.

In the single-shaft configuration (tandem compound, Figure 6.11), the HP, IP, and LP turbines and the generator are all arranged in-line on the same turbine shaft, which is operated at a rotation speed corresponding to the grid frequency of either 50 or 60 Hz (3000 or 3600 1/min, respectively). Especially in Northern America and Asia, in steam power plants with an output of about 1000 MW$_{el}$, the double-shaft arrangement is mostly applied (cross-compound). Because the grid frequency of 60 Hz is slightly higher in these countries, the last LP turbine blades are exposed to high mechanical stresses due to higher centrifugal forces. This has led in the past to a limitation of the turbine blade length in the last LP stages. For this reason, the HP and IP turbines as well as the double-flow LP turbines are arranged on separate turbine shafts rotating at different speeds, each connected to a generator. The HP and IP turbine shaft rotates at full grid frequency, whereas the LP turbine shaft rotates only at half grid frequency (the power output of the HP–IP turbine shaft is approximately 60% of the overall output of the unit). Due to significantly lower LP turbine shaft speed, assuming the same materials are used, the application of longer turbine blades in the last stages of the LP turbine and thus a higher power output are possible. Both turbine shafts are arranged parallel to each other and the IP exhaust steam is conducted through crossover pipes to the LP turbines.

6.4.2
Design Features of High-Pressure Steam Turbines

Since for small- and medium-sized steam turbines with a power output up to approximately 700–800 MW$_{el}$ in particular of Japanese origin, a combined casing for both the HP and IP turbine stages can be found, for coal-fired power plants with a single unit output up to 1000–1100 MW$_{el}$ predominantly separate casings are utilized. For the HP turbine, almost all turbine manufacturers apply a single-flow configuration,

Figure 6.12 Three-dimensional view of a Siemens high-pressure steam turbine for 600 °C application. Copyright: Siemens AG 2011.

yielding compact HP turbine designs. However, the Japanese turbine manufacturer Mitsubishi Heavy Industries uses for power plants with an output of approximately 1000 MW$_{el}$ a double-flow design for the HP turbine and all other steam turbines (cross-compound). Figures 6.12 and 6.13 show typical HP steam turbines used in current 600 °C power plants, manufactured by Siemens and Alstom.

Basically steam turbines are double-shelled with a separate inner and outer casing. The inner casing primarily serves as thermoelastic guide blade carrier and in combination with the turbine rotor and the first stage turbine blades are exposed to highest thermal stresses. Although the inner casing at the turbine inlet section is subjected to the full system pressure of up to 28–29 MPa, the mechanical stresses are much lower because of the exhaust steam pressure of approximately 6 MPa outside the inner casing. The outer casing is stressed at significantly lower temperatures, mainly by the pressure difference of the HP exhaust steam and the ambient pressure.

For outer casings of modern HP turbines, fundamentally two different types can be distinguished. While worldwide nearly all turbine manufacturers use a horizontally split HP turbine design with an axial flange and made of cast steel, the turbine maker Siemens applies a vertically split barrel-type design (also made of cast steel) so that the inner casing is fixed to the outer casing on the periphery along with a screwed cover. Major advantage of a barrel-type steam turbine is that the basic design of the outer casing is rotationally symmetric and thin-walled, which is expected to be beneficial in

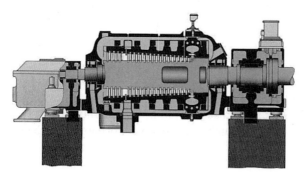

Figure 6.13 Longitudinal section of a high-pressure steam turbine of the European manufacturer Alstom. Copyright: Alstom 2011.

terms of asymmetric deformation and thermal stressing during operation. During temperature changes, particularly in thick-walled areas of the steam turbine, high thermal stresses and deformations can result, limiting the power plant's load gradient during start-up and shutdown.

In modern HP steam turbines, the live steam is admitted by the control valves directly or through short connecting pipes to the first turbine stage located downstream of the turbine inlet section. To equalize the temperature profile inside the turbine inlet section, the steam is swirled by proper measures. For this reason, in HP steam turbines of the European turbine maker Alstom, the live steam is supplied by two spiral-shaped steam admissions to the turbine inlet section, so that the steam can be mixed sufficiently. The inner casing or guide blade carrier of modern HP turbines consists of two halves normally made of 9–12% chromium steel for application in 600 °C steam power plants. Both halves are fitted into the turbine outer casing and are fixed finally with either screws or shrink rings. The application of shrink rings is a design feature of the turbine maker Alstom specifically, which in contrast to an inner casing fixed by screws has the advantage of being rotationally symmetric and thin-walled. Equivalent to the barrel-type turbine design of Siemens, this design is expected to have advantages for the power plant's start-up, shutdown, and load changes due to lower thermal stresses. If HP steam extraction for preheating of the feedwater is provided, the steam is extracted between the turbine stages and flows to the corresponding pre-heater through extraction steam pipes. Depending on the prevalent material temperatures, the outer casing is manufactured in all thermally highly stressed sections of the steam turbine from 9–12% chromium steel and, for example, thermally lower stressed areas (e.g., exhaust area) from lower alloyed chromium steel [15].

The rotating part of the steam turbine comprises the turbine rotor, including the rotor blades. The whole steam turbine blading consists of several stages, each comprising guide blades suspended in the inner casing and rotor blades fixed directly on the turbine rotor. By diverting the steam flow in the guide blades, a tangential force is induced at the rotor blades, which causes the rotation of the turbine rotor. After the application of powerful computer systems in the 1990s, the non-twisted blades were replaced by blades with a three-dimensional optimized profile. The blades are project-specific and manufactured by cutting from a single piece. For 600 °C power plants, the first-stage thermally highly stressed turbine blades are made of, for example, austenitic materials, whereas at lower steam temperatures, further downstream martensitic alloys are used instead [16]. The rotors of modern steam turbines are either forged from a single piece or are manufactured by welding different forged sections together. The latter was introduced by the turbine manufacturer Alstom in the past. The welded rotor design has the advantage of applying differently alloyed materials to the rotor sections, according to the local thermal stresses prevailing, provided that these materials can be welded satisfactorily. For 600 °C power plants, in these thermally highly stressed sections of the steam turbine, usually 9–10% chromium steels are utilized [15, 16].

Because of the high pressure inside the steam turbine casing, a special shaft sealing arrangement must be applied between the outer casing and the turbine rotor. Typically, contactless and wear-free labyrinth seals are used, which have the disadvantage of a small leakage steam mass flow through the sealing. Recently, the application of, alternative brush seal designs is being investigated further, to significantly reduce this seal leakage.

6.5
Summary and Outlook

Also in the future it is expected that a large part of the power generation will be based on coal-fired steam power plants. Compared to other fossil fuels, coal has a high content of carbon, so that this technology leads to high specific CO_2 emissions. For this reason, currently two different strategies of CO_2 reduction are considered. Besides pre- or postcombustion capture of the CO_2, the efficiency of the steam power plant must be increased significantly at the same time. In most modern steam power plants in Japan, Europe, and China, currently net efficiencies up to approximately 46% (based on lower heating value LHV) are obtained, with live steam parameters of 30 MPa/600 °C. The goal of increasing the net efficiency of coal-fired steam power plants to more than 50% can primarily only be reached by a further increase of the live steam parameters up to, for example, 35 MPa/700 °C. Regarding the 200 000 h creep strength values, the alloyed materials used today in power plants are no longer sufficient for such increased steam parameters. Thus, in the last years in Europe, various research projects (e.g., AD 700 and COMTES 700) have made efforts in the development and qualification of new materials for the application in steam generators and HP and IP steam turbines. The further increase of live steam temperatures up to 700 °C essentially affects the steam generator in the selection of materials for membrane wall tubes, final SH tubes, as well as thick-walled components like the final SH outlet header. These components are subject to the highest stresses in the steam power plant process and are exposed simultaneously to both high pressures and high temperatures. Compared to the steam generator components, the HP steam turbine is subjected to slightly lower pressures and temperatures. In comparison to the components used in the steam generator, relatively thick-walled structures for turbine rotor and casing (inner and outer casing) are used in the HP steam turbine. Particularly during load changes, these materials are stressed by high pressure and temperature gradients. Because the 9–12% chromium steels used today do not have a sufficient creep strength and oxidation resistance for steam temperatures above 600 °C, the application of nickel alloys for the highly stressed sections of both the steam generator and the HP turbine is currently being investigated.

References

1 Heitmüller, R.J. and Kather, A. (1999) Wärme- und feuerungstechnisches Konzept des Dampferzeuger für den BoA-Block Niederaußem K (in German). *VGB Kraftwerkstechnik*, **79** (5), 75–82.

2 Klebes, J. (2007) High-efficiency coal-fired power plants based on proven technology. *VGB PowerTech*, **87** (3), 80–84.

3 Bauer, F., Stamatelopoulos, G.N., Vortmeyer, N., and Bugge, J. (2003)

Driving coal-fired power plants to over 50% efficiency. *VGB PowerTech*, **83** (12), 97–100.

4 Bauer, F., Tschaffon, H., and Hourfar, D. (2008) Role of 700 °C technology for the carbon-low power supply. *VGB PowerTech*, **88** (4), 30–34.

5 Meier, H.J. (2009) Pre-engineering study for a 700 °C high-efficiency power plant. *VGB PowerTech*, **89** (10), 71–77.

6 Drbal, L.F., Boston, P.G., Westra, K.L., and Erickson, R.B. (eds) (1996) *Power Plant Engineering*, Chapman & Hall, New York.

7 Kitto, J.B. and Stultz, S.C. (eds) (2005) *Steam: Its Generation and Use*, 41st edn, The Babcock & Wilcox Company, Barberton.

8 Malek, M.A. (2005) *Power Boiler Design, Inspection, and Repair: ASME Code Simplified*, McGraw-Hill, New York.

9 VGB PowerTech e.V (2004) *Konzeptstudie Referenzkraftwerk Nordrhein-Westfalen*, VGB PowerTech Service GmbH, Essen.

10 Husemann, R.U. (2003) Development status of boiler and piping materials for increased steam conditions. *VGB PowerTech*, **83** (9), 124–128.

11 Scheffknecht, G. and Kather, A. (1997) Neue Werkstoffe im Dampferzeuger. *BWK*, **49** (11/12), 62–67.

12 Heiermann, G., Husemann, R.U., Hougaard, P., Kather, A., and Knizia, M. (1993) Steam generators for advanced steam parameters. *VGB Kraftwerkstechnik*, **73** (8), 584–594.

13 Kather, A. (1995) Verfahrenstechnische und konstruktive Auslegung moderner Braunkohle-Dampferzeuger. *VGB Kraftwerkstechnik*, **75** (9), 763–770.

14 Leyzerovich, A.S. (2008) *Steam Turbines for Modern Fossil-Fuel Power Plants*, Fairmont Press, Lilburn.

15 Kern, T.U. and Wieghardt, K. (2001) The application of high-temperature 10Cr materials in steam power plants. *VGB Kraftwerkstechnik*, **81** (5), 125–131.

16 Tremmel, A. and Hartmann, D. (2004) Efficient steam turbine technology for fossil fuel power plants in economically and ecologically driven markets. *VGB PowerTech*, **84** (11), 38–43.

7
High-Pressure Application in Enhanced Crude Oil Recovery

Philip T. Jaeger, Mohammed B. Alotaibi, and Hisham A. Nasr-El-Din

7.1
Introduction

7.1.1
Principal Phenomena in Oil and Gas Reservoirs

A brief introduction to the reservoir conditions from geological aspects will help in understanding the role of pressure in oil and gas reservoirs.

Petroleum is originally formed from insoluble organic matter called kerogen by pyrolysis under elevated temperatures up to 150 °C. Different factors contributed to the migration of fluids that are composed of alkanes and aromatics (Table 7.1). Cracking of long alkane chains into volatile components, such as methane, leads to pressure buildup in the reservoir. High reservoir temperatures (200 °C) also enhance the pressure accumulation under certain circumstances [1].

Each factor in Table 7.1 is thoroughly discussed in view of improving oil recovery techniques at high-pressure conditions.

7.1.2
Reservoir Conditions

Each petroleum reservoir has its specific properties according to the type of reservoir rock, its porosity, permeability, and conditions of temperature and pressure. Figure 7.1 shows some general relations for rock formations qualitatively.

As seen from Figure 7.1, rock porosity decreases as pressure and temperature increase along the depth. As a rule of thumb, the pressure increases at 0.1–0.25 bar/m of the reservoir depth, depending on the specific gravity of the rock matrix and whether the lithostatic or the hydrostatic pressure dominates the pressure profile. In addition, the fluid pressure depends on the phase composition of the original reservoir fluids.

Fluid saturation in reservoir rocks is divided into gas, oil, and water saturation; water refers to formation water that is believed to be the original fluid in place before

Industrial High Pressure Applications: Processes, Equipment and Safety, First Edition. Edited by Rudolf Eggers.
© 2012 Wiley-VCH Verlag GmbH & Co. KGaA. Published 2012 by Wiley-VCH Verlag GmbH & Co. KGaA.

7 High-Pressure Application in Enhanced Crude Oil Recovery

Table 7.1 Factors that influence petroleum migration in oil and gas reservoirs.

Phenomenon/property	Consequence
Overpressure due to dissolved gas	Migration, fractures
Water in formation	Buoyancy → migration upward
Capillary pressure	Migration stops
Rock porosity	Reservoir for petroleum
Rock permeability	Pathway for petroleum

being partly displaced by hydrocarbons due to capillary forces and gravity. Mathematically, the saturation is expressed in percentage of the volumetric storage capacity of a reservoir rock being occupied by the respective fluid. The respective saturation values sum up to give 100% [2]. In reservoir engineering, permeabilities are often related to the saturation giving relative values. Determination of the original hydrocarbon saturation, also referred to as the original oil in place (OOIP), is decisive for the efficiency of the exploitation of the respective reservoir.

Carbonates and sandstones are the dominating reservoir rocks. About 50% of the total oil reservoirs are found in carbonate rock formations (dolomite and limestones) [3]. In general, carbonates exhibit porosities of less than 10% and consequently less permeability than sandstones, but economical interesting production flow rates are achieved due to natural fractures that are commonly present in this type of rock. Sandstone reservoirs show a large range of texture regarding grain sizes, packing, and mineral composition, which affects the cement mineralogy. As a consequence, the porosity of sandstone formations varies to a great extent having influence on the saturation and on the permeability. Sandstone compaction leads to a 10–20% diminishing in porosity at depths of 2–3 km. Not only the texture of the solid phase but also the properties of the fluid mixture depend on the reservoir pressure, which will be discussed in the following section with more focus on the thermal

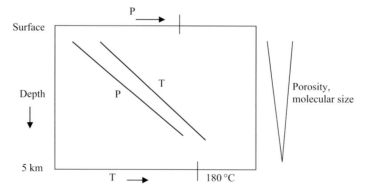

Figure 7.1 Typical conditions in hydrocarbon reservoirs depending on depth below surface.

physical properties such as phase behavior, density, viscosity, and transport properties.

7.2 Fundamentals

7.2.1 Miscibility at Elevated Pressures

As shown in Section 7.1.1, pressure buildup in a reservoir is due to the cracking of longer alkane chains into volatile components that are dissolved in the liquid phase according to the pressure generated simultaneously. Under static conditions of pressure and temperature, a vapor–liquid equilibrium is established. This equilibrium is disturbed as a result of fluid migration to new areas that show different conditions. Since migration usually occurs in the direction of decreasing pressure, solubility will be lowered until reaching the saturation pressure. According to the thermodynamic equilibrium, small-chain molecules will be released from the liquid phase mainly composed of high condensables at this point forming bubbles of a gas phase consisting of volatile hydrocarbons. This saturation condition is called bubble point and depends on the specific chemical composition of the petroleum. Above the bubble point pressure or minimum miscibility pressure (MMP), the fluid mixture forms one single phase. Below this pressure, a free gas phase will be formed and a new equilibrium state will be established between the gas phase and the liquid phase. Different methods exist for determining phase behavior, the application of which depends on the specific aim. One of the most common methods is the pressure–volume–temperature (PVT) test of the reservoir fluid [4]. This test is applied to simulate the depletion conditions of the reservoir. The apparatus used for PVT tests consists of a high-pressure view chamber with a variable but well-known volume. The reservoir fluid that is composed of one single homogeneous phase, including a dissolved gas (life oil), is placed inside the chamber at reservoir conditions. The pressure is reduced gradually by enhancing the volume at constant total mass until a second phase is formed. Volume, pressure, and temperature are recorded from which density data are retrieved. An interesting measure for reservoir engineering is the gas formation volume factor B_g, which may be determined by venting the gas formed after having reduced the pressure. Therefore, an expansion valve is opened simultaneously reducing the volume isobarically until all gas is expelled. The difference in volume resembles the gas formed at a defined pressure. Formation of the gas phase is often retarded due to nucleation effects leading to oversaturation of the liquid.

Further methods for determining the MMP are rising bubble apparatus (RBA) [5] or vanishing interfacial tension (VIT) [6]. These tests are also carried out in high-pressure view chambers preferably containing tongue-shaped windows for visualizing changing levels of the participating fluid phases. Figure 7.2 shows an

Figure 7.2 Equipment for determining the MMP from optical determination.

equipment for observation of phase behavior especially focused on determination of the MMP.

MMP tests are not limited to reservoir fluid mixtures, rather they may include fluids used for miscible or partly miscible flooding in enhanced oil recovery (EOR), which is described in more detail below. Efficiency of these processes depends very much on the miscibility of these fluids under reservoir conditions. Oil with higher contents in lighter and intermediate hydrocarbons (C7–C12) show lower miscibility pressures. Carbon dioxide is the most miscible fluid among the gases that are commonly applied in EOR. Whereas in the case of a moderately heavy oil (API 26°) complete miscibility with nitrogen is generally far from being achieved at 60 MPa and 130 °C, the same oil may mix with CO_2 in a homogeneous phase already below 20 MPa.

7.2.2
Physical Chemical Properties of Reservoir Systems at Elevated Pressures

The physical properties that affect the high-pressure processes are changed due to two reasons: (i) mechanical effect of high pressure, and (ii) increased mutual solubilities of the two adjacent phases at high pressure. The physical properties (such as density, viscosity, and interfacial tension) have an impact on the multiphase flow, phase separation, and process efficiency. Diffusivity and wetting characteristics are related to the properties mentioned, and thus also subject to alterations. Equipment available for experimental determination of physicochemical quantities at elevated pressures are also described in Chapter 14.

7.2.2.1 Density
The densities of the adjacent fluid phases at elevated pressures may vary considerably compared to atmospheric conditions due to the mechanical compressibility and

Table 7.2 Different possibilities of determining density under reservoir conditions.

Method	Working principle	Application	Remarks
PVT	Absolute volume	Crude oil	Experimental error considerable
Vibrating tube	Resonance frequency	Oil with dissolved gas	Calibration required
Interferometry	Interferometric pattern	Science	Sophisticated, single phase
Magnetic balance	Archimedes	Gas–liquid	Absolute, precise
Optical detection	Archimedes	Gas–liquid	Simple, less precise

mutual solubility. The density of both contacting phases may be measured by different methods, as listed in Table 7.2.

Occasionally, the so-called API gravity (American Petroleum Institute gravity) is used in order to differentiate between heavy and light oils. The definition is given by

$$°API = 141.5/API_{60} - 131.5$$

where API_{60} is the specific gravity at 60 °F (15.6 °C).

Crude oils are generally classified as follows:

Heavy oils	API < 20°
Normal oils	API 20–40°
Light oils	API > 40°

A density of 1 g/cc equals an API of 10°.

A common method for measuring the mixture density of liquid hydrocarbons, containing dissolved gases like methane or carbon dioxide, is the vibrating tube being adapted into a phase equilibration cycle or the Archimedes principle using a microbalance, both shown in Figure 7.3.

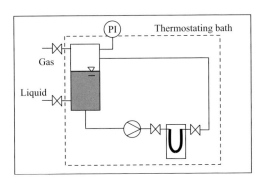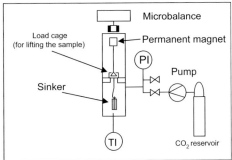

Figure 7.3 Experimental setup for determining the density of liquid–gas mixtures by the oscillating tube and the Archimedes principle under reservoir conditions described in Ref. [7].

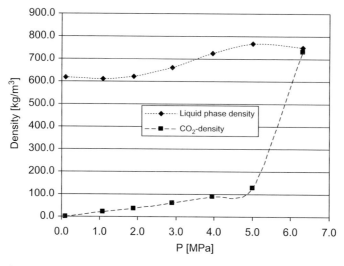

Figure 7.4 Density of n-pentane–CO_2, $T = 300$ K. Fluid phase from Ref. [8]; liquid phase: own data.

The advantage of the Archimedes principle is the possibility of measuring the density under saturation conditions. Any bubbles that are formed under these conditions will disturb the measurements in the oscillating tube. Figure 7.4 shows the density of n-pentane saturated with CO_2 as the pressure is increased.

It is clear that mixing of CO_2 with n-pentane at low pressures does not cause much variation in density. At elevated pressures, the liquid density is considerably increased even though the density of pure CO_2 is much lower. Complete miscibility is achieved at 6.3 MPa where the densities of both phases become equal.

7.2.2.2 Rheology

The viscosity is a relevant rheological property in oil recovery when injecting fluids for oil displacement. If the viscosity of the displacing fluid is too low, fingering and bypassing may occur. For determining viscosity, a number of methods exist depending on the type of fluid of interest. Table 7.3 gives an overview of possible methods and their applications.

Although crude oils may have large viscosities depending on the content in long-chain hydrocarbons especially asphaltenes, they usually still behave as Newtonian fluids. Recently, the so-called viscoelastic surfactants (VES) are increasingly studied for being applied in enhanced oil recovery in order to facilitate hydrocarbon flow toward the oil well. This class of substances shows a non-Newtonian behavior as a result of a highly concentrated solution of long-chain surfactants that are entangled within each other. [9]. In Ref. [10], the viscoelastic behavior of weak gel systems was investigated by quantifying the elastic and loss modulus and validating their behavior in view of applicability in EOR. The respective polymer solution is injected into the reservoir after conventional water flooding. In the first place, some of the remaining oil is displaced due to the higher viscosity. Within the reservoir, cross-linking is

Table 7.3 Overview of possibilities of measuring viscosity under reservoir conditions.

Method	Working principle	Application	Remarks
Rolling ball	Stokes flow of a sphere: falling time	Newtonian fluids, moderate viscosity	Experimental uncertainty
Rising bubble	Rising time	Liquids	Experimental error due to mobile interface
Capillary flow	Poiseuille flow: pressure drop	Newtonian fluids, wide range of viscosity	Simple
Oscillating crystal	Damping of oscillation	Process viscosimeter	Robust
Rotation viscosimeter	Shear flow within a channel: required torque	Newtonian/non-Newtonian fluids (e.g., polymeric solutions, VES)	Costly

initialized that induces an elastic behavior of the flooding fluid and blocks the pores. In subsequent water flooding, elevated pressures are applied until the loss modulus starts to dominate inside the larger pores forcing the main stream to take this way and displacing even more of the remaining oil.

7.2.2.3 Interfacial Tension

The interfacial tension is an important quantity having influence on the migration of liquids in small pores due to capillarity, that is, the additional pressure difference induced by curvature of fluid surfaces due to the specific wetting behavior and the interfacial tension. Since 1930s, the pendant drop method is known for determining the interfacial tension from drop shapes that are generated within a view chamber at elevated pressures. For decades, experimental values were subject to fairly high errors due to analogous photographic images and an empirical evaluation method by means of a selected plane [11]. The API still recommended the pendant drop method using this evaluation method in 1990 [12]. Later, electronic data processing allowed to digitalize drop images and thereby solve the theoretical equation of a drop profile exactly meeting the physical laws [13]. In Figure 7.5, experimental data on interfacial tension are depicted in systems showing partial or complete miscible depending on the conditions of operation.

Figure 7.5 shows a clearly decreasing interfacial tension at enhanced pressures. In the case of higher mutual miscibility, this effect is more pronounced. For instance, in n-pentane–CO_2, the interfacial tension rapidly decreases and finally vanishes at the MMP at around 6.7 MPa.

7.2.2.4 Wetting

The wetting behavior is an important issue in EOR. Next to the interfacial tension is the capillarity in small pores of the rock formations that is influenced by the wetting of the inner pore surfaces. The wetting is characterized by the so-called three-phase

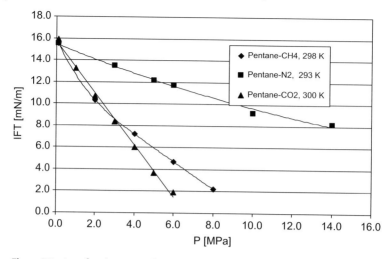

Figure 7.5 Interfacial tension of n-pentane in compressed gas atmospheres – own data.

contact angle Θ (Figure 7.6). This contact angle is associated with the interfacial tension by Young's equation [14]:

$$\sigma_{solid,fl} = \sigma_{liq,fl} \cos\theta + \sigma_{solid,liq}$$

This relationship may also be interpreted as a balance of forces in a horizontal direction between the interfacial tension $\sigma_{l,fl}$, the solid–(drop) liquid interfacial energy $\sigma_{s,l}$, and the solid–(surrounding) fluid interfacial energy $\sigma_{s,fl}$. From the surface tension and contact angle, the surface energy of solids in ambient air atmosphere can be deduced by applying Young's law to a series of experimentally determined values of contact angle and surface tension using adequate regression procedures [15]. In the condensed gas atmosphere, first attempts exist for obtaining the respective data of solid–fluid interfacial energies [16]. In general, wettability is only insignificantly affected by the pressure itself as long as no gases are present. In cases of gases being dissolved in the liquid phases, interfacial tension changes dramatically, and thus also leads to alterations in wettability. Extensive work has been published on wetting in reservoir systems mainly being restricted to investigation of single effects like temperature [17], pressure [18], and aging [19]. The effect of dissolved gases has been studied mainly in ideal systems [20]. In Ref. [7], some

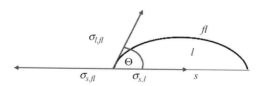

Figure 7.6 Contact angle and interfacial energies acting at the three-phase contact.

Table 7.4 Contact angle of a crude oil drop in seawater under CO_2 pressure and at 50 °C – own data.

Rock type	Pressure (MPa)	Contact angle (°)
Limestone	5.51	93
Sandstone	5.51	133
Calcium carbonate crystal (smooth)	5.51	123
	12.40	115
	19.29	112
	20.67	111

wetting data of a crude oil–seawater–CO_2 system on different types of rock surfaces are shown. Further data on contact angles are summarized in Table 7.4.

An increasing gas pressure leads to a slightly decreasing contact angle, as was also found in Ref. [16]. In cases of liquid–liquid systems and in the absence of any gases, an increasing contact angle with rising pressures was found, which may be due to the fact that the solid surface energies are almost not affected. Hence, a decreasing interfacial tension will result in an increasing contact angle fulfilling Young's equation. In general, decreasing salt concentrations result in enhanced water wetting as was found in Ref. [21].

7.2.2.5 Diffusivity

The diffusivity of gases in liquids plays a role in gas injecting systems, when displacing an oil phase or expanding the oil by gas dissolution. In Figure 7.7, the amount of CO_2 entering crude oil is depicted as a function of time.

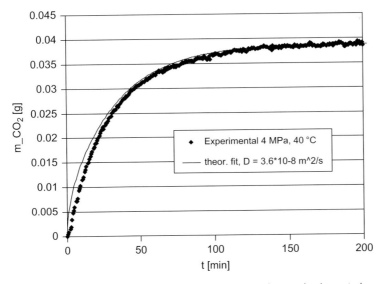

Figure 7.7 Determination of the diffusion coefficient by adapting the theoretical approach to the experimental sorption kinetics. Crude oil inside a cylindrical containment in CO_2 atmosphere, 4 MPa, 40 °C – own data.

The absorption kinetics may be described as follows [22]:

$$\left(\frac{\partial C}{\partial t}\right) = \frac{1}{r}\frac{\partial}{\partial r}\left(rD\frac{\partial C}{\partial r}\right)$$

By varying the diffusion coefficient until the theoretical approach coincides with the experimental values, the diffusion coefficient is determined. At 4 MPa and 40 °C, a diffusivity of 3.6×10^{-8} m²/s is obtained.

7.2.2.6 Permeability

Permeability is an important property of rock formations, depending to a great extent on its porosity as shown qualitatively in Figure 7.8.

The permeability results as a product of the diffusivity or mobility of the reservoir fluid and its saturation, the content of the fluid within the formation. The permeability is determined using core flooding equipment and differential pressure transducers for detecting the parameters necessary for evaluation according to Darcy's law for laminar flow:

$$\dot{M}_{CO_2} = -\varrho_{CO_2} A_{cyl} \frac{B}{\eta} \text{grad}(P)$$

where $\dot{M}_{CO_2}$ is the CO_2 mass flow, η is the dynamic viscosity, B is the bed permeability, and A_{cyl} is the cross-sectional area of the cylinder. The unit of the permeability is mainly referred to as "milli-Darcy" or "md."

In order to determine the permeability of a solid, a fluid, usually water, is passed through the solid bed or core sample located in a cylindrical autoclave and sealed toward the inner walls of the vessel in order to inhibit leakages (Figure 7.9). In a certain range of mass flow and corresponding pressure drop, a definite value of permeability can be calculated.

At porosities of 18–20%, permeabilities of 150–200 md were found for Berea sandstone rocks. In limestones of 4% porosity, permeabilities of 30 md and less have been reported [21]. As a rule of thumb, a reservoir can produce oil without stimulation in case the permeability is higher than 10 md. Above 1 md, gases may be produced, while for recovery of oil the reservoir will require stimulation below 10 md.

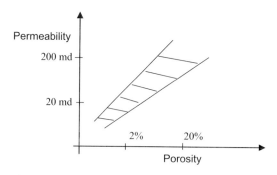

Figure 7.8 Permeability as a function of rock porosity.

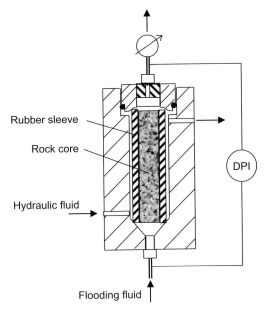

Figure 7.9 Experimental setup for determining the permeability of core samples.

7.3
Enhanced Oil Recovery

The fluid mixture in a reservoir usually contains small-chain hydrocarbons that are dissolved within the liquid phase at the elevated pressures and temperatures being present in the formation. In fact, pressure buildup is closely related to the presence of volatile compounds. When starting to exploit the reservoir, the pressure drops until reaching a certain value from where onward free gas can start to be formed from a thermodynamical viewpoint. This pressure is referred to as the bubble point pressure or minimum miscibility pressure (Section 7.2.1). Thermodynamically, the liquid phase will be oversaturated below this pressure and unless the first nuclei are formed, the gas will remain dissolved. This delay in nucleation is caused by the change in vapor pressure at a curved fluid interface described by the Kelvin equation. From the viewpoint of a gas inside a bubble, this pressure adds up to the overall system pressure and may achieve considerable values in cases of very small radii of curvature. During primary recovery, the natural pressure resulting from the expanding gas expels the oil into the wellbore, which is also named "dissolved gas drive" [23]. The gas saturation of the reservoir starts to increase until reaching a critical value because gas bubbles coalesce, starting to hamper oil production while the gas itself breaks through being driven out of the reservoir instead of the liquid oil. From this moment onward, artificial lift techniques such as pumps and compressors are required in order to bring the oil to the surface. In sum, only about 10–15% of a reservoir's original oil in place (OOIP) is typically produced during primary recovery.

Secondary recovery techniques comprise injection of water in order to displace the oil and gas injection for maintaining the pressure of the reservoir. The cumulative yield of the primary and secondary oil recovery amounts to 38–43% according to Ref. [24]. Success of this technique is mainly limited due to unfavorable wetting conditions. Especially in case of heavy oils, its high viscosity inhibits satisfactory yields merely using water for displacement. Due to the small viscosity ratio, the so-called "fingering" of the water at the water–oil interface is observed reducing the displacement effect.

Some confusion exists on the use of the expression "enhanced oil recovery" which is increasingly replaced by "improved oil recovery." Enhanced oil recovery refers to a general way of enhancing oil production by artificial means that can take place as improved primary, secondary (advanced secondary recovery(ASR)), or tertiary recovery, that is, tertiary recovery always means enhanced recovery, while the same techniques of EOR can also be applied in secondary recovery. In order to overcome the drawbacks and insufficiencies of the former recovery techniques, these were extended toward the tertiary (enhanced) oil recovery starting in the 1960s. In the meantime, about 3% of the world's oil production originates from EOR [25]. These methods offer prospects for ultimately producing more than 60% of the reservoir's original oil in place. Three major categories of EOR have been found to be commercially successful to varying degrees. The use of aqueous solutions also comprises the use of chemical agents, gas injection, and thermal treatment. The particular method applied can also be a mixture of the mentioned effects. For example, autoignition makes use of gas injection in order for combustion to take place. As a consequence of the subsequent thermal effect, the viscosity of heavy oil is lowered supporting its flow toward the wellbore.

A method for simulating the EOR process and testing achievable recoveries is the slim tube experiment. The displacing fluid and the operating conditions are preselected using results from IFT, rising bubble, and wetting experiments described above. A narrow tube is filled with milled rock (sand) that is saturated with the respective oil. The displacing fluid is passed through the tube at the selected operating conditions. A typical setup of the slim tube experiment is shown in Figure 7.10.

Another widespread method is the core flooding experiment on a core sample of normally about 40 mm diameter and up to 500 mm length. The equipment used for this type of investigation is identical with the one shown in Figure 7.9. The core sample, for example, sandstone, is previously saturated with brine. Afterward oil is passed through the rock until no water is driven out of the sample any longer, resembling the irreducible water saturation. The test consists in passing a fluid designated for reservoir flooding (e.g., brine and carbon dioxide) in order to investigate possible oil recoveries. Figure 7.11 shows performance of a test applying aqueous solutions containing different salt concentrations successively. The amount of flooding fluid is related to the total pore volume of the core sample.

Seawater injection as a secondary mode produced 76.3% OOIP at 3.6 pore volumes. Afterward, aquifer water was injected as a tertiary mode and that increased

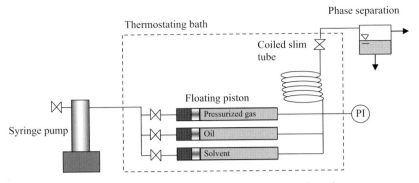

Figure 7.10 Experimental setup for determining the recovery in a slim tube.

the oil recovery to 78.4% OOIP. Dilution of aquifer water at 50 vol% showed no effect on oil recovery. Pressure drop data were not stable at the beginning because of the viscous forces and relative permeability difference. Then, results showed a consistent pressure drop across the core with time.

7.3.1
Water Flooding

After having recovered no more than 20% of the OOIP by pressure drive due to the original reservoir pressure, the simplest and cheapest way of increasing the yield is to inject an aqueous phase, as it is abundant and easy to handle. Safety requirements are not excessive and material compatibility is not an issue. The disadvantage is the relatively low effectiveness since the rock surfaces being prewetted by the oil phase

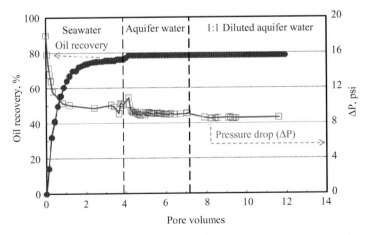

Figure 7.11 Core flooding using seawater (high salinity), aquifer water (medium salinity), and low-salinity water [21].

are unlikely to change wettability just by pure water. Hence, a significant amount of oil will still remain in the reservoir for which this technique is limited to secondary recovery giving way to more advanced techniques being discussed in the next sections. In various places, the water injected is not pure, as it contains some salts because of its origin, for example, seawater. In Saudi Arabia, low-salinity water flooding has been practiced for carbonates for more than 65 years [3]. Later investigation on the influence of the salinity on the wetting started. In Ref. [26], it was reported that low-salinity brine can improve oil recovery compared to that with higher salinities. A drawback of using aqueous systems is the formation of emulsions by inclusion of oil and consequently the need of phase separating techniques downstream of the oil well.

Water flooding has been practiced in Venezuela since the early nineties [27], but here combustion techniques were developed in order to handle oil–water emulsions directly.

7.3.2
Chemical Injection

Chemical injection can involve the use of surface active agents for lowering the interfacial tension and as a consequence changing the wetting properties of the specific rock surface toward water wet. The objective is to recover the oil by imbibition rather than drainage (Figure 7.12).

Figure 7.12a is more favorable because of the lower pressure that has to be applied and the higher yield to be obtained. On the other hand, long-chained molecules like polymers may increase the effectiveness of waterfloods by establishing a favorable viscosity. Not only the viscosity itself but also the viscoelasticity can be made use of. Small pores may be sealed by viscoelastic fluids containing a higher elasticity modulus at lower shear, forcing the main flow to pass through the larger pores. Main application of polymer flooding is the Daqing reservoir in China where production is increased by up to 12%, whereas chemical techniques account for less than 1% of US EOR production. As long as no gases are present in chemical flooding, there will not be much influence of pressure on the behavior of the used additives. In the presence of gases, especially carbon dioxide, extraction effects may lower efficiency of chemical flooding, especially nonionic surfactants will be subject to extraction.

7.3.3
Thermal Recovery

Thermal recovery involves the introduction of heat, for example, by injection of steam, to lower the viscosity or thin the heavy viscous oil and improve its ability to flow

Figure 7.12 Flooding mechanisms according to the wetting behavior. $P_i < P_d$. (a) Water wet: imbibition, P_i. (b) Oil wet: drainage, P_d. (c) Fingering.

through the reservoir. Thermal techniques account for over 50% of US EOR production, primarily in California. In steam flooding, temperatures of up to 350 °C are achieved. In order to avoid breakthrough of the displacement fluid that is the condensating water in this case, not only the oil viscosity itself but also the viscosity ratio (μ_o/μ_w) needs to be reduced [28]. In flooding with superheated steam at a temperature of 300 °C and pressure of 7.5 MPa, the MMP may eventually be passed enabling miscible flooding. Thermal recovery is especially applied to heavy oils (see below).

7.3.4
Gas Injection

Gas injection is preferred to water flooding especially in cases of low-permeability reservoirs [29]. Field tests are carried out in order to investigate the long-term effectiveness. Compared to water flooding, gas injection is superior after 3 years [30]. It has been performed in Venezuela since 1998 [27] and accounts for nearly 50% of EOR production in the United States. Gas injection techniques use gases such as natural gas, liquid petroleum gas (LPG), nitrogen, carbon dioxide, and flue gas. Depending on its miscibility with the crude oil, the mechanism ranges from miscible to immiscible displacement by expansion, lowering the viscosity for improving flow conditions and accompanying extraction of high volatiles. LPG that is usually completely miscible at moderate pressures of up to 8 MPa [31] can be used for driving crude oil from shallow reservoirs, but this technique is fairly expensive. In the case of nitrogen, mutual miscibility is negligible, making the immiscible displacement the dominating mechanism. Since the viscosity ratio between the displacement fluid and the crude oil is very low, viscous fingering will take place. Nitrogen will break through leading to insufficient yields. It was rather found to be applicable mainly to light oils of API > 35° containing volatile hydrocarbons above the MMP [32]. Applying carbon dioxide has mainly two advantages compared to nitrogen. First, the viscosity is higher and, second, CO_2 will lower the viscosity of the crude oil considerably [33], both counteracting the viscous fingering at the displacement front. Conditions of CO_2 flooding are found as 14 MPa and 80 °C [34], but can reach more than 40 MPa. Miscible gas flooding taking place at higher pressures actually comes close to a gas or supercritical (SC) fluid extraction process depending on the state of the injected fluid. However, in order to perform a real supercritical extraction process *in situ*, the pressure needs to be maintained throughout the complete production line. Reference [35] states that 78% of the OOIP were extracted from sandstone and limestone by SC CO_2 at 350 bar and 160 °C.

The advantage of using air instead of carbon dioxide, nitrogen, hydrocarbons, or flue gas is its availability. Air injection is potentially attractive for high-pressure light oil reservoirs and is being applied in the Williston Basin in the North Central United States since over 20 years [36]. When injecting air, the so-called autoignition is in principle always possible, which is described in Section 7.3.3.

In the past years, hybrid techniques have come up, for example, combining the displacement effect of an expanding solvent like propane, butane, or hexane with the temperature effect in the steam-assisted gravity drainage (ES-SAGD) being found to be more efficient than application of steam only [37]. The alkanes are directly mixed with the steam at the displacement front. The drawback of this method is the relatively high value of the alkanes being used. Furthermore, water alternating gas injection (WAG) is used in horizontal floods for compensating unfavorable gravity segregation due to the high-density difference between the gas and the oil to be removed.

Gas injection can also follow the objective of maintaining the pressure of gas condensates above the dew point for keeping condensates (heavier hydrocarbons) in the gaseous phase as long as the pressure is high enough, hence enabling recovery of these components simultaneously [38].

7.3.5
Carbon Dioxide Capture and Storage (CCS) in EOR

The use of flue gases in crude oil recovery has been mentioned in Section 7.3.4. In order to have advantage of the high miscibility of carbon dioxide and cope with the problem of eliminating large quantities of carbon dioxide from electrical power plants, development has intensified in the past years to concentrate carbon dioxide from the flue gases obtaining a condensed or supercritical phase that can be stored in smaller volumes and further used for enhanced oil recovery more efficiently [39]. The achieved purity is decisive for an efficient condensation and subsequent compression to pressures of up to 60 MPa, which is required in offshore exploitation. Flue gases from electrical power plants commonly contain considerable amounts of nitrogen, SO_2, and other trace substances that have an influence on the condensation conditions. In order to efficiently condensate carbon dioxide, noncondensables need to be separated beforehand (see also Chapter 7).

Furthermore, some components need to be eliminated, like water, since otherwise for example, compressors will be damaged due to corrosion. The corrosion issue is another main topic of investigation dedicated to develop ways of handling such large amounts of carbon dioxide.

7.3.6
Combustion

Injection of reactive gas mixtures may be used for generating a combustion front *in situ* within the reservoir. Next to the expansion and subsequent pressure increase, the enhanced temperature leads to decreasing viscosities especially in heavy oil reservoirs. The traveling high-temperature front consumes a small percentage of oil in place, while displacing and producing the rest. It is a very efficient method of enhanced oil production. For combustion, either pure air or mixtures containing methane, line oil, and other carbon sources may be applied. At reservoir temperatures of 80 °C and above, autoignition occurs. Temperatures at the fire front may reach 500 °C with pressures of around 0.8 MPa and at depths of 80–200 m.

7.4
Oil Reservoir Stimulation

Reservoir stimulation is a treatment performed to restore or enhance the productivity of a well. Stimulation treatments fall into two main groups, hydraulic fracturing treatments and matrix treatments. A fracture is a crack or surface breakage within the rock not related to foliation or cleavage in a metamorphic rock along which there has been no movement. A fracture along which there has been displacement of the contacting surfaces in a lateral way is called a fault. When walls of a fracture have moved only normal to each other, the fracture is called a joint, which usually has a more regular spacing. Fractures can enhance permeability of rocks greatly by creating new pathways and connecting pores. For that reason, fractures are induced mechanically in some reservoirs in order to boost hydrocarbon flow. Fracturing treatments are performed above the fracture pressure of the reservoir formation and create a highly conductive flow path between the reservoir and the wellbore. The required fracture pressure may reach 2–2.5 times the natural pressure in the pores (specific gravity) (see Section 7.1). The fracture width increases linearly with pressure. After pressure release, the fractures close until the closure stress is achieved at zero pore width. The viscosity of the fracture fluid also plays an important role. Whereas a higher viscosity may be an advantage in terms of fracturing, it may impede its drainage during oil production. In this sense, the use of compressed carbon dioxide as a fracturing fluid is advantageous in view of its low surface tension and consequently easy cleanup. Suspensions of a proppant like sand in water are used in order to bear the fracture openings.

Matrix treatments are performed below the reservoir fracture pressure and generally are designed to restore the natural permeability of the reservoir following damage to the near-wellbore area. Matrix acidizing in carbonates plays an important role in production enhancement in oilfields. The task is different compared to sandstone reservoirs. Negative skins can occur and in many carbonate reservoirs, natural fractures exist. Nevertheless, carbonates often show low matrix permeability. Just creating "wormholes" near the wellbore might not be sufficient. By fracture acidizing, conductive paths deeper into the formation are provided.

7.5
Heavy Oil Recovery

Initially, steam is injected into the formation for a period of time by way of an injection well and a production well. Thereafter, a combustion-supporting gas is injected through the injection well into the top of the formation to form a fluid conductive path between injection and production wells. Subsequently, steam is injected into the formation, preferably near the bottom of the formation and flows through the fluid conductive path. Heated oil adjacent to the top of the formation is produced by steam drag into the producing well or wells.

Solid hydrocarbon materials that remain in a subsurface earth formation such as the coke residue from a subterranean tar sand deposit has previously been exploited by means of a controlled oxidation process. Nowadays it is reignited and an oxygen-containing gas such as air is injected to burn the coke residue. The temperature of the sand or other formation matrix thereby increases substantially. Water is then injected into the formation to absorb heat from the hot sand or formation matrix. Hot water and/or steam is thereby generated for use in thermal oil recovery methods in the immediate vicinity, without the need for burning natural gas or other fuels that can be used more advantageously. Since the permeability of a tar sand formation is substantially greater at the conclusion of the controlled oxidation reaction than it was initially, water containing appreciable solids suspended therein as well as the minerals dissolved therein may be utilized without the danger of plugging the formation, thus eliminating the cost of water treatment as would be required in conventional steam generation practice.

Air has also been used in heavy oil reservoirs to generate heat and steam to mobilize oil in place. For light oil applications however, the *in situ* generated flue gas (85% N_2 + 15% CO_2) is the main driving force for oil displacement. The flue gas pressures up the reservoir, mobilizes, strips, and swells the oil in place, and at sufficient pressures can result in miscible and near-miscible displacement of the oil. For deeper reservoirs, the generated supercritical steam (above 22.1 MPa and 374 °C) also efficiently extracts and displaces *in situ* crude oil components.

7.6
Hydrates in Oil Recovery

In the past decades, subsea oil and gas production has reached increasingly deeper waters being associated with higher pressures and lower temperatures. At these conditions – although above the ice point of water, cage-like structures of water (host) molecules accommodate gas molecules like methane and carbon dioxide. Although similar to the strength in bonding of ice, water needs the so-called guest molecules for obtaining a stable crystalline structure as hydrates [40]. These hydrates are known to affect gas and oil recovery for almost a century mainly due to plugging of the transportation lines. A number of methods exist to prevent hydrate formation such as heating and injection of thermodynamic inhibitors like methanol. Thermodynamic inhibitors diminish the range of temperature and pressure in which hydrates are thermodynamically stable, as shown in Figure 7.13.

However, these methods become rather uneconomic at depths of up to 2000 m with the required high-throughputs of inhibitors. Consequently, it becomes important to study kinetics of hydrate formation for prediction of their growth in order to adapt the right tools like kinetic inhibitors or even depressurization. Especially, in EOR, the injector head may be heated to increase the temperature of the injected gas stream. The influence of inner wall surfaces on adhesion leading to plug formation has not been studied systematically so far.

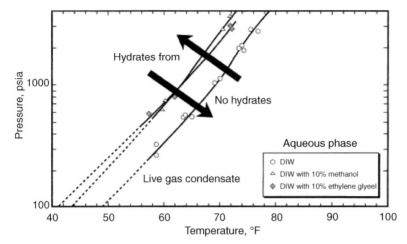

Figure 7.13 Phase diagram for CO_2 hydrates. Influence of MeOH as thermodynamic inhibitor [41]. With kind permission from SPE.

Lately, the idea came up to make use of hydrates as an energy source combined with CCS by exchanging methane in natural gas hydrates by anthropogenic carbon dioxide.

7.7
Equipment

For a detailed description of high-pressure equipment including design aspects, refer to Chapters 12 and 13. In this section, some remarks are made on specific developments in enhanced oil recovery.

7.7.1
Pumps

Compared to reciprocal pumping devices, screw pumps possess the advantage of pulsation-free conveying of multiphase fluid mixtures consisting of 100% liquid to 100% gaseous hydrocarbons. On the other hand, feasible pressure buildup amounts to only 10 MPa at flow rates of up to 4000 m^3/h (11 500 gpm). Especially, when the well head pressure declines, twin-screw pumps may be used instead of centrifugal or reciprocal pumps to achieve the pressure necessary for transporting high–viscosity heavy oils to the main pipeline and downstream treatment plants like refineries. For high-viscosity oils, the possibility of steam heating exists: heavy oil and bitumen (API 8–9) sands recovery is carried out by *in situ* steam-assisted gravity drainage (SAGD). Recently, electric submersible pumps have been developed that withstand temperatures up to 200 °C [42].

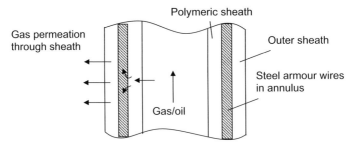

Figure 7.14 Principal design of a flexible tube.

7.7.2
Pipes

Especially in offshore oil production, common steel tubings are increasingly replaced by the so-called flexible pipes, which reveal a higher tolerance to movements caused by ocean streams. The principal idea is to separate required resistance to mechanical pressure from liquid and gas tightness by different wall layers. An inner polymeric pressure sheath is responsible for conducting the injection or production fluid inside the tubing. Surrounding the inner sheath, the steel reinforcement wires are wound up in a diagonal manner in order to withstand the mechanical stress while maintaining the flexibility of the tubing. An outer sheath prevents seawater from entering the annulus in-between both sheaths, hence protecting the steel armour wires from corrosion. The principal design of a flexible tube is shown in Figure 7.14. For a detailed description of the design and working principle, refer to Ref. [43] who also describes the permeation process of hydrocarbons through the polymeric sheath material, as shown in Figure 7.14.

Since gas permeation cannot be completely inhibited, criteria have to be fulfilled to ensure a long-term stability of the pipes against fatigue failure due to changing material properties. These criteria are given in ISO 13628:2006 and were adopted by the API specifications 17 J.

7.7.3
Seals

In many applications, elastomers are used as sealing materials in contact with typical gases present in oil and gas recovery like NBR, FKM (Viton®), EPDM, or recently also TFEP (AFLAS®), a copolymer of tetrafluoroethylene (TFE) and propylene (P) that can be used up to 200 °C. Like in the case of the flexible pipes described above, the gas diffuses into the sealing material to some extent leading to loss of flows into the atmosphere. Especially H_2S and CO_2 dissolve to a considerable extent resulting in high permeation coefficients [44].

In case of emergency shutdown and rapid depressurization of high-pressure equipment used in oil recovery, the gas that is dissolved inside elastomeric seals will

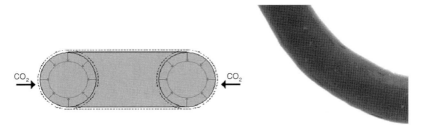

Figure 7.15 Penetration of CO_2 into seals and result of posterior explosive decompression.

not have enough time to leave by diffusion, but will rather form gas bubbles with a high internal pressure. The elastomer does not withstand this pressure and will give way to the enclosed gas by disruption, which is known as explosive decompression illustrated in Figure 7.15.

This problem of gas permeation through seals can partly be counteracted by use of thermoplastic sealing materials with higher diffusion resistance like polyamide or teflone (especially used for dynamic sealing) or by constructive solutions for enhancing the shielding effect and constrain dimensional changes thereby diminishing gas dissolution. Test requirements that have to be met for qualification of sealing materials are given in the Norsok Standard M-710.

7.7.4
Separators

Because of the enhanced oil recovery, especially using aqueous surfactant solution, an emulsion is obtained that needs to be separated for obtaining the pure hydrocarbon phase. For the working principle and design of common separators, refer to Ref. [45]. An interesting technology has come up in recent years for cleaning crude

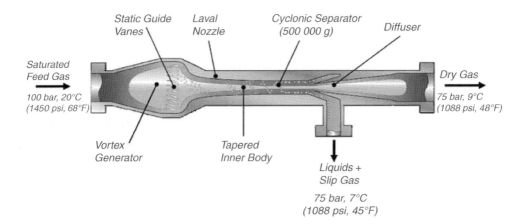

Figure 7.16 Principal design of a Twister separator. With permission from Twister B.V.

natural gases, making use of a temporary pressure and temperature drop resulting from an isentropic expansion in a Laval nozzle [46]. The hydrocarbon–water mixture is accelerated to supersonic velocities reducing the hydrostatic pressure. Simultaneously, a swirl is induced by which tiny droplets of water or condensed hydrocarbons are centrifuged toward the inner wall, from where the separated liquid is collected like in a cyclone and retrieved through openings in the circumference. The working principle is illustrated in Figure 7.16. These so-called Twisters have passed the prototype state and are now reaching qualification for subsea applications, for example, offshore Brasil. Working pressures amount up to >15 MPa, but the pressure is rather limited by the application than due to the apparatus itself.

References

1 Allen, P.A. and Allen, J.R. (2005) Chapter 10, in *Basin Analysis*, 2nd edn, Blackwell Publishing, Oxford.
2 Dandekar, A.Y. (2006) *Petroleum Reservoir Rock and Fluid Properties*, Taylor & Francis, Boca Raton.
3 Okasha, T.M. and Al.-Shiwaish, A.-J.A. (2009) Effect of brine salinity on interfacial tension in Arab-D carbonate reservoir, Saudi Arabia. Proceedings of the SPE Middle East Oil and Gas Show and Conference, 2009, MEOS, pp. 236–244.
4 Danesh, A. (1998) *PVT and Phase Behaviour of Petroleum Reservoir Fluids*, Elsevier.
5 Elsharkawy, A.M. (1996) Measuring CO_2 minimum miscibility pressures: slim tube or rising bubble method. *Energy Fuels*, **10** (2), 443–449.
6 Rao, D.N. and Lee, J.I. (2003) Determination of gas-oil miscibility conditions by interfacial tension measurements. *Journal of Colloid and Interface Science*, **262**, 474–482.
7 Jaeger, P., Alotaibi, M., and Nasr-El-Din, H.A. (2010) Influence of compressed carbon dioxide on capillarity of reservoir systems. *Journal of Chemical and Engineering Data*, **55**, 5246–5251.
8 Kunz, O., Klimeck, R., Wagner, W., and Jaeschke, M. (2004) The GERG-2004 wide-range equation of state for natural gases and other mixtures. *VDI Fortschrittber.*, **6**, 557.
9 Houa, J., Liub, Z., Zhangc, S., Yuea, X., and Yangc, J. (2005) The role of viscoelasticity of alkali/surfactant/polymer solutions in enhanced oil recovery. *Journal of Petroleum Science and Engineering*, **47** (3–4), 219–235.
10 Wang, W., Liu, Y., and Gu, Y. (2003) Application of a novel polymer system in chemical enhanced oil recovery. *Colloid and Polymer Science*, **281**, 1046–1054.
11 Andreas, J.M., Hauser, E.A., and Tucker, W.B. (1938) Boundary tension by pendant drop. *Journal of Physical Chemistry*, **42**, 1001–1019.
12 API, Recommended Practices for Laboratory Evaluation of Surface Active Agents for Well Stimulation, API Recommended Practice 42, 2nd edn, Reaffirmed February 1992.
13 Song, B. and Springer, J. (1996) Determination of interfacial tension from the profile of a pendant drop using computer-aided image processing 1: theoretical. *Journal of Colloid and Interface Science*, **184**, 64–76.
14 Young, T. (1805) An essay on the cohesion of fluids. *Philosophical Transactions of the Royal Society of London*, **95**, 65–87.
15 Girifalco, L.A. and Good, R.J. (1960) A theory for estimation of surface and interfacial energies III estimation of surface energies of solids from contact angle data. *Journal of Physical Chemistry*, **64**, 561–565.
16 Jaeger, P. and Pietsch, A. (2009) Characterization of reservoir systems at elevated pressures. *Journal of Petroleum Science and Engineering*, **64** (1–4), 20–24.

17 Nasr-El-Din, H.A., Al-Othman, A., Taylor, K.C., and Al-Ghamdi, A. (2004) Surface tension of acid stimulating fluids at high temperatures. *Journal of Petroleum Science and Engineering*, **43** (1–2), 57–73.

18 Hansen, G., Hamouda, A.A., and Denoyel, R. (2000) The effect of pressure on contact angles and wettability in the mica/water/n-decane system and the calcite + stearic acid/water/n-decane system. *Colloids and Surfaces A: Physicochemical and Engineering Aspects*, **172**, 7–16.

19 Rao, D.N. (2002) Measurements of dynamic contact angles in solid–liquid–liquid systems at elevated pressures and temperatures. *Colloids and Surfaces A: Physicochemical and Engineering Aspects*, **206** (1–3), 203–216.

20 Sutjiadi-Sia, Y., Jaeger, P., and Eggers, R. (2008) Interfacial phenomena of aqueous systems in dense carbon dioxide. *Journal of Supercritical Fluids*, **46**, 272–279.

21 Alotaibi, M.B., Azmy, R.M., and Nasr-El-Din, H.A. (2010) A comprehensive EOR study using low salinity water in sandstone reservoirs. 2010 SPE Improved Oil Recovery Symposium, Oklahoma, April 24–28, 2010.

22 Crank, J. (1975) *The Mathematics of Diffusion*, 2nd edn, Oxford University Press.

23 Allen, M.B., Behie, G.A., and Trangenstein, J.A. (1988) Lecture Notes in Engineering, in *Multiphase Flow in Porous Media*, vol. **34**, Springer.

24 OTA (1978) Enhanced Oil Recovery Potential in the United States, Office of Technology Assessment.

25 Taber, J.J., Martin, F.D., and Seright, R.S. (1997) EOR screening criteria revisited – Part 1. Introduction to screening criteria and enhanced recovery field projects. *Journal SPE Reservoir Engineering*, **12** (3), 189–198.

26 Zhang, Y., Xie, X., and Morrow, N.R. (2007) Waterflood performance by injection of brine with different salinity for reservoir cores. Proceedings of the SPE Annual Technical Conference and Exhibition, pp. 1217–1228.

27 Figuera, L.A., Khan, K.A., Quijada, D., and Maestrazzi, W.J., EOR evaluation method for highly heterogeneous and complex reservoirs 2007. 69th European Association of Geoscientists and Engineers Conference and Exhibition 2007: Securing the Future, Society of Petroleum Engineers, pp. 1291–1301.

28 Mollaei, A., Maini, B., and Jalilavi, M. (2007) Investigation of steam flooding in naturally fractured reservoirs. International Petroleum Technology Conference, 2007, pp. 44–56.

29 Zhang, S. (2009) NMR study on porous flow mechanisms in low permeability reservoirs with CO_2 flooding. *Journal of Shenzhen University Science and Engineering*, **26** (3), 228–233.

30 Chukwudeme, E.A., and Hamouda, A.A. (2009) Enhanced oil recovery (EOR) by miscible CO_2 and water flooding of asphaltenic and non-asphaltenic oils. *Energies*, **2** (3), 714–737.

31 Bath, P.G.H. (1989) Status report on miscible/immiscible displacement by carbon dioxide. *Journal of Petroleum Science and Engineering*, **2**, 103–117.

32 Vahidi, A. and Zargar, Gh. (2007) Sensitivity analysis of important parameters affecting minimum miscibility pressure (MMP) of nitrogen injection into conventional oil reservoirs. SPE/EAGE Reservoir Characterization and Simulation Conference 2007, Society of Petroleum Engineers, pp. 247–257.

33 Holm, L.W. and Josendal, V.A. (1974) Mechanisms of oil displacement by carbon dioxide. *Journal of Petroleum Technology*, **26**, 1427–1438.

34 Hamouda, A.A., Chukwudeme, E.A., and Mirza, D. (2009) Investigating the effect of CO_2 flooding on asphaltenic oil recovery and reservoir wettability. *Energy and Fuels*, **23** (2), 1118–1127.

35 Al-Marzouqi, A.H., Zekri, A.Y., Azzam, A.A., and Dowaidar, A. (2009) Hydrocarbon recovery from porous media using supercritical fluid extraction. *Journal of Porous Media*, **12** (6), 489–500.

36 De Zwart, A.M., Van Batenburg, D.W., Blom, C.P.A., Tsolakidis, A., Glandt, C.A., and Boerrigter, P. (2008) The modeling challenge of high pressure air injection. *Proceedings – SPE Symposium on Improved Oil Recovery*, **3**, 1204–1216.

37 Gates, I.D. (2007) Oil phase viscosity behaviour in expanding-solvent steam-assisted gravity drainage. *Journal of Petroleum Science and Engineering*, **59** (1–2), 123–134.

38 Ayala, L.F. and Ertekin, T. (2007) Neuro-simulation analysis of pressure maintenance operations in gas condensate reservoirs. *Journal of Petroleum Science and Engineering*, **58** (1–2), 207–226.

39 Carroll, J.J. (2010) *Acid Gas Injection and Carbon Dioxide Sequestration*, John Wiley & Sons, Inc., New York.

40 Sloan, E.D. and Koh, C.A. (2008) *Clathrate Hydrates of Natural Gases*, 3rd edn, CRC Press, Boca Raton.

41 Notz, P.K., Bumgardner, S.B., Schaneman, B.D., and Todd, J.L. (1996) Application of kinetic inhibitors to gas hydrate problems. *SPE Production & Facilities*, **11** (4), 256–260.

42 Gaviria, F., Santos, R., Rivas, O., and Luy, Y. (2007) Pushing the boundaries of artificial lift applications: SAGD ESP installations in Canada. Proceedings – SPE Annual Technical Conference and Exhibition 2007, pp. 2063–2077.

43 Last, S., Groves, S., Rigaud, J., Taravel-Condat, C., Wedel-Heinen, J., Clements, R., and Buchner, S. (2002) Comparison of models to predict the annulus conditions of flexible pipe. OTC Conference, Houston, TX, 2002, Paper No. 14065.

44 Jaeger, P., Eggers, R., and Buchner, S., Sorption Kinetics of High Pressure Gases in Polymeric Tubing Materials, Proceedings of the 25th Int. Conf. on Offshore Mechanics and Arctic Eng., 2006 Hamburg, OMAE2006–92393, pp. 581–584.

45 Abdel-Aal, H.K., Aggour, M., and Fahim, M.A. (2003) *Petroleum and Gas Field Processing*, CRC Press.

46 Kalikmanov, V., Betting, M., Bruining, J., and Smeulders, D. (2007) New Developments in nucleation theory and their impact on natural gas separation. SPE Annual Technical Conference 2007, 110736-PP.

8
Supercritical Processes
Rudolf Eggers and Eduard Lack

8.1
Introduction

Although the first phenomenological studies of the solvent power of supercritical gases go back to the time period from 1861 until 1910, it was not until the 1960s that a strong development started in industrial application of the "gas extraction" effect, reported for the first time in 1879 by Hannay and Hoghart [1] when they detected the solvent power of fluids at temperatures and pressures higher than their critical data. Besides the Solexol process for the purification of plant and fish oils [2], the invention of Zosel [3] and the first international conference on supercritical processes in 1978 in Essen, Germany [4] helped the idea of gas extraction to be accepted. Since that time, a huge number of scientific articles have been published and a broad diversity of possible technical applications have been investigated. The racy development is clearly registered in Figure 8.1, illustrating the number of new patent applications per year in supercritical processing (Schütz, E. (2010), Databank on SCF Patents, personal communication).

Although Figure 8.1 reveals a maximum in 2002, there is a constant trend to innovative processes using supercritical processes [5]. Currently, the main interest is focused on the production and refining of pharmaceuticals, including the generation of particles, further the treatment of polymers [6], chemical reactions in the field of renewable energies and waste water treatment, and even on the use of supercritical gas as a supporting medium in enhanced recovery of oil and gas [7].

Table 8.1 gives an overview of compounds that have been proposed for use as supercritical fluids in ascending order related to the critical temperature T_c. For reasons of low costs, inflammability, low critical temperature, and solvent-free production, the food industry was the first to start a commercial decaffeinating process using CO_2 as the solvent [8] followed by hop extraction [9] and many other processing of natural material [10]. Although supercritical CO_2 is being used most frequently in food industry due to the products being total free of solvent residues, supercritical water has been proved to be an effective reaction medium especially for wastewater treatment [11].

Industrial High Pressure Applications: Processes, Equipment and Safety, First Edition. Edited by Rudolf Eggers.
© 2012 Wiley-VCH Verlag GmbH & Co. KGaA. Published 2012 by Wiley-VCH Verlag GmbH & Co. KGaA.

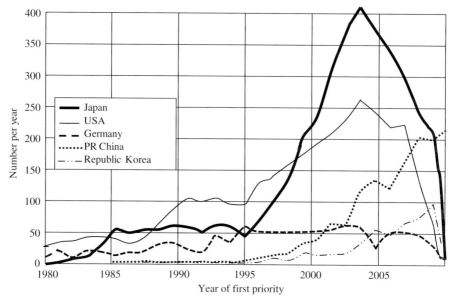

Figure 8.1 Supercritical processes: new patents per year, data from Ref. (Schütz, E. (2010), Databank on SCF Patents, personal communication).

Table 8.2 visualizes a classification of supercritical processes regarding the solid or liquid state of the feed material to be treated. Solids are processed batchwise in high-pressure vessels, whereas liquids are moved continuously via countercurrent columns either as a falling film or as a droplet spray. Most important is the formation of a phase boundary between the feed material and the supercritical fluid in order to

Table 8.1 Survey on critical conditions of several fluids, ordered by critical temperature T_c.

Solvent	Chemical formula	Molar mass (g/mol)	T_c (K)	P_c (MPa)
Methane	CH_4	16.04	190.4	4.60
Ethylene	C_2H_4	28.05	282.4	5.04
Carbon dioxide	CO_2	44.01	304.1	7.38
Ethane	C_2H_6	30.07	305.3	4.87
Ethyne	C_2H_2	26.04	308.3	6.14
Propylene	$(C_3H_6)_n$	42.08	364.9	4.60
Propane	C_3H_8	44.09	369.8	4.25
Dimethyl ether	C_2H_6O	46.07	400.0	5.24
Ammonia	NH_3	17.03	405.6	11.3
Diethyl ether	$C_4H_{10}O$	74.12	466.7	3.64
Acetone	C_3H_6O	58.08	508.1	4.70
Methanol	CH_4O	32.04	512.6	8.09
Toluene	$C_6H_5CH_3$	92.14	591.8	41.1
Water	H_2O	18.02	647.3	22.1

Table 8.2 Classification of supercritical processes.

Material	Phase boundary	Pretreatment	Process	Example
Solids	s/f s/f → l/f l/f	Pelletizing Wetting/swelling Destroying of cell walls	High-pressure vessel Batch operation	Hop resins Decaffeination Plant oils
Liquids	l/f falling film	Filtration	Countercurrent continuous operation	Column fraction of minor components from plant oils
	l/f drop	Preaeration	High-pressure spray continuous operation	Phospholipids

generate an effective mass transfer area. Aiming a sufficient accessibility for the supercritical fluid to the valuable components of the feed material, different methods of pretreatment are available. In case of solid feed, it is decisive to have short and open diffusion paths enabled by mechanical, hydrothermal, or biological methods of conditioning. Liquids are often to be purified from disturbing sediments by filter elements or centrifuges. Moreover, the solubility of the supercritical fluid in some liquids supports the diffusive mass transport by decreasing the viscosity of the liquid.

In summary, the knowledge of both the phase equilibrium data and the mass transfer kinetics is essential for the specification of supercritical processes. As an example, Figure 8.2 shows the well-known data for solubility of seed oil in CO_2 and the corresponding kinetic profile of oil extraction from differently prepared

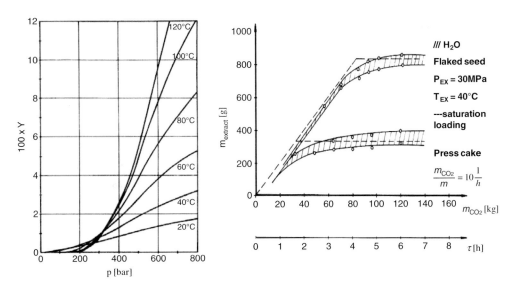

Figure 8.2 Equilibrium data and mass transfer kinetics [8].

oilseed [12]. Caused by strongly increasing solubility data at very high pressures, the common pressure range in supercritical fluid extraction (SFE) processes has been increased from 30 MPa to 100 MPa recently [13].

Supercritical processes for solids and liquids are dealt with in detail in Sections 7.2 and 7.3 and shown in Table 8.2.

8.2
Processing of Solid Material

There are several basic principles that influence extraction processes based on supercritical fluids. It is of advantage if soluble substances can be obtained at moderate conditions. Contrary to organic solvents, the solubility of substances can – to some extent – be adjusted by means of pressure/temperature variation, which also have influence on the phase equilibrium. Essential for economic processing is the knowledge of mass transfer. The influence of pressure and temperature on yield is demonstrated in Figure 8.3 as a typical example [14]. During the first period of time, the linear increase in yield shows the influence of solubility, that is, the higher pressures/temperatures lead to faster extraction, whereas in principle elevated pressures cause higher densities and elevated temperatures cause increase in vapor pressure. The influence of vapor pressure at higher pressures and temperature is more effective than the decreased fluid density. After removal of soluble substances from the surface of the feed material, further extraction is mainly controlled by time-consuming diffusion. To reach acceptable results, the raw material has to be pretreated in order to remove diffusion barriers on the one hand and to shorten diffusion distances on the other.

As the mass transfer depends on the mean concentration difference of solubles within the fluid, in the solid matrix and the fluid flow the correct selected mass flow

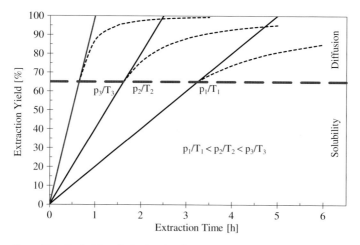

Figure 8.3 Typical trend of extraction lines.

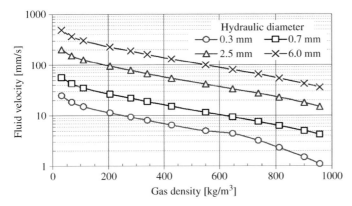

Figure 8.4 Fluidized bed velocities dependent on gas density and particle diameter.

rate is decisive for an optimized extraction process. Influence of the mass transfer is further given by the interfacial area that is higher on the smaller particle sizes. A compromise must be found between particle size and flow velocity to avoid channeling within the extractor (see Figure 8.4).

The solubility in supercritical CO_2 (nonpolar solvent) depends on chemical structure, say polarity of the substance to be obtained, the higher the polarity the less is the equilibrium concentration. To some extent, the polarity of CO_2 can be changed by adding an entrainer and thereby increasing the solvating power. In some cases – like diffusion-controlled processes for decaffeination of coffee (further discussed on page 185) and tea or cleaning of cork (see on page 193) – the use of matrix modifiers is appropriate. The feed is moisturized with water to some extent for swelling the structure and accelerating the mass transfer. Additional water-saturated supercritical CO_2 is applied to avoid drying out of feedstock and to enhance the solubility. The disadvantage of using cosolvents is that one gets solvent-diluted extracts and corresponding residues in the spent raw material.

Separation of dissolved substances from the fluid is performed by *precipitation*. Most applications use *an isenthalpic decrease of pressure and temperature* to reduce the fluid density and, consequently, the solvent power of the fluid. After pressure reduction, the extraction dense gas is heated so that it reaches the gaseous state. In this phase, no solvent power is present for any substances and therefore nearly complete separation of the extracted substances takes place. The dissolved substances precipitate and can be discharged at the bottom of the separator. The solvent cycle is closed by recycling the recompressed CO_2, either cooled in case of a compressor process or condensed by using the pump process.

In some cases, the solubility can be reduced by changing the temperature and keeping the pressure, obtaining a nearly isobaric process that requires less energy. However, for this kind of separation, the solubility behavior of the dissolved substance has to be taken into account.

Absorption is a further method for separation of the extracted substances. This method requires a high solubility of the dissolved substances in the absorption media, without any effect on the extract on the one hand and preferably no transfer of said media into the dense solvent on the other. Separation of absorbent and extract should also be easily possible. The decaffeination process is a perfect example, water with a low solubility in CO_2 is used as matrix modifier and entrainer for extraction and as an absorbent in the separation step. The caffeine in the water solution can be concentrated and recovered by crystallization. Another separation method is *adsorption* of solubles on adsorbents like activated charcoal or resins, but one has to consider that in most cases the adsorbed substances are hardly to recover. Industrial examples of such an application are the decontamination of rice – see on page 189 – or cleaning of cork – see on page 193 – and decaffeination of coffee or tea as described on page 14.

Both separation possibilities described above utilize an isobaric process allowing processing under optimized conditions.

Separation by adding a substance that lowers the solubility – like nitrogen – into the separation unit causing precipitation of the dissolved substances is also a theoretical option, but such an *antisolvent* has to be removed from the dense CO_2 before recycling. Another method under investigation is the usage of membranes that seems possible in case molecular weight differences between solvent and extract are high enough [15].

From the above, it is obvious that the design engineer has to evaluate optimized extraction conditions and to select the best way of separation in order to obtain a viable process. In the following sections, various design possibilities are described in more detail.

8.2.1
Isobaric Process

To optimize the energy consumption, an isobaric process should be applied, especially for plants requiring very high specific CO_2 mass flow rates of more than 70 kg CO_2/kg feed. Especially for decaffeination plants, it is of advantage because flow rates of 100 kg/kg–150 kg/kg for tea and 200 kg/kg–250 kg/kg for green coffee beans are required. Separation by means of pressure reduction would require enormous amount of energy for condensation and evaporation, and consequently application on industrial scale is far beyond the economical area. Both possibilities – adsorption on activated charcoal or absorption by means of a washing column – are already used on industrial scale (see on page 185).

8.2.2
Single or Cascade Operation with Multistep Separation

Most multipurpose plants are equipped with two or more extractors switched in cascade mode, which enables to come as close as possible to phase equilibrium and are equipped with a multiple separation system, operating at different reduced pressures for fractionation. Such design is mainly used for the extraction of spices and herbs, allowing concentration of pungent substances like piperine from pepper or capsaicin from chili in a first step and the corresponding volatile oil afterward in one additional separator. In case that the valuables are highly concentrated in the raw material and also relatively highly soluble in the dense gas, installation of only two extractors is sufficient as nearly equilibrium is already obtained after passing just the first extractor or later on after the second extractor is switched on line.

8.2.3
Cascade Operation and Multistep Separation

Except for decaffeination plant of green coffee beans in Texas that uses a discontinuous operated lock system for the input and output, until now all other high-pressure extraction plants for solid materials operate as batch process. Although a continuous process for solid feed material was investigated in different research studies, using screw presses or moving cells, no other commercial use is utilized maybe due to very high investment and related risks or results are not convincing enough [16].

To obtain a quasi-continuous operation, a series of extractors (Figure 8.5) is operated in quasi-continuous countercurrent operation. Fresh CO_2 is always switched to the extractor that is longest in line and consequently a sufficient concentration difference is maintained and the highest possible loading with soluble substances is ensured.

8.2.4
Extractable Substances

In principle, solid material extraction can be divided into two different types, the selective and the total.

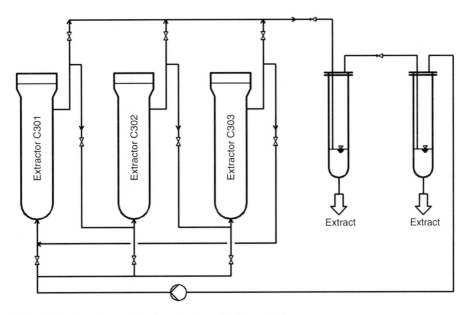

Figure 8.5 Cascade operation for extraction of solid materials.

8.2.4.1 Selective Extraction

Selective extraction shall remove certain substances from the feed and, maintaining the properties, the extract could only be a by-product. Examples are as follows:

- Decaffeination of green coffee beans
- Decaffeination of black and green tea
- Removal of nicotine from tobacco
- De-alcoholization of wine and beer
- Removal of undesired substances like TCA from cork and plant protectives from spices, herbs, or cereals
- Removal of organic solvents from solids

Essential for such processes is a very high selectivity for substances to be removed without influencing flavor, appearance, color, shape, smell, or size, because the treated insert represents the final product.

8.2.4.2 Total Extraction

Total extraction has the target to obtain the highest possible yield of soluble substances, whereas a single separation results in a so-called total extract, but multiple separations allow fractioned products. Typical examples are the extraction of spices, herbs, and hops. One likes to obtain the exhaustive lead substances as α-acids from hops or piperine from pepper, or gingeroles from ginger, or carotinoids from paprika and so on on the one hand and the corresponding aromas on the other. The extraction residue from the extractor is mostly without any value, but can sometimes be used as fertilizer basis or animal food addition.

A specific case is a total extraction, but with a selective removal of one undesired soluble component, which can be separated by adsorption within the high-pressure area, either by an insert in the extractor or by one additional vessel.

8.2.5
Pretreatment of Raw Materials

As far as possible, one should use the same pretreatment methods than that used for conventional extraction. Raw materials are supplied and stored under dried conditions to avoid degradation, not to lose valuable extractor volumes, and to limit the water concentration in extracts. Spices and herbs should have moisture contents between 8% and 12%, coffee beans between 35% and 45%, tea between 20% and 30%, fruits between 13% and 20%, and cereals about 12%.

A certain content is required in order to avoid shrinking of the matrix and, consequently, hindrance in diffusion. For example, a reduced moisture content of 5% instead of 12% decrease the yield by the extraction of calendula from about 7.5% down to 5.5%.

For total extraction, the raw materials are ground down to particle sizes of 0.3 mm–0.8 mm. Smaller particles would have a better mass transfer, but reduce the fluidized bed velocity – larger diameters and higher costs, or result in channeling.

Some materials like chamomile with active components on the surface do not need to be milled. Grinding for most spices is executed by means of a hammer mill; leaves, roots, or barks preferably with cutting mills; and for some seeds and cereals, roller mills are used to destroy the structure.

The bulk density is decisive for the plant capacity and economy and, therefore, raw materials with a low bulk density should if possible be pelletized, for hops one get an increase from about $150\,kg/m^3$ to $500\,kg/m^3$ and even more.

Another kind of pretreatment for materials requiring swelling of the cell structure is moisturizing, which depends on the type to be used, for example, for coffee up to 35% or 45%, for black tea to about 25%, for green tea up to 40%, for pesticide reduction of rice up to 20%, and for vanilla beans up to 30%, which would otherwise increase the volume during extraction and cause difficulties for emptying. For the latter mentioned treatment, it is to be considered that the moisture concentration during extraction remains constant, which can be maintained by using saturated CO_2. Furthermore, drying afterward is absolutely necessary, avoiding fungi infection.

8.2.6
Design Criteria

For the design of the process and equipment, the following has to be considered and experimentally verified on lab and pilot scale [17]:

- Specific basic data and behavior of feed material and extract
- Thermodynamic conditions for extraction and separation
- Fluid dynamics, like fluidized bed velocities

The above-mentioned factors not only determine the design pressures and temperatures but also the dimensions:

- Corresponding mass and heat transfer
- Energy optimization by means of a T,s diagram
- Design of pumps, cooling, heating, and piping system
- Selection of appropriate separation system
- Selection and sizing of proper plant equipment

Besides the desired plant capacity, raw material and product specification are also necessary to get knowledge of local conditions like climate, altitude, and earthquake factor in order to design adequate building and service units, especially cooling machines and electrical drives.

The right selection and preparation of the raw material are the determining factors for obtaining high-quality products.

Depending on the final product requirement, processing conditions, pressure, temperature, mass flow, and pretreatment of raw materials are influenced and fractionated separation or further downstream treatment as concentration, removal of moisture, or purification may be required.

Figure 8.6 Bottom closure (Archive Natex).

8.2.7
Design with the Use of Basket

Materials like green coffee beans or rice that are free flowing throughout processing can be relatively easily filled from top and discharged from bottom, provided adequate design as shown in Figure 8.6 is used. Raw materials with low bulk densities and no tendency to agglomerate allow pneumatic transport to and from extractors.

Most feed materials must be grinded and consequently have a certain particle size distribution. Fine particles (less than 0.2 mm) tend to float upstream and can in case of higher concentration block the filter, creating possibly very fast an increased pressure drop, forming a *filter cake* or in worst case one *big pellet* and the extraction process has to be stopped. In spite of a careful selected fluid velocity, channeling of the fluid (Fig. 8.7) or larger strong agglomerates cannot be avoided for materials in

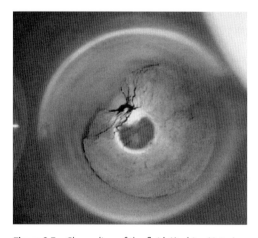

Figure 8.7 Channeling of the fluid (Archive Natex).

powder form like egg yolk or those that lose their structure like roasted sesame seeds during extraction. Raw materials with higher fiber content, like ginger, vanilla, and black and green tea, swell during extraction (volume increase up to 30%) and therefore limit the bed height. From the foregoing discussion, it is evident that in most cases baskets are to be used, because the above-mentioned failures can be handled outside, extraction can be continued without larger interruptions and changing of feed, and spent material can be executed in few minutes.

8.2.8
Thermodynamic Conditions

Contrary to liquid solvents, dense gases allow by changing the thermodynamic conditions to influence the solubility and to process different product qualities and by means of multiple separation, a fractionation of the extracts. Determination of possible loadings and corresponding solvent flow rates is important for a viable process. Such in mind and considering a corresponding selectivity, it is obvious to look for highest possible loading during extraction and to perform separation at the minimum for maximum precipitation within the foreseen area and to avoid carryover or recirculate extracted substances. The level of solubility in CO_2 depends on polarity of the substance and the density of the fluid.

Economy and technical possibilities determine and limit to some extent the extraction pressure, but for industrial plants a trend to higher pressures can be recognized in recent years.

For already known systems, modifications based on the old semiempirical equation of state from Van der Waals have been developed, a successful one from Redlich and Kwong or the one from Soave. A variation specifically describing the properties in the critical and supercritical area is given by the equation of Peng and Robinson [18]. Derivations from the association laws for unknown solute properties are used to describe experimental results, the one from Chrastil [19] is the most popular one and useful for the calculation of equilibrium distribution.

8.2.9
Mass Transfer

Modeling supercritical extraction of solid material is based on the mass balances that are relevant for the internal diffusive transport of the extract within the solid matter and the external convective transport of the extract from the solid surface to the solvent fluid.

Internal transport for spherical particles

$$\varepsilon_p \frac{\partial y_p}{\partial t} = D_{\text{eff}} \cdot \frac{1}{r^2} \frac{\partial}{\partial r}\left(r^2 \frac{\partial y_p}{\partial r}\right) - \varrho_p \cdot \frac{\partial x_p}{\partial t}$$

where

D_{eff} is the effective diffusivity (m²/s)
y_p is the pore fluid concentration (kg/m³)
x_p is the particle phase concentration expressed as mass ratio
ε_p is the particle porosity
r is the particle radial coordinate (m)
ϱ_p is the particle density (kg/m³)

The most important parameters in mass transfer modeling are the effective diffusion coefficient D_{eff} and the mass transfer rate $\partial x_p / \partial t$, which is linked to the mass transfer coefficient k in the balance equation for external transport. As for the calculation of effective diffusion coefficients, simple equations have been proposed in literature [10].

External transport

$$\frac{\partial c}{\partial t} = D_{ax} \frac{\partial^2 c}{\partial z^2} - u \frac{\partial c}{\partial z} + \frac{k \cdot a_s}{\varepsilon}(c^* - c)$$

where

c is the bulk fluid concentration (kg/m³)
c^* is the saturated concentration at the phase boundary (kg/m³)
a_s is the specific surface aria (m²/m³)
ε is the porosity of solid bed
k is the mass transfer coefficient (m/s)

The coupling condition is given by the mass change of solid that is caused by mass transfer to the fluid.

Coupling condition

$$\frac{k \cdot a_s}{\varepsilon}(c^* - c) = -\varrho_p \cdot \frac{\partial x_p}{\partial t}$$

Concerning different simplifications, several solutions of these governing equations have been derived [10, 15, 18, 20, 22]. As an example, the axial diffusion D_{ax} is neglected in case of no back mixing of the fluid flow (plug flow). Until now the problem remains that these models neglect both, the change of the bed porosity and the particle diameter. It is known that especially solid beds of natural material tend to increase their porosity along the extraction time [23]. However, the most important transport parameters in the differential balance equations are the mass transfer coefficient k and the specific mass transfer area a_s.

As the velocities within the porous structure of the solid bed are relatively low (1 mm/s–10 mm/s), the mass transfer by forced convection may be superposed by natural convection.

The target of this investigation is to find the optimized specific solvent mass flow rate in order to obtain the highest extraction efficiency. The result of the investigation should indicate the amount of CO_2 per kilogram of raw material and hour, the extraction time, and the length-to-diameter ratio of the extractor.

In order to choose optimal thermodynamic conditions, a compromise between solubility and investment is to be found. The mass flow rate is the determining factor for energy consumption and investment cost, and therefore most attention must be paid for the optimization, which allows adequate evaluating of extraction cycles and sizing of entire equipment.

Mass flow rates can highly differ between individual raw materials and applied conditions.

Examples:

Hops extracted by supercritical CO_2	30–40 kg CO_2/kg
Hops extracted by subcritical CO_2	50–70 kg CO_2/kg
Decaffeination of green coffee beans	200–270 kg CO_2/kg
Decaffeination of black tea	250–400 kg CO_2/kg
Decaffeination of green tea	350–600 kg CO_2/kg
Most spices and herbs	25–70 kg CO_2/kg

Figure 8.3 shows at the beginning of extraction a more or less uniform mass transfer with time because substances on and close to surface, called leaching zone, are directly in contact with the solvent and pass through the limit layer directly into it. Afterward the substance has to diffuse from the core to the surface in order to be dissolved. The diffusion time depends on the corresponding distribution ratio of the substance within the solid matrix and if adsorbed in it or not.

The following transport phenomena in natural solid material can be assumed:

- The raw material absorbs the fluid, swelling the cells, and extends the pores, increasing the decisive mobility of substance and solvent.
- The substance dissolves in the solvent and diffuses to surface layer and pass through it.
- The substance passing the surface layer is removed by upstreaming CO_2. Diffusion velocity depends on prevailing substance concentration difference (within cell structure and CO_2).

Examples

- Dense gases dissolve up to 40% in vegetable oils, reduce the viscosity, and increase diffusion.
- Moisture content in decaffeination processes is essential, because water and CO_2 break up the weak binding of caffeine to organic acids and caffeine, dissolved in water, and can diffuse to the surface.
- Moisture content in CO_2 and raw material like oilseeds influence the extraction. Loading of triglycerides in CO_2 decrease with increasing water content and is about 40% neglectable.

- The yield for medical herbs is 20%–30% reduced if the moisture content is dropped from 8%–12% down to 2%–5%.

From the above, it can be summarized that diffusion can be influenced by the following:

- Destruction of plant cells and short diffusion paths
- Breaking cell walls by enzymatic treatment or pelletizing
- Swelling of the cell structure
- Breaking of possible bondings by means of pH change or increased moisture

8.2.10
Hydrodynamics

Size and its distribution, form, and density of solid particles are the determining factors of the hydrodynamic condition within the extractor. Practical experience in pilot and industrial scale ask for particle sizes between 0.3 mm and 0.8 mm. Finer particles like dust, in spite of its smaller flow velocities, can be fluidized and be collected on filters, which in most cases create a dramatic and fast pressure drop, with the consequence that in worst cases the basket can be damaged, if not a corresponding design like pump shut off at a certain pressure difference is considered or carryover of raw materials into downstream equipment happens. Besides fluidization, one must also count with channeling, which can be watched, for example, for materials like powders from paprika, egg yolk, algae, and cacao. Pelletizing, which is possible for paprika and egg yolk, can eliminate such problems due to enlarged particle size and bulk density.

In each case, the fluidized bed velocities, during extraction, pressurizing, and depressurizing, have to be determined taking into account dependence on the given particle size and on the solvent density. The maximum allowable fluid velocity, about 90% of the one from fluidization, is taken to calculate the extractor diameter.

8.2.11
Energy Optimization

Contrary to conventional extraction processes, dense gas extraction enables more freedom in designing the optimum solvent cycle by means of the T,s diagram. Depending on solubilities and corresponding selected thermodynamic conditions, the CO_2 circulation system can be driven by means of either a pump or a compressor, whichever needs less consumption of energy. Eggers [24] and Lack [18] compared both processes by means of a T,s diagram.

8.2.12
Pump Process

Liquid CO_2 from a storage tank is subcooled or prepressurized by means of an inline pump (avoiding cavitations), pressurized and heated up to extraction conditions.

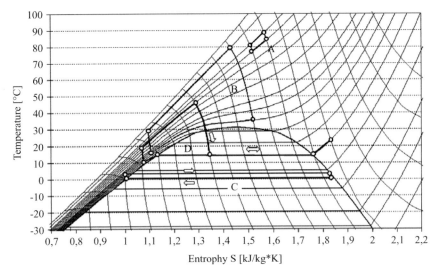

Figure 8.8 Pump process.

As shown in Figure 8.8 various possibilities for separation are possible. Cycle A shows an isobaric process (besides pressure drop), which is used for decaffeination of green coffee beans (see on page 185). Cycle B demonstrates separation at supercritical separation, not used on industrial scale so far. Cycle C shows a liquid CO_2 process, applied on industrial scale for processing of volatile oils and hops (see on pages 185). Cycle D shows the most applied method used for most commercial plants, whereas separation is performed by expansion and phase change of precipitation, the gaseous CO_2 condensed and recycled via the CO_2 storage tank. Condensation after gaseous separation requires highest energy demand. In commercial plants, the pumping mode is used mainly for extraction under supercritical conditions using extraction pressures between 170 bar and 480 bar, extraction temperatures between 40 °C and 80 °C, and fractioned separation at about 90 bar–120 bar and 45 bar–60 bar, which are within an economical area.

8.2.13
Compressor Process

In Figure 8.9, cycle A represents again a more or less an isobaric process, remaining within a gaseous phase, whereas separation can be either by temperature change or using an adsorbent. Cycle B shows a process performed throughout under supercritical conditions with separation due to pressure and temperature change. A variation is applied for purification of cork (see on page 193), using subcritical conditions for separation. Cycle C is again a liquid extraction system with the difference in the pump mode, that is, the gaseous CO_2 after the separator is compressed first and afterward condensed for extraction. Cycle D represents an economic possibility for the extraction of highly soluble substances in CO_2, requiring

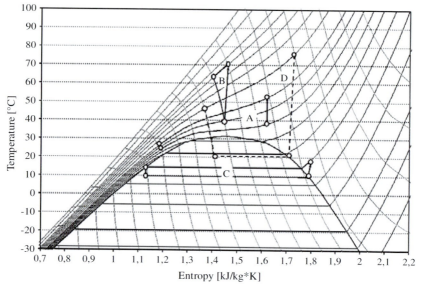

Figure 8.9 Compressor process.

consequently lower extraction pressures between 80 bar and 200 bar and low separation pressures and temperatures for recovery of highly volatile substances. Highest energy demand is required for the compressors, which in case of higher compression rates need multistage design with intermediate cooling.

8.2.14 Applications of Supercritical Extraction of Solids

8.2.14.1 Decaffeination of Green Coffee Beans

[3] described in 1980 different possibilities for the decaffeination of green beans, a process that was the first to be applied on large industrial scale. ...baric process (Figure 8.10) operates at pressures 160 bar–220 bar, using a ...current column for caffeine separation by means of demineralized water at ...atures between 70 °C and 90 °C. Depending on caffeine concentration, 3 l–5 l ...r per kilogram of coffee is required.

...tions of the above-described process use activated charcoal instead of the ... for the removal of caffeine. The adsorbent is placed either in a separate vessel ...d with the coffee beans and placed within the extractor. Depending on the type of coffee, the extraction time is between 6 h and 8 h and about 1 kg of activated charcoal is needed for 3 kg of coffee beans. Recovery of caffeine is not possible, but reactivation of charcoal at 600 °C is possible with about 30% of loss.

8.2.14.2 Production of Hops Extract

Hop is not a stable product and has to be dried immediately after harvesting. During its storage, the α-acid content decrease with time, not as fast when stored under

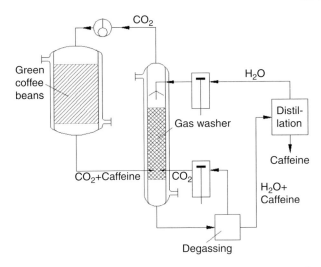

Figure 8.10 Zosel's suggestion for the decaffeination of green coffee beans [3].

cooled condition; however, because of its higher stability, corresponding extracts are mainly used to bitter the beer.

Since 1980, the formerly used solvent methylene chloride has been substituted step by step by CO_2 processes, covering nowadays over 30% of the hops production. Such plants are located in Europe, United States, and Australia; while the first two countries use supercritical CO_2, the latter use a process involving liquid CO_2. Due to less solvent power, liquid extraction requires more than 50 kg of CO_2/kg hops in order to obtain the demanded yield of 95 wt% for α-acid, although hardly getting any hard resins.

The processing at supercritical condition allows more or less exhaustive extraction of the bitter substances and most of resins, requiring – due to higher solubilities (up to 3.2 wt%) – less mass flow rate and shorter extraction time. Because the advantages justify the higher investment, all modern plants use such conditions, that is, between 200 bar and 450 bar and a temperature range of 40 bar–65 bar. Industrial plants use three or four extractors, each designed with volumes in the range of $2\,m^3$–$6\,m^3$.

Independent of design, hops has to be pelletized, increasing the bulk density from about $125\,kg/m^3$ to $500\,kg/m^3$ and reducing consequently the extraction volume in the same extent.

Hop extracts have the advantage of easier and cheaper storage and transport due to the remarkably reduced volume; it can be easily standardized with regard to α-acid content and, consequently, is very easy to dose and maintain the beer quality.

A comparison between beers brewed with CO_2 extracts, hop pellets, or dichloromethane shows no difference in bitterness and similar long-term stability, but foam values are better if CO_2 extracts are used, because precipitation of protein is reduced due to lack of tannins, which forms complexes with protein. During wort boiling, a bit of aroma hop pellets are added to increase the polyphenol content and to compensate the slightly less full-bodied note, obtaining thereby an improved stability after maturation.

8.2.14.3 Extraction of Spices and Herbs

Spices and herbs are mainly used to improve the taste of food. Parts of used plants can be seeds, leaves, flowers and parts of flowers, barks, roots, or fruits and are used fresh or dried for transportation and/or storage and available as whole, broken, ground, or rubbed shape on the market. For many different applications, such conditions do not meet the requirement with regard to the necessary low microbiological contamination, consistent quality, and longer storage stability [9, 22].

Total extracts from plants, the so-called oleoresins, are obtained by extraction and contain both the essential oils and the nonvolatile substances. Before processing, spices or herbs are selected with regard to quality and in most cases are dried and grinded independent of used solvent standardized afterward. As advantages of oleoresins, the consistent flavor and aroma, spice equivalence, long-lasting stability with low storage space, sterility, and easy handling and dosing can be mentioned. Several organic solvents can be used, taking into account the prevailing regulations and corresponding residual levels in the oleoresins and spent raw materials. Thermal stress on certain valuable compounds and consequently required low temperatures for extraction and evaporation can be a limitation for quite some solvents. Undesired high boiling components can remain in the final product and influence the quality [25, 26]. Because of the well-known advantages of dense CO_2, supercritical extraction of spices and herbs has been well established.

Spices and herbs are processed in medium-sized extraction plants, equipped with two or three extractors, with payload volumes of 200 l–1000 l, and in most cases two separators and are designed according to their product mix between 300 bar and 550 bar.

Figure 8.11 shows a principal flow sheet of a multipurpose plant, equipped with three extractors, two separators, and a CO_2 circulation system with heat exchangers and a CO_2 storage tank. Raw materials containing higher valued substances – used as intermediate products for cosmetics or nutraceuticals – can be extracted with smaller extraction units, while most common spices/herbs like chili, coriander, pepper, paprika, nutmeg, or maces need larger extraction volumes to be viable. Because components in raw materials like pigments, pungent substances, special fatty oils, or antioxidants have a rather low solubility, such units should be designed in most cases for 550 bar, but higher pressures up to 800 bar could be necessary for sufficient yield of valuables.

The feed is normally placed in multiple baskets for easy product handling and the plant has to be designed for frequent and simple cleaning.

8.2.14.4 Extraction of Essential Oils

Volatile oils produced by hydrodistillation are called essential oils, a complex mixture of hydrocarbons as terpenes, sesquiterpenes, and diterpenes and oxygenated ones like alcohols, esters, aldehydes, ketones, ethers, and phenols. Substances with sulfur or nitrogen can also be found in the mixture. Hydrolization of esters to corresponding alcohols during distillation is a negative aspect.

CO_2 extraction of volatile oils can be executed by liquid, subcritical, or supercritical conditions, but in any case within a gentle pressure and temperature range in order to avoid coextraction of undesired components (waxes) or to create thermal stress to

8.2 Processing of Solid Material

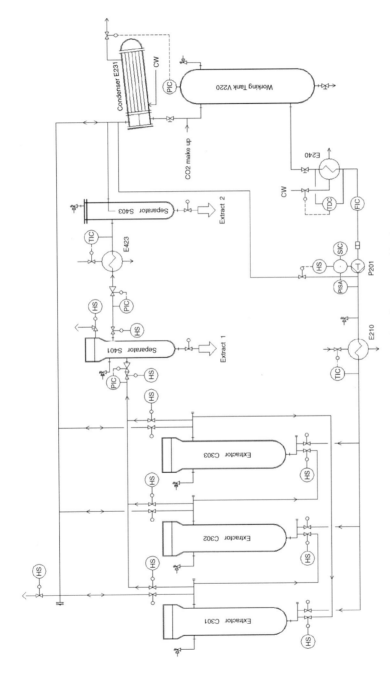

Figure 8.11 Flow sheet of a multipurpose supercritical fluid extraction plant for spices and herbs.

heat sensitive components. Volatile components have a higher solubility in dense CO_2 and, consequently, fractionated extraction of terpene hydrocarbons and oxygenated components is hardly possible.

The extraction of volatile oils and related products has already been intensively reported in the literature [27], and most of them are better produced by steam distillation. For example, compared to steam distillation, CO_2 extraction has an advantage of the production of wood oils from sandalwood, agar wood, hinoki wood, and vetiver roots. These wood oils have a high boiling point and therefore need high temperatures when they are produced by steam distillation, which leads frequently to a burned flavor of the essential oil. Furthermore, the mass transfer of the oil through the wood chips is strongly diffusion controlled, which is a minor problem during CO_2 extraction. These oils have a high market price, which is a further advantage of CO_2 extraction.

Sandalwood oil has very good fixative properties and finds application in many perfumes [28]. The highest yields of sandalwood oil can be obtained from trees and their roots at the age of 50 years–60 years. The classical steam distillation of sandalwood needs very long time, more than 24 h, and reaches a yield of 3.0%–4.0%, while supercritical processing allows a yield up to 5.5%.

8.2.14.5 Production of Natural Antioxidants

Oxidation of lipids is the reason for unpleasant flavor and smell in oil- or fat-containing food and oxidation in human bodies can cause pathological changes. Antioxidants neutralize and scavenge free radicals and are used to minimize lipid peroxidation.

Extracts and leaves of rosemary and sage have the most effective antioxidant activities, but some are recognized in oregano, clove, thyme, pepper, and allspice. Supercritical CO_2 extraction enables enrichment of carnosolic acid from rosemary up to concentrations of 25 wt%–30 wt%. Figure 8.12 shows that nearly exhaustive extraction of carnosolic acid is possible at elevated pressures. Stability tests on lard (based on peroxide numbers) demonstrated that CO_2 extracts had higher lifetimes compared to BHA (butylhydroxyanisole), up to about 2.5 times higher in case of 700 bar extracts.

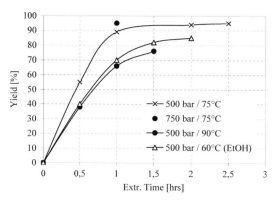

Figure 8.12 Yield in carnosolic acid.

8.2.14.6 Production of High-Value Fatty Oils

Production of oil from commodity oilseeds (sunflower, soybeans, and rape) does not appear economically viable, because of the higher costs of high-pressure batch processes. But there can be areas in which supercritical fluid extraction can be useful especially for the extraction of high-value oils.

These can be, for example, special gourmet oils (almond, apricot, avocado, grape seed, hazelnut, and walnut) or oils used for pharmaceutical and cosmetic applications (corn germ, wheat germ, evening primrose, and borage). If the deoiled residual can also be used for food or cosmetic applications, the economy of the process increases substantially.

A successful implementation for a commercial application was the erection of a supercritical CO_2 extraction plant for roasted sesame oil. This development of a CO_2 supercritical fluid extraction process for roasted sesame seeds can serve as an example for the viable production of such special oils. Sesame seeds have a relatively low tocopherol content, but contain other very effective antioxidants that stabilize the oil on the one hand and make it very tasty on the other. Roasted premium sesame oils are very popular in Asia, particularly in Korea and Japan.

The most important antioxidants in sesame are sesamoline and sesamine, which are well soluble in CO_2 and, consequently, nearly completely present in the extracted oil. The active substance is sesamole, which is formed continuously from sesamoline during aging.

8.2.14.7 Extraction of γ-Linolenic Acid

A further successful example is the extraction of essential polyunsaturated fatty acids, like γ-linolenic acid, which can be viably extracted with CO_2. The ω-6,9-double bonds, γ-linolenic acid, and $C_{18:3}$ (6c, 9c, 12c) are the important physiological active compounds of this oil and therefore of particular interest.

CO_2 extracted oils are of high grade, clear, light yellow, and with nearly the same fatty acid distribution as solvent extracts, but have the advantage that the phospholipids are hardly extracted.

Due to the high content of polyunsaturated fatty acids, high extraction pressures do not show advantages and pressures of about 500 bar are sufficient. Like evening primrose, many other raw materials are very interesting for the production of γ-linolenic acid.

8.2.14.8 Cleaning and Decontamination of Cereals Like Rice

Residues of plant protective agents can remain in cereals and have to be avoided or removed as most of them are considered to have a negative effect on human organism. Such harmful substances fall in most countries under legal regulations and have to be controlled.

Rice, grown mainly in the Southern Hemisphere and tropical countries, is one of the basic foods all over the world and the production in necessary quantities is difficult without the use of pesticides.

For the development of an extraction process based on CO_2 as solvent, the solubility of pesticides – mainly selected representative carbamates – has to be evaluated, because their residual concentration must be very low. Such an investigation is preferably performed first on fortified inert matrixes like sand and afterward under preoptimized extraction conditions with contaminated rice [29]. As a result, the authors mentioned that pressure and temperature are not significant parameters and moderate conditions can be applied. The moisture content of the rice has to be kept constant throughout the extraction and should be in a range of 10 wt%–15 wt%, because higher levels give an increasing loss of efficiency. Based on the research data, a patent was applied and granted [30] and by means of a corresponding scale-up, a large industrial plant (Figure 8.13) was set up in Taiwan.

In summary, supercritical CO_2 is an appropriate solvent for a moderate removal of the investigated group of pesticides from rice. Water has to be used as a matrix or CO_2 modifier with an optimum water content of around 10%–15%, and an amount of 7 up to 15 kg CO_2/kg rice provides an acceptable effect for pesticide reduction, but the so-called *grown pesticides* located within the plant cells and incorporated during growing from the soil are hardly removed especially in case of such a polarity.

The plant owners claim that CO_2 extraction process has added advantage that the treatment of brown rice removes the surface fat, which extends the shelf life, because this fat becomes easily oxidized and usually contains most of the pesticides. Furthermore, the cooking time of the brown rice is reduced, the rice is softened, and retains all vitamins and trace elements, that is, maintain a higher nutrition value, which is not the case for white, polished rice.

8.2.14.9 Impregnation of Wood and Polymers

Very often wood is exposed to environmental impacts like water, wind, sunlight, or even biological attacks like insects, rot, or fungus. Such attacks cause big damages

Figure 8.13 Rice purification plant (Archive Natex).

and result in costly renovation requirements comparable to corrosion problems with iron constructions (Superwood™ process at Hampen company profile, 2003).

In order to prevent such damages, surfaces are sealed and/or biocides are deposited in the wood. Surface sealing is typically done by painting with substances containing biocides, UV blockers, and materials that prevent moisture penetration. The treatment with biocides is normally done by depositing the active ingredients in the wood by vacuum or pressure.

Larger structural timber or wood mainly for outdoor uses like poles or garden fences are pressure impregnated with different metal salts.

Smaller structural wood for doors and windows is mainly impregnated with different biocides solved in organic solvents using the vacuum method.

Impregnation of wood with the help of supercritical carbon dioxide has been investigated during the past two decades [31]. Very early it was detected that supercritical carbon dioxide penetrates completely through the wood together with the solved biocides. But there is a gradient in concentration of biocides depending on thermodynamic conditions and the kind of wood. It seems that it is not the result of incomplete penetration of the wood by the supercritical CO_2/biocides mixture. This gradient can be caused by different adsorption and desorption processes, by re-extraction of biocides from the wood during depressurization or the biocides get filtered from the carbon dioxide as the solution fills up the wood matrix.

Although a lot of knowledge was not yet available, the company Supertrae A/S in Hampen, Denmark started wood impregnation by means of supercritical CO_2 in 2002 (Figure 8.14). They started directly with an industrial size plant that caused some problems at the beginning. But this investment induced strong development efforts in the wood industry.

The actual impregnation process can be described as follows:

- The wood is placed in the impregnation vessel.
- The required amount of active ingredients is placed in a mixing vessel.
- CO_2 flows into the cycle and pressure and temperature are adjusted to the desired impregnation conditions, where the active ingredients are dissolved in the CO_2.
- CO_2 together with the active ingredient is circulated through the impregnation vessel for a certain time to ensure an even distribution of the active ingredients in the wood.
- The vessel is depressurized and any excess active ingredients are separated in a separation vessel, so that CO_2 and active ingredients can be recycled.

Numerous examinations have been carried out, which document that the process does not affect the mechanical characteristics of wood and does not change the moisture content. As the product is not on the market since a very long time, the experience with the durability is still limited.

Tests in Denmark and in Malaysia show that the provided protection is equal to or better than the protection of products treated with competing processes.

Wood impregnation is not only limited to avoidance of fungi and bacteria attacks but can also be used to increase the fire and water resistance, to dye the wood especially for furniture and floors and also for the hardening of wood surfaces.

Figure 8.14 An impregnation plant for wood (Courtesy of Superwood).

8.2.14.10 Cleaning of Cork

For centuries cork stoppers have been used to seal wine bottles and in the western society, people expect the typical sound "blob" when opening a bottle of wine. Cork is a natural polymer foam based on the substance suberine and grown as bark of the oak tree (*Quercus suber*).

The average chemical composition of cork is as follows:

- Suberine (45%) – the main component of the cell walls and responsible for the resilience of cork.
- Lignin (27%) – the binding compound.
- Polysaccharides (12%) – component of the cell walls.
- Tannins (5%) and ceroids (5%).

Secondary metabolites produced by infections of fungi or bacteria can damage the quality of cork. The worst one is trichloroanisole (2,4,6-TCA) that is primarily responsible for the typical "cork taint" and causes drastic losses in the wine industry. Good wine specialists can recognize a TCA content in wine of about 2 ng/l, while usual consumers notice concentrations between 5 ng/l and 10 ng/l. The cork industry made a lot of efforts to solve this problem because it became very serious during the recent years. A number of processes were developed mainly using overheated steam to reduce the TCA concentration in the cork. But even with the best of such plants, a TCA reduction of only 70% can be reached.

One of the leading cork producers, Sabaté, began the development of a cork cleaning process with supercritical gases in 1997 [32, 33]. The tests were very successful and, consequently, the process was patented. With supercritical gas, it was possible to reduce the TCA content below the detectable limit, which is at the moment around 0.2 ng/l. Furthermore, CO_2 has special properties regarding reduction of pesticides and inhibition of fungus growth.

In 2003, NATEX Prozesstechnologie started investigations for scale-up of an industrial process. With a production capacity of 2500 tons per year, it corresponds to 500 million cork stoppers. In order to optimize the energy consumption, a careful comparison between the pump and the compressor process was executed. The evaluation showed economic advantages for the compressor process and in consequence design engineering was adjusted accordingly.

Because of the successful process implementation and the strong demand on the market, the capacity of the plant had to be more than doubled and the corresponding unit set in operation at the end of 2010 [34].

8.2.14.11 Economics – Especially Investment Cost for Multipurpose Plants

Figure 8.15 shows the relative investment cost for the high-pressure part of different sizes of supercritical fluid extraction plants, not including costs for infrastructure and erection. A plant with two extractors, each with a payload volume of 850 l – representing an economical size for the extraction of spices and herbs – with an investment of about €3 500 000 – was selected as basis for the comparison.

The influence of different separation conditions on processing costs is demonstrated by means of Figure 8.16, representing a decaffeination process for tea.

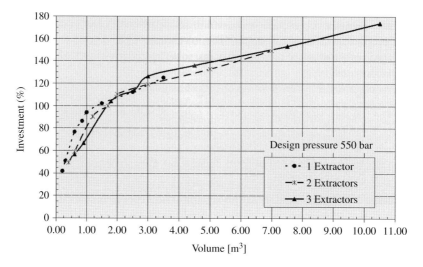

Figure 8.15 Investment cost for multipurpose plants designed for 550 bar.

Processing costs are more or less equal for the nonisobaric process using an external adsorbent. Larger plants with a total volume of above 4.4 m^3 should for economical reasons use a washing column, which also allows additional caffeine recovery.

8.3
Processing of Liquids

The processing of liquids with supercritical fluids mainly depends on their flow ability. Liquid feed materials of low viscosity like water-based solutions – for example, ethanol in water – are operated in high-pressure countercurrent columns that are

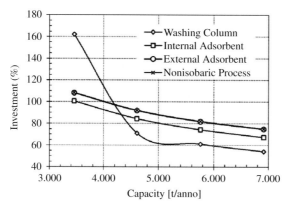

Figure 8.16 Processing costs depending on separation methods.

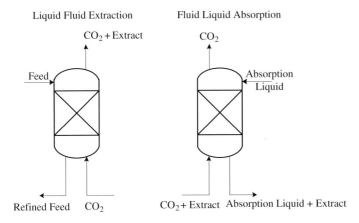

Figure 8.17 High-pressure column processes with supercritical fluids.

equipped with sieve plates or structured packing or even filling bodies in order to obtain thin falling films with low mass transfer resistance and high mass transfer area. These high-pressure columns are used for both the separation of distinct components from a liquid mixture by extraction and the regeneration of the loaded supercritical fluid by an absorption fluid (Figure 8.17). Generally, supercritical processing of liquids enables continuous operation. Furthermore, the regeneration of the supercritical solvent by an absorption liquid offers the advantage of an isobaric process circle. For example, the decaffeinating of concentrated liquid coffee extracts is possible in a circle containing an extraction column and an absorption tower [35].

Although a continuous and isobaric operated high-pressure column process seems to be beneficial, no absorption liquid with suitable distribution coefficients is often available. An incomplete purification of the supercritical fluid leads to strongly increased solvent to feed ratios. Thus, the extracted components are separated by pressure reduction or by adsorption like in supercritical processes for solid material. An example is given in Figure 8.18 for the deacidification of vegetable oil [36].

With the aim of designing a high-pressure column process, some basic information is needed on the following:

- Phase equilibrium: distribution coefficients and selectivity coefficients
- Mass transfer kinetics: mass transfer coefficients and mass transfer areas
- Hydrodynamic behavior of the countercurrent fluid flow: flooding points and static and dynamic holdup
- Interfacial phenomena: interfacial tension and wetting angles

First, the distribution of the extract component between the feed and the contacting supercritical fluid is given by the phase equilibrium. The distribution coefficient relates the extract concentration in the supercritical phase to the extract concentration in the corresponding liquid phase at equilibrium conditions. Distribution coefficients have been evaluated for ternary systems in a broad range of

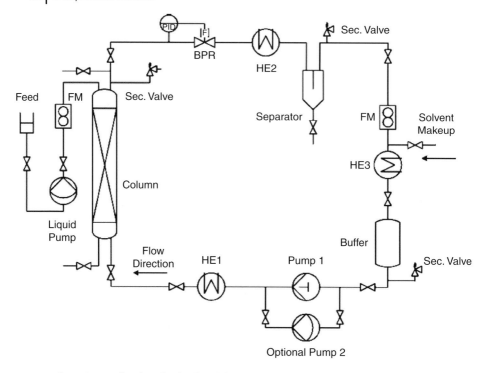

Figure 8.18 Pilot plant for the deacidification of vegetable oil in a countercurrent column process [35]. HE: heat exchanger, BPR: back pressure regulator.

pressures and temperatures [15, 37, 38]. It is evident that an effective countercurrent separation requires high distribution coefficients at least greater one. Often the liquid feed is a mixture of various components having a binary solubility in the supercritical fluid. The design of these processes additionally needs the selectivity coefficient giving the relation of the concentration of a distinct component to the concentration of other soluble components from the mixture. Although the distribution coefficient depends on the concentration of the feed solution to be separated, the design of the minimum solvent to feed ratio is approximately given by the inverse number of the distribution coefficients. As shown by Treybal [39], this is true when the distribution coefficient as the slope in a concentration diagram of the corresponding phases is a constant figure. However, the result is valid for an infinitely high column. Thus, the design is carried out with a step construction between the equilibrium line and an operation line using a solvent to feed ratio higher than the minimum value.

Aiming separation of liquid mixtures with supercritical fluids, the governing mass transfer resistance normally exists in the liquid phase. Furthermore, the solvent-to-liquid ratios are high. Thus, the favored mode of operation is to run the supercritical solvent as the continuous phase. The liquid phase forms out thin films, rivulets, and droplets. The application of usual mass transfer equation $Sh = f(Re, Sc)$ as given in Table 2.6, enables the calculation of mass transfer coefficients. However, in high-pressure countercurrent columns, one has to consider mutual mass transport, because the supercritical solvent dissolutes in the liquid changing its material properties like density and viscosity. The density slightly increases [40] and the viscosity decreases [41]. Besides the changing material properties of the liquid phase, the main problem in calculating the rate of mass transfer along the high-pressure column is caused by an inaccurate knowledge of the mass transfer area between the liquid and the supercritical phase. The interfacial tension drops down at high pressures and in consequence the liquid film may become decomposed [42]. The difficulty in precise calculation of mass transfer is often overcome by determining the product of mass transfer coefficient and mass transfer area with the help of experimental results. However, the measurement of interfacial tension and contact angles at high pressures relieves the estimation of the wetted area in a high-pressure column [42]. Generally, the wetting of oil-based liquid feed mixtures is better compared with the aqueous liquids (see Figure 2.9). In case of protein-containing feed mixtures, foam generation may be a risk in separating the liquid from a supercritical fluid.

The hydrodynamic behavior of the countercurrent flow is of high importance because of the density difference between the two phases at high pressures. Thus, with increasing gas density, the risk of flooding the column reveals. In order to design a safe liquid dynamics of a supercritical countercurrent column process, Stockfleth [43] developed an equation nondependent on geometrical data that enabled the prediction of the flooding point.

$$\sqrt{Fr} = \frac{0.4222}{1 + 1.1457 \cdot \sqrt{\Phi}}$$

With the Froude number of the gas phase related to the hydraulic diameter of the column and Φ as a flow parameter,

$$Fr_g = \left(\frac{u_l}{\varepsilon}\right)^2 \frac{\varrho_l}{g d_h \Delta \varrho}$$

$$\Phi = \frac{u_l}{u_g \left(\varrho_l/\varrho_g\right)^{0.5}}$$

$$d_h = \frac{4\varepsilon}{a_s}$$

where

u_g is the supercritical gas velocity (m/s)
u_l is the liquid film velocity (m/s)
ϱ_g is the gas density (kg/m³)
ϱ_l is the liquid density (kg/m³)
$\Delta \varrho$ is the density difference $\varrho_l - \varrho_g$
ε is the free volume ratio of the column packing
d_h is the hydraulic diameter
a_s is the specific area of filling bodies or structured packing (1/m)

In addition, the efficiency in the high-pressure separation process depends on the hydraulic capacity of the column. Besides the solvent to feed ratio, the liquid holdup is an important figure in order to evaluate the performance of the column packings and the pressure drop of the fluid flow. Due to the enhanced density of the gas phase, the influence of buoyancy is high and the knowledge of the dynamic holdup below the loading point is most important. The dynamic holdup denotes the ratio of the liquid volume in operation to the bed volume of the packing. It is determined by the total drainage of the liquid phase from the column after stopping the steady-state fluid flow. In addition, the static holdup gives the ratio of the remaining volume of the liquid to the bed volume. Especially for high-pressure countercurrent columns that are operated with supercritical gases, this ratio is small and of less importance. The reason is a low amount of liquid adhering to the surfaces of the packings due to low interfacial tension at high pressures.

$$h_{dyn} = \frac{\text{dynamic liquid flow}}{\text{bed flow volume}}$$

In the high-pressure columns, the density of the supercritical gas has to be considered. Thus, the dynamic holdup relates to the dimensional numbers of the Froude number and the Reynolds number of the liquid phase.

$$h_{dyn} = C_1 \left(\frac{Fr_l^2}{Re_l}\right)^{C_2}$$

where

$$Fr_l = \frac{u_l^2 a_s \varrho_l}{g \Delta \varrho}$$

$$Re_l = \frac{u_l \varrho_l}{\eta_l a_s}$$

Zacchi et al. [36] evaluated in order to find out a reliable relation for the dynamic holdup. Figure 8.19 demonstrates a linear regression in double logarithmic scale:

$$h_{dyn} = 2.72 \left(\frac{Fr_l^2}{Re_l}\right)^{0.20}$$

The high-pressure spray processes are schematically assembled in Figure 8.20. These processes using supercritical fluids are of increasing importance for the following advantages:

- High surface area for heat and mass transfer in countercurrent spray extraction or two-phase spray extraction. An example is the deoiling of raw lecithin [44].
- Homogeneous precipitation of particles from supercritical solvents (rapid expansion of supercritical fluids (RESS)). Applications are microparticles for pharmaceutics [45].

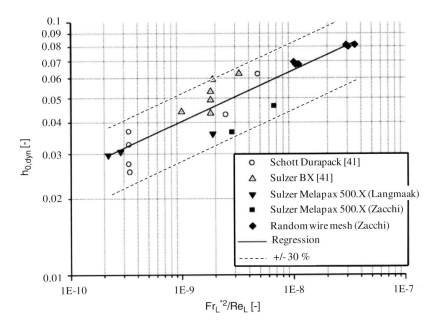

Figure 8.19 Dynamic holdup of different packings in high-pressure columns [36].

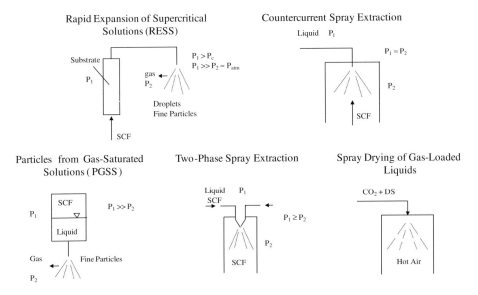

Figure 8.20 High-pressure spray processes.

- Homogeneous precipitation of particles from gas-saturated solutions (particle generation from saturated solutions (PGSS)). Applications are the powder production from molten polymers or fat compounds [46].
- Spray drying of liquid solutions saturated with dense gas and containing high contents of dry substances (ds). An example is the production of instant products [47].

Although continuous processing of liquids is possible, it is restricted to liquids with low viscosity. Known industrial applications are the caffeine recovery from corresponding loaded CO_2 streams (see Figure 8.10), processing of lipids, and the recovery of organic solvents from wastewater.

Nevertheless, some supercritical processes for treatment of viscous material have been developed.

Brunner and Peter [48] comprehensively performed research on liquid/liquid fluid extraction, using packed columns and adding entrainers to enhance the solubility as well as to reduce the viscosity. A particular problematic case of deoiling of lecithin was investigated by Weidner et al. [49].

Stahl et al. [8] developed the so-called *high-pressure nozzle extraction*. This process uses a two-phase nozzle with two concentric stainless steel capillaries of different diameters and fitted into each other, whereas the viscous feed was in the center and the dense gas in the outer annular space. Intense intermixing is ensured due the high resulting velocities and thereby caused turbulences. Tests for deoiling of lecithin performed at 900 bar and 90 °C resulted in light colored uneven structured powder with a residual content of oil below 3.5 wt%, without bitter substances and carotenoids. Eggers et al. [50] proposed the extraction of substances from sprayed particles within a high-pressure vessel and the precipitating of the dissolved substances afterward in a separator at a reduced pressure.

For viscous materials, the viscosity can strongly increase during extraction and due to lack of mixing substances that are extracted just from the surface, which is hardened, forms agglomerates, hinders further diffusion, and consequently reduces the yield. Therefore, the following criteria have to be considered for the high-pressure extraction of viscous materials:

- The supercritical gas shall have steady and easy access to soluble substances.
- Soluble substances shall have a steadily homogeneous radial distribution within a thin film.
- Mass transfer shall be increased by short diffusion distances.
- Product agglomerations shall be prevented/destructed.
- The surface shall be as large as possible, renewing itself continuously.

Such requirements were carefully considered by the development of the so-called *thin film extraction* (TFE) process [51].

A flow sheet for the design of the prototype pilot plant for the thin film extraction process is shown in Figure 8.21. The design pressure and temperature are 70 MPa and 150 °C, respectively.

- The extractor is equipped with two separate heating jackets, each with its own electrical heating circuit.

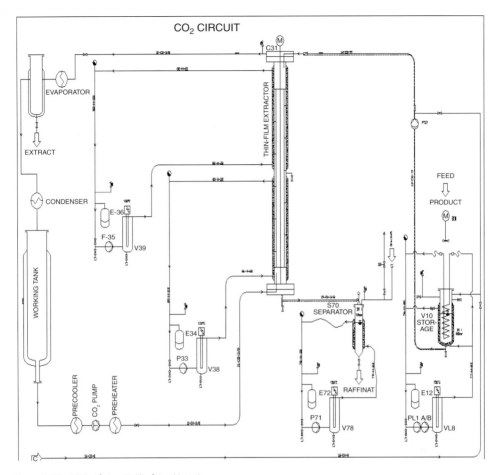

Figure 8.21 TFE pilot unit (Archive Natex).

- The wiper shaft is fixed with a specific ball bearing at the lower end, insensitive to dirt and CO_2 resistant.
- Above the top end of the wiper, a facility for uniform material distribution is placed.
- The wiper is shaped in such a way that it forces the film downward on the one hand and the flow of the dense gas upward along the rolls on the other.
- The wiper is driven via a magnetic clutch and a frequency changer controls the revolution.
- The CO_2 flow is limited to the roller space only and short path flow is prevented by specific internals.
- A nozzle is foreseen to take samples.

For the continuous processing of viscous materials or dispersions containing solids, the new thin film extraction process could be an excellent separation method and can serve – in cases of heat-sensitive or high-boiling substances in the substrate – as an alternative solution to short path distillation. If processing compounds with close vapor pressures and molecular steam distillation fails, the TFE process could be an appropriate solution, because separation depends on differences in polarity of the substances concerned. Another possible application is, according to the miscibility gap at applied conditions, the recovery of alcohol- and CO_2-soluble fractions from the extract solutions.

8.4
Future Trends

8.4.1
Drying of Aerogels

Aerogels have already been known for 70 years, but most of them are produced from inorganic materials like silica or metal oxides. It is also possible to produce aerogels from cellulosic materials. Aerogels belong to a group of materials with a high porous structure, a low solid content, and a very low bulk density. Therefore, cellulose, an easily available and renewable material, was chosen as an interesting candidate for the production of aerogels [52–54]. After gelation, the aerogel bodies were immersed in a solvent that is suitable for *supercritical drying* by means of a solvent change.

The substitution/displacement of solvent molecules located within the solid matrix by dense CO_2 is the basis for drying of highly porous materials. Supercritical carbon dioxide is not miscible with water, but completely miscible with most organic solvents and, therefore, very suitable for such a process. The specific volume of the CO_2–solvent mixture changes when the phase is changed from liquid to gas within the porous structure. In case part of the liquid remains in the structure, the surface tension at the solid–liquid interface creates capillary forces that destroy the structure, that is, delicate structures tend to break up and porous structures collapse.

Supercritical drying is a very delicate process in order to obtain a high-quality Aerocell with a desired pore structure. Several standard drying procedures were developed to minimize shrinkage of the samples and to preserve the pore structure of the wet gels. Static as well as dynamic extraction and a combination of both were investigated for optimization of the drying process. The solvent concentration in the CO_2 was measured online at the outlet of the separator in order to indicate the end of the drying process.

Currently, the supercritical drying process is the best solution to avoid such capillary forces. With the supercritical fluid drying, it is possible to move from the liquid phase via the supercritical fluid directly to the gas phase without touching the two-phase area. With optimized drying parameters, capillary forces can be avoided and nanoporous structures are maintained [55–57].

The supercritical drying process is carried out batchwise (see Figure 8.22). The material to be dried is inserted into a high-pressure vessel equipped with a quick-acting closure at the top. The samples are soaked in the organic solvent before the

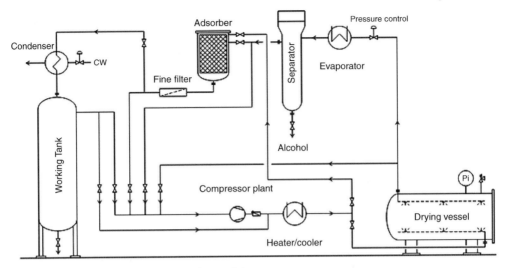

Figure 8.22 Flow sheet of an supercritical aerogel drying process.

extractor is closed. CO_2 from a storage vessel is pressurized up to 9 MPa–20 MPa (depending on the wet gel and the sample size) and heated up to 40 °C before entering into the extractor, where the displacement of solvent by CO_2 happens. It is necessary that CO_2 and organic solvent are in a homogeneous mixture. After the drying conditions are reached, the samples remain in the extractor without any CO_2 flow and the CO_2 can penetrate into the samples. After a certain time, CO_2 and the organic solvent form a homogeneous supercritical mixture inside and outside the samples.

It is extremely important that any surface tension between the organic solvent, CO_2, and the samples is avoided at any time during the whole process. Afterward the CO_2 flow is started again, which dilutes the organic solvent–CO_2 mixture. The loaded CO_2 leaving the extractor is depressurized in order to separate the organic solvent from the now gaseous CO_2. The CO_2 flow is maintained until the organic solvent is completely removed from the samples. In the final step, the extractor is depressurized to atmospheric pressure according to a certain pressure–time gradient. After opening of the extractor, the dried material can be unloaded.

8.4.2
Treating of Microorganisms

For the treatment of microorganisms, high-pressure technology offers two different processes:

Hydrostatic high-pressure process up to 4000 bar, which is discussed in Chapter 9. And as an alternative, microorganisms are processed with supercritical fluids.

The idea to use CO_2 as the supercritical fluid for inactivation of bacteria is already quite old. The first more successful results were published in 1987 [58] describing the treatment of *Escherichia coli, Staphylococcus aureus*, Baker's yeast, and spores of *Aspergillus niger* in a batch equipment at operation pressures up to 20 MPa and at 35 °C. The publication showed a sufficient reduction of microbes by many orders of magnitude not only for yeasts but also for spores. Wei *et al.* [59] published interesting results about microbial inactivation by treating pathogenic microbes in pure water at different operating pressures. Since this time, the important role of water has been recognized and the best results can be reached if the microbes are in an aqueous solution.

It was found that bacterial inactivation is achieved by CO_2 adsorption in the liquid phase, although the detailed procedure of the inactivation is not clear.

Finally, the intensive investigations in the past 10 years led supercritical CO_2 treatment of food to a promising stage by achieving cold pasteurization and/or sterilization of liquids. Now much more knowledge about the mechanisms of microbial inactivation is available.

Some hypotheses explain the peculiar action of CO_2 on microorganisms [60, 61]. In brief, dense CO_2 is claimed to cause the following:

- Cytoplasmatic pH decrease (acidification)
- Explosive cell rupture due to high internal pressure under pressure reduction
- Modification of cell membrane and extraction of cell wall lipids
- Inactivation of key enzymes for cell metabolism
- Extraction of intracellular substances

In summary, supercritical CO_2 treatment seems to be optimal for aqueous solutions, especially for fruit juices and vitamin-containing liquids. As liquids are mainly processed at moderate pressure, a continuous CO_2-operated plant for industrial use is possible.

8.4.3
Use of Supercritical Fluids for the Generation of Renewable Energy

Microalgae contain lipids and fatty acids as membrane components, storage products, metabolites, and sources of energy. Algae fatty acids and oils have a range of potential applications. Algal oils possess characteristics similar to those of fish and vegetable oils, and can thus be considered as potential substitutes for the products of fossil oil [62].

Direct extraction of microalgae lipids appears to be a more efficient methodology for obtaining energy from these organisms than the fermentation of algal biomass to produce either methane or ethanol. The lipid and fatty acid contents of microalgae vary in accordance with culture conditions.

As microalgae are growing very fast and harvesting can be done every few days, they are a very significant raw material for the production of different products, such as astaxanthine, carotenoids, lipids for bulk chemicals or biodiesel, protein for animal food, raw material for biogas production, and finally as natural process to convert carbon dioxide to oxygen by means of sun energy.

8.4.4
Gas-Assisted High-Pressure Extraction

Besides the attempts to generate polymer foams from CO_2-loaded melts [6], some other possible applications of gas-assisted processes are currently under investigation. Recently, processes that use the solubility of liquids for supercritical fluids are of increasing interest. The common background of these processes is the beneficial changes of thermophysical liquid data by dissolving of supercritical fluids (see Chapter 2). As such interfacial tension and dynamic viscosity drop improving the flowability and especially the drainage of liquids through porous systems. In consequence, the pressure drop decreases and the accessibility of the liquids under pressure to narrow capillary systems increases. For example, high-pressure fluids assist the enhanced recovery of oil and gas from the partial exhausted reservoirs [7], see Chapter 7, by reducing the wetting forces of the remaining oil in capillary systems. Another promising process is the gas-assisted pressing (GAP) [63, 64] of oilseed. A continuously working cage screw press is equipped with high-pressure nozzles injecting liquid or supercritical carbon dioxide into the compressed solid matter at distinct positions along the axis of the press (Figure 8.23).

There are some simulated running effects of liquid injection: dissolution of CO_2 in oil, enhanced specific volume of the CO_2-loaded oil, and displacing of oil by CO_2. These effects elucidate diverse benefits: first, the oil yield increases, and the temperatures of the drained oil and even of the deoiled press cake drop down by the Joule–Thompson effect of the rapid degassing CO_2. Thus, the quality of both the

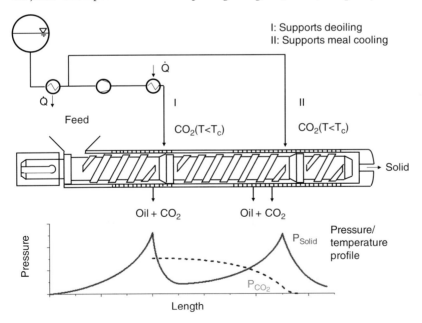

Figure 8.23 Gas-assisted pressing of oilseed.

oil and the protein-containing press cake will be improved by a reduced thermal impact. Similarly, a new gas-assisted process has been introduced aiming at cleaning the very small channels and ducts in turbine blades [65]. These channels, needed for inner cooling of the turbine blades, have to be deliberated from wax fillings that protect the surfaces after the manufacturing process. The gas-assisted cleaning process injects periodically supercritical CO_2 into the small channels. Dissolution of CO_2 and displacement of wax content are pushed out of the channels. A further gas-assisted process is the cleaning of textiles with a mixture of supercritical CO_2 and some interfacial active agents [66]. The low viscous fluid flow dissolves in lipid-based impurities and it lowers the adhesion of the dirty components at the textile surfaces. At the end, the disturbing particles are flushed away by the dense fluid. Other applications like absorption of supercritical fluids in ionic liquids are getting increasing interest. Also, the breaking of emulsion systems by means of supercritical fluids reveals successful steps [67]. Obviously, the development of new processes using supercritical fluids is by far not at its end.

References

1 Hannay, J.B. and Hoghart, J. (1879) On the solubility of solids in gases. *Proceedings of the Royal Society of London*, **29**, 324.
2 Passino, H.J. (1949) The solexol process. *Industrial & Engineering Chemistry Research*, **41**, 280.
3 Zosel, K. (1964) Patent 646641.
4 Schneider, G.M., Stahl, E., and Wilke, G. (eds) (1980) *Extraction with Supercritical Gases*, Verlag Chemie, Weinheim.
5 Brunner, G. (2010) Clean processing with supercritical fluids. Paper presented at the 12th European Meeting on Supercritical Fluids, May 9–12, 2010, Graz, Austria.
6 Maartje, F.K. and Meyer, T. (eds) (2005) *Supercritical Dioxide in Polymer Reaction Engineering*, Wiley-WCH Verlag GmbH, Weinheim.
7 Jaeger, P., Abaibi, M., and Nasr-El-Din, H. (2010) Influence of compressed carbon dioxide on the capillarity of the gas-crude oil-reservoir water systems. *Journal of Chemical & Engineering Data*, **55**, 5246–5251.
8 Stahl, E., Quirin, K.-W., and Gerard, D. (1988) *Dense Gases for Extraction and Refining*, Springer, Heidelberg.
9 Hubert, P. and Vitzthum, O. (1978) *Angewandte Chemie*, **90**, 756.
10 King, M.B. and Bott, T.R. (1993) *Extraction of Natural Products Using Near-Critical Solvents*, Blackie Academic & Professional, Glasgow.
11 Modell, M. (1982) US Patent No. 4,338,199.
12 Eggers, R. (1996) Supercritical fluid extraction of oilseeds/lipids in natural products, in *Supercritical Fluid Technology in Oil and Lipid Chemistry* (eds J.M. King and G.R. List), AOCS Press.
13 Steinhagen, V., Lütge, C., and Knez, Z. (2011) Ultra high pressure supercritical carbon dioxide extraction and fractionation of plant material. Paper presented at the 13th European Meeting of Supercritical Fluids, October 9–12, 2011, Hague, The Netherlands.
14 Bertucco, G. and Vetter, G. (eds) (2001) *High Pressure Process Technology: Fundamentals and Applications*, vol. **9**, Elsevier, pp. 351–403.
15 Brunner, G. (1994) *Gas Extraction: An Introduction to Fundamentals of Supercritical Fluids and the Application to Separation Processes*, Steinkopff, Darmstadt.
16 Eggers, R. and Hagen, R. (1986) Zum Stand der kontinuierlichen Extraktion von Feststoffen mit überkritischen Gasen, *Fette, Seifen. Anstrichmittel*, **9**, 344–351.

17 Eggers, R. and Tschiersch, R. (1979) *Extraction with Supercritical Gases: Development and Design of Plants for High-Pressure Extraction of Natural Products*, Verlag Chemie, Weinheim.

18 Lack, E. (1985) Kriterien zur Auslegung von Anlagen für die Hochdruckextraktion von Naturstoffen. PhD thesis. Technical University of Graz.

19 Chrastil, J. (1982) Solubility of solids and liquids in supercritical gases. *Journal of Physical Chemistry*, **86**, 3016.

20 Sovovà, H. (1994) Rate of the vegetable oil extraction with supercritical CO_2: I. Modeling of extraction curves. *Chemical Engineering Science*, **49**, 409–414.

21 Sovovà, H. (2005) Mathematical model for supercritical fluid extraction of natural products and extraction curve evaluation. *Journal of Supercritical Fluids*, **33**, 35–52.

22 Zizovic, I., Stamenić, M., Ivanović, J., Orlović, A., Ristić, M., Djordjević, S., Petrović, S.D., and Skala, D. (2007) Supercritical extraction of sesquiterpenes from valerian root. *Journal of Supercritical Fluids*, **43**, 249–258.

23 Stamenić, M., Zicović, I., Eggers, R., Jaeger, P., and Rój, E. Skala, D. (2010) Supercritical carbon dioxide extraction of Hop pellets. Paper presented at the 12th European Meeting on Supercritical Fluids, May 9–12, 2010, Graz, Austria.

24 Eggers, R. (1981) Energetische Optimisierung der Hochdruckextraktion. *Chemie Ingenieur Technik*, **53**, 551.

25 Farrell, K.T. (1990) *Spices Condiments and Seasonings*, 2nd edn, Van Nostrand, Reinhold.

26 Starmans, D.A.J. and Nijhuis, H.H. (1996) Extraction of secondary metabolites from plant material: a review. *Trends in Food Science and Technology*, **7**, 191.

27 Reverchon, E. (1997) Supercritical fluid extraction and fractionation of essential oil and related products. *Journal of Supercritical Fluids*, **10**, 1–37.

28 Mukhopadhyay, M. (2000) *Natural Extracts Using Supercritical Carbon Dioxide*, CRC Press, p. 155.

29 Bertucco, A. and Vetter, G. (eds) (2001) *High Pressure Process Technology: Fundamentals and Applications*, Elsevier, vol. **9**, p. 566–569.

30 Lack, Lang, Glanz, Seidlitz, Trukses, Liaw (1997) Verfahren zur Extraktion von Pflanzenschutzmitteln und/oder Reduzierung von unerwünschten Begleitstoffen aus Getreide, Patent No. 408,599, Austria.

31 Kjellow, A.W. and Henriksen, O. (2009) Supercritical wood impregnation. *Journal of Supercritical Fluids*, **50**, 297–304.

32 Taylor, M.K., Young, T.M., Butzke, C.E., and Ebeler, S.E. (2000) Supercritical fluid extraction of 2,4,6-trichloroanisole from cork stoppers. *Journal of Agricultural and Food Chemistry*, **48**, 2208–2211.

33 Felsvang, K., Iversen, S., Larsen, T., Lüthje, V., Aracil, J.M., Seidlitz, H., Lang, F., and Lack, E. (2005) The world's first supercritical plant for cork cleaning. Paper presented at the 7th International Symposium on Supercritical Fluids, Orlando, FL.

34 Lack, E., Seidlitz, H., Bakali, M., and Zobel, R. (2009) Cork treatment: a new industrial application of supercritical fluids. Paper presented at the 9th International Symposium of Supercritical Fluids, May 18–20, 2009, Arcachon, France.

35 Marckman, H. (2005) Überkritische Extraktion von aufkonzentrierten Kaffelösungen in Hochdruckkolonnen. PhD thesis. Technical University of Hamburg.

36 Zacchi, P., Bastida, S.C., Jaeger, P., Cocero, M.J., and Eggers, R. (2008) Countercurrent de-acidification of vegetable oils using supercritical CO_2, holdup and RTD experiments. *Journal of Supercritical Fluids*, **45**, 238–244.

37 Dohrn, R. (1994) *Berechnungen von Phasengleichgewichten*, Vieweg Verlag, Braunschweig.

38 Mc Hugh, M. and Krukonis, V. (1994) *Supercritical Fluid Extraction*, Butterworth-Heinemann, Boston.

39 Treybal, R.E. (1987) *Mass Transfer Operations*, McGraw Hill, New York.

40 Tegetmeier, A., Dittmar, D., Fredenhagen, A., and Eggers, R. (2000) Density and volume of water and triglyceride mixtures in contact with carbon dioxide. *Chemical Engineering and Processing*, **39**, 399–405.

41 Hobbie, M. (2005) Bildung von Tropfen in Verdichteten Gasen und Stationäre Umströmung fluider Partikel bei Drücken bis zu 50 MPa. PhD thesis. Technical University of Hamburg.

42 Jaeger, P.T. (1997) Grenzflächen und Stofftransport in verfahrenstechnischen Prozessen am Beispiel der Hochdruck-Gegenstromfraktionierung mit überkritischem Kohlendioxid. PhD thesis. Technical University of Hamburg.

43 Stockfleth, R. (2002) *Fluiddynamik in Hochdruckgegenstromkolonnen für die Gasextraktion*, Fortschritt-Berichte VDI, Maintal.

44 Wagner, H. (1999) Extraktion hochviskoser Medien in einer Turbulenten Zweiphasenströmung unter erhöhten Drücken am Beispiel der Entölung von Sojalecithin. PhD thesis. Technical University of Hamburg.

45 Türk, M. (2010) Supercritical fluids as novel particle formation media: 1. Application to the formation of organic and inorganic materials. Paper presented at the Proceedings of the 12th European Meeting on Supercritical Fluids, Graz, Austria.

46 Martin, A., Pham, H.M., Kilzer, A., and Weidner, E. (2010) Particles from gas saturated solutions (PGSS): drying process. Fundamentals and application to micronization of polyethylene glycol. Paper presented at the Proceedings of the 12th European Meeting on Supercritical Fluids, Graz, Austria.

47 Hassenklöver, E. and Eggers, R. (2008) Atomization of solid suspensions with dissolved inert gases. Paper presented at the ILASS Proceedings, September 8–10, 2008, Como, Italy.

48 Brunner, G. and Peter, S. (1981) Zum Stand der Hochdruckextraktion mit komprimierten Gasen. *Chemie Ingenieur Technik*, **53**, 529.

49 Weidner, E., Peter, S., Schneider, M., and Ziegelitz, R. (1987) Die Viskosität koexistierenden Phasen bei der überkritischen Fluidextraktion. *Chemical Engineering & Technology*, **10**, 37–42.

50 Eggers, R., Wagner, H., and Jaeger, P. (1996) Extraction of Spray-Particles with Supercritical Fluids. Proceedings of the 3rd International Symposium on High Pressure Chemical Engineering, October 7–9, 1996, Zürich, Switzerland, 247.

51 Lack, Seidlitz (2003) Verfahren und Vorrichtung zur Extraktion von Stoffen aus Flüssigkeiten und Feststoffdispersionen, PCT/AT 03/00246.

52 Innerlohinger, J., Weber, H.K., and Kraft, G. (2006) Aerocellulose: Aerogels and aerogel-like materials made from cellulose. *Macromolecular Symposia*, **244** (1), 126–135.

53 Gavillon, R. (2007) Preparation et Caractérisation de Matériau Cellulosique Ultra Poreux. PhD thesis. Sophia Antipolis.

54 Lack, E., Seidlitz, H., Sova, M., and Lang jun, F. (2007) Supercritical drying of cellulose aerogels. Paper presented at the 5th International Symposium on High Pressure and Chemical Engineering, June 24–27, 2007, Segovia, Spain.

55 Fischer, F., Rigacci, A., Pirard, R., Berthon-Farby, S., and Achard, P. (2006) *Cellulose-based aerogels. Polymer*, **47**, 7636.

56 Smirnova, I. and Arlt, W. (2003) Synthesis of silica aerogels: influence of the supercritical CO_2 on the sol–gel process. *Journal of Sol-Gel Science and Technology*, **28** (2), 175–180.

57 Mehling, T., Smirnova, I., Guenter, U., and Neubert, R.H. (2009) Polysaccharide-based aerogels as drug carriers. *Journal of Non-Crystalline Solids*, **355**, 2472–2479.

58 Kamahira, M., Taniguchi, M., and Kobayashi, T. (1987) Sterilisation of microorganisms with supercritical carbon dioxide. *Agricultural Biological Chemistry*, **51**, 407–412.

59 Wei, C.I., Balaban, M.O., Fernando, S.Y., and Peplow, A.J. (1991) Bacterial Effect of High Pressure CO2 Treatment of Foods spiked with Listeria or Salmonella. *Journal of Food Protection*, **54**, 189.

60 Spilimbergo, S. and Bertucco, A. (2003) Non-thermal bacteria inactivation with dense CO_2. *Biotechnology and Bioengineering*, **84** (6), 627–638.

61 Spilimbergo, S., Elvassore, N., and Bertucco, A. (2002) Microbial inactivation by high-pressure. *Supercritical Fluids*, **22** (1), 55–63.

62 Chisti, Y. (2007) Biodiesel from microalgae. *Biotechnology Advances*, **25**, 294–306.
63 Venter, M.J. (2006) Gas assisted mechanical expression of cocoa nibs. PhD thesis. University of Twente, The Netherlands.
64 Voges, S., Eggers, R., and Pitsch, A. (2008) Gas assisted oilseed pressing. *Separation and Purification Technology*, **63**, p. 1–14.
65 Pietsch, A. and Unger, M. (2010) High pressure cleaning of turbine blades with supercritical CO_2: a new pressure process comes into operation. Paper presented at the Proceedings of the 48th EHPRG International Conference, July 25–29, 2010, Sweden.
66 van Roosmalen, M.J.E., Woerlee, G.F., and Witkamp, G.J. (2004) Surfactants for particulate soil removal in dry-cleaning with high-pressure carbon dioxide. *Journal of Supercritical Fluids*, **30**, 97–109.
67 Alex, M. (2010) Spaltung von Emulsionen mit verdichteten Gasen. PhD thesis. Ruhr-Universität, Bochum.

9
Impact of High-Pressure on Enzymes
Leszek Kulisiewicz, Andreas Wierschem, Cornelia Rauh, and Antonio Delgado

9.1
Introduction

Many proteins are enzymes that catalyze various biochemical reactions. Enzymes are naturally present in foods. Some cause unwanted processes leading to deterioration of food quality, changing the taste, appearance, aroma, and so on [1–3]. Hence, besides reduction of viable microorganisms, inactivation of unwanted enzymes is the primary role of food preservation. Inactivation of microorganisms and enzymes is most commonly carried out by means of thermal processing, which, along with beneficial improvement of food safety and shelf life, often entails significant losses in nutritional and organoleptic properties of the products. Therefore, innovative technologies are of interest, which provide pasteurization with minimum deterioration of food quality. In the last two decades, high-pressure processing (HPP) is being considered as an alternative to thermal treatment [1–13]. Nowadays, the effect of pressures up to around 1 GPa is studied, although industrial equipment is usually operated at somewhat lower pressure levels [4, 14].

The feature of pressure treatment that has attracted the attention of food technologists is the ability to kill pathogenic microorganisms, even at low temperatures. Bacteria, yeasts, moulds, and viruses can be inactivated by means of high-pressure treatment [4, 5, 8]. Obviously, the process parameters, such as pressure and temperature level, rate of pressure change, treatment time, mode of pressurization (single or repeated cycles), influence inactivation. This kind of "cold pasteurization" allows retention of heat-sensitive nutrients, flavor, and color. Besides reduction of viable microorganism cells, high pressure enables destabilization and inactivation of certain enzymes present in foods [1–7]. In this way, the stability and shelf life of the product can be extended and unwanted changes in taste or appearance catalyzed by enzymes can be avoided. However, the reactions of particular enzymes to pressure treatment are diverse. Besides the process parameters, they also depend on the physicochemistry of the environment [1–3]. Enzymes, being proteins, undergo pressure-induced disruption of the sensitive equilibrium between the molecular interactions stabilizing the native three-dimensional protein structure. This can result

Industrial High Pressure Applications: Processes, Equipment and Safety, First Edition. Edited by Rudolf Eggers.
© 2012 Wiley-VCH Verlag GmbH & Co. KGaA. Published 2012 by Wiley-VCH Verlag GmbH & Co. KGaA.

in unfolding, denaturation, aggregation, or gelification [7, 10–12]. Enzymatic activity of proteins may increase or decrease due to change of conformation and functionality. The change may be complete or partial, reversible or irreversible [1–3, 15].

If the processing goal is the enhancement of the food product's shelf stability, the desired effect is usually the irreversible inactivation of the quality-deteriorating enzymes. However, pressure can also be applied to enhance the action of enzymes [2–16]. Pressure-tolerant enzymes are required, for example, in the area of enzyme-catalyzed synthesis involving organic or volatile solvents. Increase of pressure leads to increase of the boiling point of organic solvents and so allows high-efficiency enzyme-catalyzed synthesis of, for example, drugs and flavors at elevated temperatures and pressures [16]. Furthermore, pressure can influence the enantioselectivity of enzymes, that is, their ability to catalyze asymmetric synthesis of chiral substances, which is of high importance particularly in pharmaceutical industry [17–19].

Hence, enzyme activity is not always unwanted. On the contrary, many industrial applications of enzymatic catalysts are known, including food processing (e.g., brewing, dairy, and starch industry) and chemical and pharmaceutical industry [9, 16–20]. However, this chapter is focused on the case of high-pressure processing of foods aimed at the prolongation of shelf life. Hence, the emphasis is placed on the inactivation of enzymes related to quality of food products.

9.2
Influence of Pressure on Biomatter

The enzymatic activity depends, among other factors, not only on the chemical environment, for example, pH value, water content, and presence of salts and sugars [2, 21–23], but also on the thermodynamic state given by temperature and pressure. Obviously, a change in pressure leads to shift of the equilibrium of the system. The effect of pressure on biomatter is governed by Le Chatelier's principle, which predicts that pressure increase shifts equilibrium toward the state occupying a smaller volume. In other words, processes associated with a negative change in volume are favored by pressure application. This behavior can be quantitatively described by Eq. (9.1) provided by Planck:

$$\left(\frac{\partial \ln K}{\partial P}\right)_T = -\frac{\Delta V}{RT} \tag{9.1}$$

In this expression, K is the equilibrium constant for a given chemical equilibrium and the reaction volume ΔV is the difference between the volumes of the final and initial states in any particular process, for example, due to a change in molecule packing, electrostatic action of charges, solvation, and hydration effects. From Eq. (9.1), it is clear that the sign of the reaction volume ΔV imposes the direction of equilibrium shifts induced by pressure. If ΔV is negative, the equilibrium shifts toward the transformed state; if ΔV is positive, the equilibrium shifts toward the initial state. Accordingly, for $\Delta V = 0$, pressure does not influence the reaction equilibrium [10, 11, 24].

Similarly, pressure influences the rate of reactions, accelerating reactions with negative activation volume $\Delta V^{\#}$ and decelerating reactions with positive activation volume, according to Eq. (9.2):

$$\left(\frac{\partial \ln k}{\partial P}\right)_T = -\frac{\Delta V^{\#}}{RT} \quad (9.2)$$

where k is the rate constant. An overview of the activation volumes of various organic reactions is provided by van Eldik et al. [25].

These principles allow a fundamental understanding of pressure effects on biomolecules such as enzymes. Enzymes are stabilized by interactions such as covalent, electrostatic, hydrophobic, and hydrogen bonds involved at different structure levels. The key to explaining the overall effect of pressure on any particular macromolecule is the combination of the changes in all bonds. These changes are governed by the volumes of formation or breaking of the interactions [10, 11, 24].

Covalent bonds are strong intramolecular interactions due to sharing of electron pairs between atoms. The bond energy is very high (on the order of several hundreds of kiloJoule/mole), whereas the reaction volume ΔV for the exchanges in bonds or bond angle changes is close to zero [24]. Therefore, covalent bonds are generally pressure insensitive, at least up to pressures of 1–1.5 GPa [10, 11]. Electrostatic interactions are caused by Coulomb forces acting on charged ions. Formation of an ion in aqueous solution results in electrostriction, that is, compression of the nearby water dipoles due to action of the ion's electrostatic field. Dissociation of a neutral molecule into two ions results in volume decrease ΔV on the order of -20 ml/mol [10, 11]. Thus, according to Le Chatelier's principle, electrostatic bonds in biomolecules become weaker at high pressures, while ionization is a process promoted by pressure. This is the explanation for the pressure-induced decrease in the pH value of water and aqueous solutions, which is due to dissociation of water molecules. A pressure increase from ambient conditions to 100 MPa shifts the pH value of water by 0.3 [24]. Similarly, the pH value of acids decreases with pressure, for example, by about 0.2 per 100 MPa for acetic acid and 0.4 per 100 MPa for phosphoric acid [26]. Hydrogen bonds are interactions involving a hydrogen atom bonded to an electronegative atom and attracted by another electronegative atom. Formation of hydrogen bonds in biomolecules is accompanied by small or vanishing values of reaction volume ΔV, which may be positive or negative [24, 25]. Masson and Tonello [27] report on reaction volumes in the range of -4 to 1 ml/mol. Accordingly, some studies [10, 11, 26] indicate that hydrogen bonds are slightly stabilized by high pressure, while others show the opposite result [24, 28]. Finally, hydrophobic interactions arise due to repelling of hydrophobic groups from the bulk of the water. The formation of hydrophobic contacts requires a positive reaction volume ΔV in the range of approximately

1–20 ml/mol, depending on the system [11]. Therefore, hydrophobic contacts are disfavored at elevated pressures. However, exposure of hydrophobic groups to water leads to generation of a hydrophobic solvation layer that is assumed to be more densely packed than that generated when exposed to the bulk water [29]. Hence, the exposure of hydrophobic residues occurring, for example, during the unfolding of globular proteins is favored by pressure increase [24].

Enzymes are generally globular proteins fold into a three-dimensional structure, which is considered at four levels [9, 12]. The primary structure is given by the sequence of amino acid residues connected by covalent peptide bonds. The secondary structure refers to regularly repeating ordered configurations of local segments of the protein molecule, for example, α-helices and β-sheets, which are stabilized by hydrogen bonds. The remaining parts of the molecular chain generally adopt random coil conformations. The tertiary structure is the overall shape of a single protein molecule defined through the spatial relationship of the secondary structures to one another. The tertiary structure is stabilized by a combination of hydrogen bonds, electrostatic interactions, and covalent disulfide bonds and also through formation of a hydrophobic core [9, 12]. Globular proteins have a roughly spherical shape, often with a hydrophobic core due to complex fold of the molecular chain. The quaternary structure is the arrangement of oligomeric complexes involving more than one protein subunit. Protein complexes or aggregates are stabilized by noncovalent and disulfide bonds [9, 12, 24].

Depending on the kind of interactions bonding the structure element of a protein molecule, pressure increase leads to its stabilization or destabilization. The covalent bonds interconnecting the primary structure of proteins are not influenced by pressure [10, 12, 24]. Hence, the sequence of amino acids does not change due to pressure treatment. However, as discussed above, noncovalent bonds undergo significant changes leading to reversible and irreversible modifications of the secondary, tertiary, and quaternary structures [11]. Generally, macroscopically observed denaturation of proteins can be described with a characteristic elliptic diagram in the pressure–temperature domain [9, 12, 20], shown schematically in Figure 9.1. The native conformation is found in a limited range of pressure and temperature and exceeding the limit values leads to denaturation. Irreversible denaturation is preceded by an intermediate state of reversible unfolding [12, 16]. Hence, at low temperatures (point A in Figure 9.1), pressure-induced denaturation already takes place at relatively low pressures, whereas moderate pressures lead to increased stability of proteins against thermal denaturation (point B in Figure 9.1).

However, the elliptic phase diagram can be observed only for proteins in solution. If the water content is low and proteins are in dry state, their pressure stability is significantly increased [12]. This appears to be related to the decreased mobility and flexibility of enzyme molecules at low water content that prevents unfolding and denaturation [2]. Furthermore, also the pH value influences the protein's behavior. Zipp and Kauzmann [30] studied pressure–temperature denaturation of myoglobin in a wide pH range and observed decreased stability of the protein at extreme pH values. Hence, the size of the ellipse shown in Figure 9.1 is reduced at very high and very low pH [12].

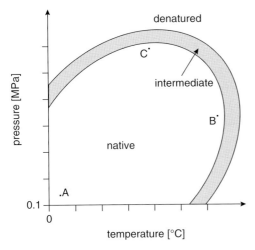

Figure 9.1 Schematic representation of the elliptic phase diagram of proteins in pressure–temperature domain.

9.3
Influence of Pressure on the Kinetics of Enzyme Inactivation

Generally, the structure of enzyme molecules and their function can be correlated. Denaturation leads to changes at the active site, resulting in loss of the catalytic activity [2, 31]. Experimental observations appear to confirm this prediction. Based on the measurements of the kinetics of enzyme denaturation, isokineticity curves can be drawn by connecting the points with the same denaturation rate constants in the p–T domain. As a result, isokineticity curves with a shape similar to the elliptic phase diagram (see Figure 9.1) are obtained [1–4, 12, 20]. It is then possible that the increase of temperature stabilizes enzymes against pressure-induced inactivation (left from point C in Figure 9.1) or destabilizes it against pressure (right from point C in Figure 9.1), depending on the initial temperature of the process and location of the isokineticity–ellipse of the particular enzyme relative to the temperature and pressure axes. In the same way, pressure increase can enhance enzyme activity at elevated temperatures (down from point B in Figure 9.1) or reduce its activity (up from point B in Figure 9.1).

The temperature–pressure-induced inactivation can be described by a general nth order scheme [2, 30]:

$$N \underset{}{\overset{k_{rev}}{\longleftrightarrow}} I \overset{k_{ir}}{\longrightarrow} D \tag{9.3}$$

where N, I, and D denote the native, intermediate (reversibly unfolded), and denatured states of the enzyme, respectively. The measured apparent rate constant k can be expressed in terms of the rate constants for the respective reversible and irreversible denaturation steps k_{rev} and k_{ir} [2] as

$$k = \frac{k_{ir}}{1 + k_{rev}} \tag{9.4}$$

At sufficiently high temperatures and pressures, concentration of the native form is much smaller than that of the reversibly unfolded intermediate form and $k_{rev} \ll 1$ (compare with Figure 9.1). Hence, the influence of the reversible unfolding equilibrium constant k_{rev} is negligible and $k = k_{ir}$ [2, 32].

In fact, enzyme inactivation often appears to follow first-order kinetics according to a scheme:

$$N \xrightarrow{k} D \tag{9.5}$$

The first-order kinetics was observed to be suitable for description of inactivation of *Bacillus subtilis* α-amylase [33], lipoxygenase from green pea, green bean and soy bean [34–37], polyphenoloxydase from avocado, white grapes, strawberry [38–40], polygalacturonase from tomato [41], peroxidase from strawberry [42], and plasmin and alkaline phosphatase from bovine milk [43, 44]. This happens independent of the fact that the inactivation itself is a complex process involving a number of steps and possibly more than one molecule [45]. In case of first-order kinetics ($n = 1$), the general model describing the decrease of enzyme activity A as a function of time t

$$\left(\frac{dA}{dt}\right) = -k \cdot A^n \tag{9.6}$$

can be transformed to

$$A = A_0 \exp(-k \cdot t) \tag{9.7}$$

where A_0 is the initial enzyme activity [2]. Alternatively, the inactivation process can be expressed in terms of thermal death time [46] or pressure death time [47, 48] with the observed decimal reduction time $D = 1/k \ln(10)$ defined as the time needed to reduce the initial enzyme activity to 10% of its original value at a given temperature or pressure [2]:

$$A = A_0 \cdot 10^{-t/D} \tag{9.8}$$

Nevertheless, inactivation of enzymes is a process involving several events, as mentioned above. Therefore, simple first-order inactivation kinetics is rather a simplification that cannot be expected sufficient in all cases. For example, Buckow and coworkers report on kinetics order of 1.4 for β-amylase from barley [49], 1.6 for β-glucanase from barley [50], 2.1 for α-amylase from barley [51], and 2.2 for polyphenoloxydase from apple [52] and Rademacher and Hinrichs [53] found inactivation kinetics order of 1.5 for alkaline phosphatase in bovine milk.

Furthermore, the kinetics of inactivation processes often shows a biphasic or multiphasic behavior, which is attributed to occurrence of isozyme fractions with different stabilities [2, 20]. It is then convenient to assume a discrete number of distinct isozymes according to a scheme (here for biphasic case):

$$N_S \leftrightarrow I_S \xrightarrow{k_S} D_S \\ N_L \leftrightarrow I_L \xrightarrow{k_L} D_L \tag{9.9}$$

with indices S and L referring to the stable and labile isozyme fractions, respectively. The enzymatic activity can be treated additively:

$$A = A_{0_S} \exp(-k_S \cdot t) + A_{0_L} \exp(-k_L \cdot t) \tag{9.10}$$

where A_{0S} and A_{0L} are the initial activities of both isozymes. Biphasic kinetic models of pressure-induced inactivation were proposed, for example, for glucoamylase [54], lipoxygenase from tomato [55] and pectinmethylesterase from banana [56], carrot [57], grapefruit [58, 59], and orange [47, 59–61].

Also, other alternative models were proposed to explain inactivation behavior. The consecutive step model refers to a number of intermediate steps in the overall inactivation

$$N \longleftrightarrow I \xrightarrow{k_1} D_1 \xrightarrow{k_2} D_2 \longrightarrow D \tag{9.11}$$

due to respective irreversible (first-order) reaction steps of conversion of the native enzyme to intermediate conformations with reduced activity [2].

Another model, the fractional conversion model, [2, 3], can be applied to describe the presence of residual active enzyme fraction resistant to pressure/temperature treatment in addition to a fraction following a first-order inactivation according to

$$N + RF \longleftrightarrow I + RF \xrightarrow{k} D \tag{9.12}$$

where RF denotes the fraction resistant to the treatment.

The temperature and pressure dependency of the inactivation rate constants can be described, respectively, by means of Arrhenius equation at constant pressure

$$k = k_{T_0} \exp\left(\frac{-E^{\#}}{R}\left(\frac{1}{T} - \frac{1}{T_0}\right)\right) \tag{9.13}$$

and Eyring equation at constant temperature [2]:

$$k = k_{p_0} \exp\left(\frac{-V^{\#}}{RT}(p - p_0)\right) \tag{9.14}$$

The inactivation rate constants at reference temperature T_0 and pressure p_0 are given by k_{T_0} and k_{p_0}, respectively. $E^{\#}$ is the activation energy and $V^{\#}$ is the activation volume, both derived from the corresponding slopes of the semilogarithmic plots [2]. Alternatively, for the first-order kinetics, the thermal death time and pressure death time approach can be applied (compare with Eq. (9.8)) with D_{T_0} and D_{p_0} respectively being the decimal reduction times at reference temperature and pressure [2, 46–48].

$$D = D_{T_0} \cdot 10^{(T_0 - T)/z_T} \tag{9.15}$$

$$D = D_{p_0} \cdot 10^{(p_0-p)/z_p} \tag{9.16}$$

The values z_T and z_p are the temperature and pressure increase necessary to obtain a reduction of the decimal reduction time to 10% of its initial level and are determined from the negative reciprocal slopes of the semilogarithmic plots of D as function of temperature or pressure, respectively [2].

In an attempt to develop mathematical models describing the combined pressure–temperature dependency of the inactivation rate constants in (9.13) and (9.14) or decimal reduction times in (9.15) and (9.16), the kinetics of enzyme inactivation is measured experimentally. The experimental conditions are aimed at isobaric and isothermal cases, which are not straightforward as discussed in the next section. The models usually require the dependencies on pressure and temperature of the parameters k_{T_0}, k_{p_0}, $E^{\#}$, $V^{\#}$, D_{T_0}, D_{p_0}, z_T, and z_p (see (9.13)–(9.16)) [34–37, 61–63]. The suitable functions (e.g., polynomial, exponential, and potential) are usually determined by curve fitting. The clear disadvantage of this semiempirical approach is that the resulting model equations are valid only in the experimentally verified pressure and temperature ranges. Extrapolation from the experimentally studied domain is usually not possible [2]. Model formula describing the combined pressure–temperature impact on the kinetics of inactivation of particular enzymes can be found in databases dedicated to this issue [1, 2, 20]. In particular cases (e.g., B. subtilis α-amylase), the data on the pressure–temperature behavior of enzymes are sufficiently precise to enable application of these enzymes (in combination with suitable solvent) as pressure–temperature–time integrating indicators for control of the impact of pressure–temperature treatment [64, 65].

Essential factors in inactivation behavior are the origin of the enzyme and the milieu conditions (liquid solvent, juice, puree, and solid tissue). For example, threshold pressure for inactivation of pectinmethylesterase has been shown to vary from 150 to 1200 MPa, depending on the source and the medium in which the treatment is carried out [47, 59, 66–70]. Pressure treatment can cause some damage to cell structure, for example, by influencing membrane functions, which in turn would cause release and contact of enzyme and its substrate. As a result, apparent activation of the catalytic activity can be observed, like in the case of polyphenoloxidase [71]. Obviously, the catalytic activity of enzymes depends not only on the state of the enzymes itself but also on the impact of pressure on the catalyzed reaction, particularly the sign and magnitude of reaction volume [16, 49, 50, 72]. It has been shown that enzymatic reactions can be reduced or enhanced [73–75], for example, due to alteration of the rate-limiting structures of the substrate molecules, conformational change, or gelatinization [72].

9.4
Technological Aspects

As already stated, the inactivation of enzymes depends not only on the hydrostatic pressure but also on a number of other parameters such as pressure-holding time, temperature, pH value, and other interactions with the ambient media. These dependencies may influence significantly the result in industrial applications. Heterogeneous distribution of temperature or pH value may induce nonuniform spatiotemporal enzyme inactivation, which is an undesired effect, since it results in over- and underprocessed fractions.

Industrial inactivation is generally carried out as a batch process. The treated food is often packaged and then placed into an autoclave made from steel. After filling the autoclave with a liquid, generally water, which serves as a pressure-transmitting medium, the autoclave is pressurized. Subsequently, the process continues with a phase of constant high-pressure or periodic pressure cycles. Finally, the pressure is released, the product is recovered, and a new treatment may start.

As shown by Delgado and Hartmann, the high-pressure treatment in autoclaves is intrinsically an inhomogeneous process [76]. This is essentially due to the heating caused by compression. For an adiabatic process, this reads

$$\frac{\partial T_{\text{ad}}}{\partial p} = \frac{\alpha_p}{\varrho c_p} T \tag{9.17}$$

where T, p, α_P, ϱ, c_P, and the subscript ad indicate temperature, pressure, isobaric thermal expansion coefficient, density, specific heat capacity and adiabatic process, respectively. Since α_P, ϱ, and c_P are material parameters, the temperature increase differs for the materials involved. This has far-reaching consequences. The filled pressure vessel represents a nonuniform system consisting of pressure-transmitting medium, steel vessel wall, product, and packaging material. Hence, a heterogeneous temperature field can arise in the vessel during pressurizing and pressure-holding time, despite an initial uniform temperature distribution [77–89]. Subsequently, heat transfer processes set in. Due to the pressure–temperature dependency of enzyme inactivation, heterogeneous temperature distributions can influence the uniformity of the process. This holds even in the basic system consisting of the autoclave and a pressure-transmitting medium. The temperature increase of the autoclave walls made from stainless steel is weaker than that of the pressure-transmitting medium; this creates temperature differences between the container walls and the pressure-transmitting liquid [13]. Heat conduction through the container walls cools down the neighboring liquid and causes thermal heterogeneities within the vessel, as has been pointed out by Delgado and Hartmann [76].

Various groups have studied velocity fields and temperature distributions in the pressure-transmitting medium in the autoclaves [76, 87, 90–95]. Delgado and Hartmann could show numerically that the horizontal temperature gradient yields an unstable density configuration in the neighborhood of the cool sidewalls [76]. Due to gravity, the cooler liquid flows downward in the vicinity of the sidewalls. In large industrial containers, this results in a hydrodynamic boundary layer [92]. The maximum down-flow velocity in such a case as determined numerically is shown in Figure 9.2. Rauh [92] showed that in water and water–saccharose solutions, the corresponding thermal boundary layer is thinner than the hydrodynamic one. Hence, hotter liquid close to the cool near-wall liquid is also dragged downward,

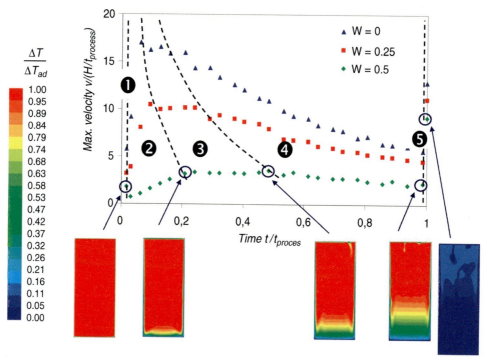

Figure 9.2 Maximum velocity in a cylinder–piston system as a function of time during pressure treatment. Triangles, squares, and diamonds indicate the use of water and water–saccharose solutions with a mass fraction of 0.25 and 0.5, respectively. For selected time instances, the distribution of the temperature increase in the autoclave is visualized in the insets underneath the diagram [92].

seen as hot fingers in the lower part of the insets depicted in Figure 9.2. As time proceeds during pressure holding, the same figure shows that natural convection produces a stable temperature and density stratification in the pressure vessel. Traces of the stratification may remain even after pressure release [92].

While the cold container bottom yields a stably stratified configuration right from the start, the unstable density stratification at the cool top of the container may result in a Rayleigh–Bénard instability [93]. In the third and fourth inset of Figure 9.2, the instability is observed as cold liquid draining down from the top into the bulk.

Besides natural convection during pressure holding, forced convection may also arise during the pressure ramp. In systems where pressurizing is achieved by injection of pressure-transmitting liquid through a tube into the vessel, the incoming liquid is usually colder than that in the vessel [96]. Because of higher contact area, the heat transfer between inflowing liquid and piping is more efficient than in the autoclave. Here, a free jet enters the vessel with high flow velocities and induced thermal heterogeneities can be large. Hartmann [81] and Hartmann and coworkers showed the strong effect of the forced convection on the temperature and on the velocity field in the vessel [90, 94, 97]. Recently, Song et al. gave evidence of turbulent

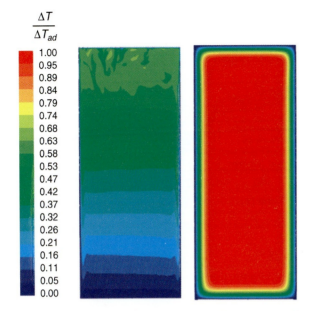

Figure 9.3 Comparison of the temperature fields at the end of the pressure-holding phase in water (left) and in a model liquid with the same material parameters as water except for a 20 000 times higher viscosity [101].

inflow and relaminarization in liquids with sufficiently strong pressure dependence of the viscosity [98–100].

Forced and natural convection enhance heat transfer through the container walls and a temperature decrease in the bulk. As shown in Figure 9.2, the intensity of the convection strongly depends on the viscosity of the pressure-transmitting liquid, which increases with the saccharose concentration. The effect of viscosity has been further studied in detail by Hartmann et al. [94], Baars et al. [95], and Hartmann and Delgado [97]. Rauh et al. showed that natural convection could be suppressed in pressure-transmitting liquids of sufficiently high viscosity, as illustrated in Figure 9.3 [101].

To validate the numerical results, Delgado and coworkers also studied the temperature field experimentally. They showed that numerical and experimental results coincide within measurement uncertainty, as indicated in Figure 9.4. Detailed information about the two-dimensional temperature and velocity fields inside the autoclave has been obtained by using liquid crystals as temperature indicators in a particle image velocimeter [86, 87, 102, 103]. The liquid crystals color code the local temperature. Figure 9.5 shows the stratification of the temperature field in the autoclave, where blue indicates higher and red the lower temperatures, respectively.

The aforementioned heterogeneous temperature field in the autoclave and its temporal evolution may significantly affect the inactivation process of enzymes. For enzymes diluted in pressure-transmitting media, Rauh et al. have numerically studied inactivation and its heterogeneity in the presence of the spatiotemporal

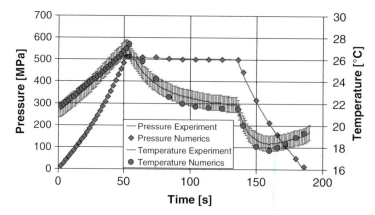

Figure 9.4 Comparison of numerical and experimental results for the temperature at a certain location in the pressure chamber [97].

temperature field [101]. For this purpose, they modeled the inactivation kinetics of several enzymes in water as a first-order reaction:

$$\frac{\partial A}{\partial t} + \vec{v} \cdot \nabla A = -k_p(1+\alpha)A \tag{9.18}$$

Here, A and $\vec{v}$ indicate the enzyme activity and the local velocity of the liquid medium, respectively. As shown, they split the inactivation rate into a pressure-sensitive one for the given starting temperature k_p and a temperature-sensitive one at high static pressure α [95].

Figure 9.6 shows showcases of different outcomes for the enzyme activity at the end of the pressure-holding phase. At a starting temperature of 293.15 K, the activity

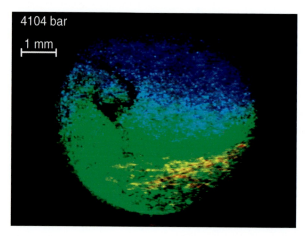

Figure 9.5 Temperature stratification in the autoclave as visualized by liquid crystal thermometry. Blue indicates higher and red lower temperatures, respectively [87].

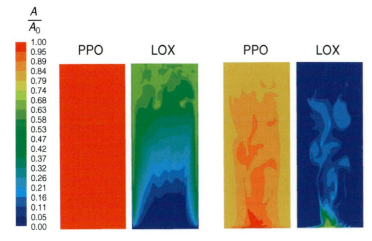

Figure 9.6 Activity of polyphenoloxidase (PPO) and lipoxygenase (LOX) at the end of the pressure-holding phase. Starting temperature on the left: 293.15 K, on the right: 323.15 K [101].

of polyphenoloxidase from avocado modeled according to Ref. [38] is rather pressure insensitive. Lipoxygenase [104], on the other hand, reacts sensitively to the high-pressure treatment. Due to the temperature sensitivity, the inactivation is heterogeneous, reflecting roughly the footprint of the temperature stratification in the autoclave. In this particular case, the inactivation process is more efficient at lower temperature.

At higher temperatures, the pressure sensitivities of the inactivation of both enzymes converge and thermal sensitivity rises [38, 101, 104]. As shown for a starting temperature of 323.15 K in Figure 9.6, both enzymes depict qualitatively the same activity heterogeneities. Strong deviations between activity and temperature heterogeneities are apparent. Using numerical particle tracking, Kitsubun [105] has shown that the discrepancies between activity and temperature heterogeneities are due to the convective transport of the enzymes during the pressure treatment: Enzymes that are at the end of the holding phase in a warm region may have been during most of the time in rather cold regions with different inactivation sensitivity. Figure 9.7 shows an example of the sensitivity under initial conditions for the inactivation of two enzymes even of neighboring enzymes [105]. As a consequence, the enzymes experience varying mechanical and thermal histories, which originate different inactivation. Kitsubun found variations in the residual activity of *B. subtilis* α-amylase in the range of 0.59–0.72. This example underlines that the actual state depends on the entire treatment history of the process, as outlined in Ref. [106].

Several authors have studied the role of the heat losses through the container walls [94, 101, 107, 108]. Hartmann *et al.* [94] showed that overall inactivation rates of *Bacillus subtillis* α-amylase increase with thermal insulation. They point out that this is due to higher mean temperature in the vessel, which favors inactivation. Inhibiting heat losses through the container walls also reduces effectively thermal and thus

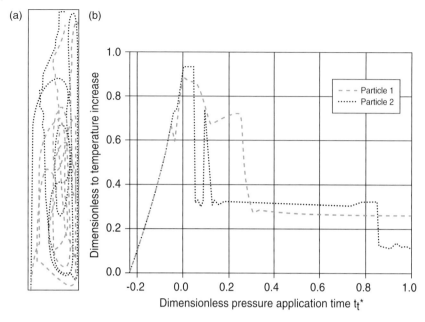

Figure 9.7 Particle tracks of two different enzymes (a) lead to different treatment histories due to nonuniform temperature fields (b) [105].

inactivation inhomogeneities, as reported by Hartmann et al. [94] and de Heij et al. [108]. Rauh et al. [101] studied the impact of preheating the container walls on the temperature of the pressure-transmitting medium after compression and compared it with the effect of a higher starting temperature and of the suppression of convection by using a highly viscous liquid. With the preheated wall, they found a much higher degree of uniformity for the temperature field than with the other methods. Preheating the walls had also a considerable impact on enzyme inactivation: For all the enzymes studied, that is, polyphenoloxidase, lipoxygenase, β-glucanase, and B. subtilis α-amylase, they obtained the highest inactivation uniformity and the lowest overall enzyme activity.

The size of the autoclaves may also have a significant effect on the temperature distribution and hence on inactivation [97, 105, 109]. Hartmann and Delgado showed that overall inactivation rates of B. subtilis α-amylase increase with the container size [97, 109]. They point out that this is due to the reduced effect of heat losses and higher mean temperature in the vessel, analogous to thermal insulation of the container walls. Kitsubun [105] focused on the role of autoclave size on inactivation heterogeneity. He found that in small vessels, effective heat transfer through the container walls reduces thermal and thereby activity heterogeneities. On the other extreme, in large chambers, the heat transfer through the walls is inefficient for the large bulk volume, which results again in small heterogeneities even without turbulence mixing. Hence, maximum heterogeneities due to thermal effects are to occur in midsize containers [105].

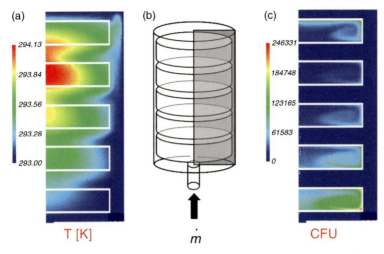

Figure 9.8 Temperature (a) and inactivation distribution (b) in packaged food [82].

The effect of packaging on thermal heterogeneities and also on inactivation has been studied by Hartmann et al. [82, 109, 110]. Besides serving as a mechanical barrier to prevent mixing and large-scale convection, they find, in line with the effect of the container walls, that the package material serves as a heat barrier. Thus, elevated temperature levels are maintained through a longer part of the pressure treatment in better insulating packaging material resulting in lower inactivation heterogeneity [109, 110]. Nevertheless, Hartmann et al. [82] reported variations of inactivation efficiency depending on the relative position of the packaging inside the autoclave (Figure 9.8).

Inhomogeneous inactivation is not restricted to liquids. Due to different compressibility of the materials, the temperature rise during compression alters with the material distribution inside the vessel. Under these conditions, heat conduction inside solids or gels may also result in heterogeneous inactivation. Several authors have studied the temperature distribution in solid biomatter [79, 91, 104, 111–113]. Denys et al. studied the inactivation of B. subtilis α-amylase immobilized in agar gel [111]. In the gel matrix, the heterogeneities are due to heat conduction. Similar results were obtained by the same group for heat conduction problems in other materials [104]. Taking into account the temperature distribution and convection in the pressure-transmitting medium, Otero et al. reported considerable thermal heterogeneities in solid foods. Besides heat conduction, solid food structure may also shelter its interior from the high pressure applied in the autoclave. Minerich and Labuza reported on indications of a pressure drop of about 9 MPa in ham during high-pressure processing, which could result in underprocessing the interior of the product during inactivation [114].

Process timescale is another parameter that affects enzyme heterogeneity. Quasi-isothermal processing and thus improved uniformity could be achieved in principle in processes that are either much slower or faster than the heat transfer processes involved. Slower processes could be achieved, for instance, with slow compression rates (<0.1 MPa/s). However, in this case, process time would exceed the time required for industrial throughput [89, 96]. On the other extreme, with pressure ramps on the order of 60–600 MPa/s [115–117], nearly adiabatic conditions can be achieved with fast pressure ramp and release periods in short-time high-pressure processes [101]. For these processes, Ardia *et al.* reported enhanced microbial inactivation and product quality compared to conventional high-pressure processes [115].

The fact that thermal heterogeneities may have strong effects on the process quality in high-pressure treatment has raised the demand for appropriate quality indicators. Since enzyme inactivation is sensitive to pressure, temperature, and process time, Hendrickx and coworkers identified this process as a possible candidate for these indicators [64, 65]. Using solvent engineering, they were able to shift the inactivation window in the processing range of interest, thus allowing a wider dynamic range of the sensor. The detectors may be placed at the position of lowest impact of a specific treatment to monitor process quality in the autoclave [65].

9.5
Summary

Pressure is a fundamental state variable. It profoundly influences the thermodynamic state of material systems. Obviously, a change in pressure leads to shifting of the equilibrium of the system. As a result, complex biomolecules, such as enzymes, exhibit distortion of structure and functionality. Application of pressure treatment to enzymes may influence their catalytic activity, either increasing or decreasing it until complete inactivation in regard to the substrate. The pressure-induced change can be reversible or irreversible, partial or complete. Numerous models in the form of kinetic equations with parameters fitted to experimental data can be found in the literature. However, large variability of the enzyme behavior and respective model equations must be noticed, depending on the origin of the particular enzyme population and medium in which the pressure–temperature treatment is carried out.

Nevertheless, information on the influence of time, temperature, and pressure on the enzyme inactivation given in the form of a kinetic model is not sufficient for process engineering purposes. In order to investigate the impact of treatment in a particular technical application, it is necessary to take into account the macroscopic aspects of the process, such as the heat and mass transport effects, size, and form of the pressure vessel, product packaging, and the specific boundary and initial conditions. Inevitable temperature inhomogeneities and convective motion accompanying the pressurization result in an inhomogeneous treatment of particular molecules embedded in liquid or solid product. As a result, the global impact of pressure on the enzymatic activity depends on the entire treatment history of all individual molecules.

References

1 Van Buggenhout, S., Messagie, I., Van der Plancken, I., and Hendrickx, M. (2006) Influence of high-pressure-low-temperature treatment on fruit and vegetable quality related enzymes. *European Food Research and Technology*, **223**, 475–485.

2 Ludikhuyze, L., Van Loey, A. Indrawati, I., Smout, C., and Hendrickx, M. (2003) Effects of combined pressure and temperature on enzymes related to quality of fruits and vegetables: from kinetic information to process engineering aspects. *Critical Reviews in Food Science and Nutrition*, **43**, 527–586.

3 Hendrickx, M., Ludikhuyze, L., van den Broeck, I., and Weemaes, C. (1998) Effects of high pressure on enzymes related to food quality. *Trends in Food Science and Technology*, **9**, 197–203.

4 Knorr, D. and Mathys, A. (2008) Ultrahochdrucktechnik für innovative Behandlungsverfahren von Lebensmitteln. *Chemie Ingenieur Technik*, **80**, 1069–1080.

5 Guerrero-Beltrán, J.A., Barbosa-Cánovas, G.V., and Swanson, B.G. (2005) High hydrostatic pressure processing of fruit and vegetable products. *Food Reviews International*, **21**, 411–425.

6 Raso, J. and Barbosa-Cánovas, G.V. (2003) Nonthermal preservation of foods using combined processing techniques. *Critical Reviews in Food Science and Nutrition*, **43**, 265–285.

7 Cheftel, J.C. (1992) Effects of high hydrostatic pressure on food constituents: an overview, in *High Pressure and Biotechnology* (eds C. Balny, R. Hayashi, K. Heremans, and P. Masson), John Libbey Eurotext, pp. 195–209.

8 Cheftel, J.C. (1995) Review: high pressure microbial inactivation and food preservation. *Food Science and Technology International*, **1**, 75–90.

9 Hinrichs, J. (2000) Ultrahochdruckbehandlung von Lebensmitteln mit Schwerpunkt Milch und Milchprodukte – Phänomene, Kinetik und Methodik. Düsseldorf, VDI Verlag.

10 Mozhaev, V.V., Heremans, K., Frank, J., Masson, P., and Balny, C. (1996) High pressure effects on protein structure and function. *Proteins: Structure, Function and Genetics*, **24**, 81–91.

11 Balny, C. (2004) Pressure effects on weak interactions in biological systems. *Journal of Physics: Condensed Matter*, **16**, S1245–S1253.

12 Smeller, L. (2002) Pressure–temperature phase diagrams of biomolecules. *Biochimica et Biophysica Acta*, **1995**, 11–29.

13 Ting, E., Balasubramaniam, V.M., and Raghubeer, E. (2002) Determining thermal effects in high-pressure processing. *Food Technology*, **56**, 31–35.

14 Hernando Sáiz, A., Tárrago Mingo, S., Purroy Balda, F., and Tonello Samson, C. (2008) Advances in design for successful commercial high pressure food processing. *Food Australia*, **60**, 154–156.

15 Northrop, D.B. (2002) Effects of high pressure on enzymatic activity. *Biochimica et Biophysica Acta*, **1595**, 71–79.

16 Eisenmenger, M.J. and Reyes-De-Corcuera, J.I. (2009) High pressure enhancement of enzymes: a review. *Enzyme and Microbial Technology*, **45**, 331–347.

17 Harada, T., Kubota, Y., Kamitanaka, T., Nakamura, K., and Matsuda, T. (2009) A novel method for enzymatic asymmetric reduction of ketones in a supercritical carbon dioxide/water biphasic system. *Tetrahedron Letters*, **50**, 4934–4936.

18 Matsuda, T., Marukado, R., Mukouyama, M., Harada, T., and Nakamura, K. (2008) Asymmetric reduction of ketones by *Geotrichum*

candidum: immobilization and application to reactions using supercritical carbon dioxide. *Tetrahedron Asymmetry*, **19**, 2272–2275.

19 Matsuda, T., Kanamaru, R., Watanabe, K., Kamitanaka, T., Harada, T., and Nakamura, K. (2003) Control of enantioselectivity of lipase-catalyzed esterification in supercritical carbon dioxide by tuning the pressure and temperature. *Tetrahedron Asymmetry*, **14**, 2087–2091.

20 Buckow, R. and Heinz, V. (2008) High pressure processing: a database of kinetic information. *Chemie Ingenieur Technik*, **80**, 1081–1095.

21 De Cordt, S., Hendrickx, M., Maesmans, G., and Tobback, P. (1994) The influence of polyalcohols and carbohydrates on the thermostability of α-amylase. *Biotechnology and Bioengineering*, **43**, 107–114.

22 Volkin, D.B., Staubli, A., Langer, R., and Klibanov, A.M. (1991) Enzyme thermoinactivation in anhydrous organic solvents. *Biotechnology and Bioengineering*, **37**, 843–853.

23 Masson, P. and Balny, C. (2005) Linear and non-linear pressure dependence of enzyme catalytic parameters. *Biochimica et Biophysica Acta*, **1724**, 440–450.

24 Gross, M. and Jaenicke, R. (1994) Proteins under pressure: the influence of high hydrostatic pressure on structure, function and assembly of proteins and protein complexes. *European Journal of Biochemistry*, **221**, 617–630.

25 van Eldik, R., Asano, T., and le Noble, W.J. (1989) Activation and reaction volumes in solution. *Chemical Reviews*, **89**, 549–688.

26 Heremans, K. (1995) High pressure effects on biomolecules, in *High Pressure Processing of Foods* (eds D.A. Ledward, D.E. Johnston, R.G. Earnshaw, and A.P.M. Hasting), Nottingham University Press, pp. 81–97.

27 Masson, P. and Tonello, C. (2000) Potential applications of high pressures in pharmaceutical science and medicine. *High Pressure Research*, **19**, 223–231.

28 Zhang, S.H., Casalini, R., Runt, J., and Roland, C.M. (2003) Pressure effects on the segmental dynamics of hydrogen-bonded polymer blends. *Macromolecules*, **36**, 9917–9923.

29 Kauzmann, W. (1959) Some factors in the interpretation of protein denaturation. *Advances in Protein Chemistry*, **14**, 1–67.

30 Zipp, A. and Kauzmann, W. (1973) Pressure denaturation of metmyoglobin. *Biochemistry*, **12**, 4217–4228.

31 Tanford, C. (1968) Protein denaturation. *Advances in Protein Chemistry*, **23**, 121–282.

32 Zale, S.E. and Klibanov, A.M. (1983) On the role of reversible denaturation (unfolding) in the irreversible thermal inactivation of enzymes. *Biotechnology and Bioengineering*, **25**, 2221–2230.

33 Ludikhuyze, L.R., Van Den Broeck, I., Weemaes, C.A., Herremans, C.H., Van Impe, J.F., Hendrickx, M.E., and Tobback, P.P. (1997) Kinetics for isobaric–isothermal inactivation of *Bacillus subtilis* α-amylase. *Biotechnology Progress*, **13**, 532–538.

34 Indrawati, I., Van Loey, A.M., Ludikhuyze, L.R., and Hendrickx, M.E. (2000) Kinetics of pressure inactivation at subzero and elevated temperature of lipoxygenase in crude green bean (*Phaseolus vulgaris* L.) extract. *Biotechnology Progress*, **16**, 109–115.

35 Indrawati, I., Ludikhuyze, L.R., Van Loey, A.M., and Hendrickx, M.E. (2000) Lipoxygenase inactivation in green beans (*Phaseolus vulgaris* L.) due to high pressure treatment at subzero and elevated temperatures. *Journal of Agricultural and Food Chemistry*, **48**, 1850–1859.

36 Indrawati, I., Van Loey, A.M., Ludikhuyze, L.R., and Hendrickx, M.E. (2001) Pressure–temperature inactivation of lipoxygenase in green peas (*Pisum sativum*): a kinetic study. *Journal of Food Science*, **66**, 686–693.

37 Indrawati, I., Van Loey, A.M., Ludikhuyze, L.R., and Hendrickx, M.E. (1999) Soybean lipoxygenase inactivation by pressure at subzero and elevated temperatures. *Journal of*

38. Weemaes, C.A., Ludikhuyze, L.R., Van Den Broeck, I., and Hendrickx, M.E. (1998) Effect of pH on pressure and thermal inactivation of avocado polyphenol oxidase: a kinetic study. *Journal of Agricultural and Food Chemistry*, **46**, 2785–2792.

39. Rapeanu, G., Van Loey, A., Smout, C., and Hendrickx, M. (2005) Thermal and high-pressure inactivation kinetics of polyphenol oxidase in Victoria grape must. *Journal of Agricultural and Food Chemistry*, **53**, 2988–2994.

40. Dalmadi, I., Rapeanu, G., Van Loey, A., Smout, C., and Hendrickx, M. (2006) Characterization and inactivation by thermal and pressure processing of strawberry (*Fragaria ananassa*) polyphenol oxidase: a kinetic study. *Journal of Food Biochemistry*, **30**, 56–76.

41. Fachin, D., Smout, C., Verlent, I., Nguyen, B.L., Van Loey, A.M., and Hendrickx, M.E. (2004) Inactivation kinetics of purified tomato polygalacturonase by thermal and high-pressure processing. *Journal of Agricultural and Food Chemistry*, **52**, 2697–2703.

42. Terefe, N.S., Yang, Y.H., Knoerzer, K., Buckow, R., and Versteeg, C. (2009) High pressure and thermal inactivation kinetics of polyphenol oxidase and peroxidase in strawberry puree. *Innovative Food Science and Emerging Technologies*, **11**, 52–60.

43. Ludikhuyze, L., Claeys, W., and Hendrickx, M. (2000) Combined pressure–temperature inactivation of alkaline phosphatase in bovine milk: a kinetic study. *Journal of Food Science*, **65**, 155–160.

44. Borda, D., Van Loey, A., Smout, C., and Hendrickx, M. (2004) Mathematical models for combined high pressure and thermal plasmin inactivation kinetics in two model systems. *Journal of Dairy Science*, **87**, 4042–4049.

45. Lencki, R.W., Arul, J., and Neufeld, R.J. (1992) Effect of subunit dissociation, denaturation, aggregation, coagulation, and decomposition on enzyme inactivation kinetics: I. First-order behaviour. *Biotechnology and Bioengineering*, **40**, 1421–1426.

46. Bigelow, W.D. (1921) The logarithmic nature of thermal death time curves. *Journal of Infectional Diseases*, **29**, 528–536.

47. Basak, S. and Ramaswamy, H.S. (1996) Ultra high pressure treatment of orange juice: a kinetic study on inactivation of pectin methyl esterase. *Food Research International*, **29**, 601–607.

48. Mussa, D.M. and Ramaswamy, H.S. (1997) Ultra high pressure pasteurization of milk: kinetics of microbial destruction and changes in physico-chemical characteristics. *LWT – Food Science and Technology*, **30**, 551–557.

49. Heinz, V., Buckow, R., and Knorr, D. (2005) Catalytic activity of β-amylase from barley in different pressure/temperature domains. *Biotechnology Progress*, **21**, 1632–1638.

50. Buckow, R., Heinz, V., and Knorr, D. (2005) Effect of high hydrostatic pressure–temperature combinations on the activity of β-glucanase from barley malt. *Journal of the Institute of Brewing*, **111**, 282–289.

51. Buckow, R., Weiss, U., Heinz, V., and Knorr, D. (2007) Stability and catalytic activity of α-amylase from barley malt at different pressure–temperature conditions. *Biotechnology and Bioengineering*, **97**, 1–11.

52. Buckow, R., Weiss, U., and Knorr, D. (2009) Inactivation kinetics of apple polyphenol oxidase in different pressure–temperature domains. *Innovative Food Science and Emerging Technologies*, **10**, 441–448.

53. Rademacher, B. and Hinrichs, J. (2006) Effects of high pressure treatment on indigenous enzymes in bovine milk: reaction kinetics, inactivation and potential application. *International Dairy Journal*, **16**, 655–661.

54. Buckow, R., Heinz, V., and Knorr, D. (2005) Two fractional model for evaluating the activity of glucoamylase from *Aspergillus niger* under combined pressure and temperature conditions.

Food and Bioproducts Processing, **83**, 220–228.

55 Rodrigo, D., Jolie, R., Van Loey, A., and Hendrickx, M. (2006) Combined thermal and high pressure inactivation kinetics of tomato lipoxygenase. *European Food Research and Technology*, **222**, 636–642.

56 Ly-Nguyen, B., Van Loey, A.M., Smout, C., Verlent, I., Duvetter, T., and Hendrickx, M.E. (2003) Effect of mild-heat and high-pressure processing on banana pectin methylesterase: a kinetic study. *Journal of Agricultural and Food Chemistry*, **51**, 7974–7979.

57 Ly-Nguyen, B., Van Loey, A.M., Smout, C., Özcan, S.E., Fachin, D., Verlent, I. Vu Truong, S., and Hendrickx, M.E. (2003) Mild-heat and high-pressure inactivation of carrot pectin methylesterase: a kinetic study. *Journal of Food Science*, **68**, 1377–1383.

58 Guiavarch, Y., Segovia, O., Hendrickx, M., and Van Loey, A. (2005) Purification, characterization, thermal and high-pressure inactivation of a pectin methylesterase from white grapefruit (*Citrus paradisi*). *Innovative Food Science and Emerging Technologies*, **6**, 363–371.

59 Goodner, J.K., Braddock, R.J., and Parish, M.E. (1998) Inactivation of pectinesterase in orange and grapefruit juices by high pressure. *Journal of Agricultural and Food Chemistry*, **46**, 1997–2000.

60 Polydera, A.C., Galanou, E., Stoforos, N.G., and Taoukis, P.S. (2004) Inactivation kinetics of pectin methylesterase of greek Navel orange juice as a function of high hydrostatic pressure and temperature process conditions. *Journal of Food Engineering*, **62**, 291–298.

61 Van Den Broeck, I., Ludikhuyze, L.R., Van Loey, A.M., and Hendrickx, M.E. (2000) Inactivation of orange pectinesterase by combined high-pressure and -temperature treatments: a kinetic study. *Journal of Agricultural and Food Chemistry*, **48**, 1960–1970.

62 Katsaros, G.I., Katapodis, P., and Taoukis, P.S. (2009) High hydrostatic pressure inactivation kinetics of the plant proteases ficin and papain. *Journal of Food Engineering*, **91**, 42–48.

63 Weemaes, C., Ludikhuyze, L., Van Den Broeck, I., and Hendrickx, M. (1998) High pressure inactivation of polyphenoloxidases. *Journal of Food Science*, **63**, 873–877.

64 van der Plancken, I., Grauwet, T., Oey, I., van Loey, A., and Hendrickx, M. (2008) Impact evaluation of high pressure treatment on foods: considerations on the development of pressure–temperature–time integrators (pTTIs). *Trends in Food Science & Technology*, **19**, 337–348.

65 Grauwet, T., van der Plancken, I., Vervoort, L., Hendrickx, M., and van Loey, A. (2009) Investigating the potential of *Bacillus subtilis* α-amylase as a pressure–temperature–time indicator for high hydrostatic pressure pasteurization processes. *Biotechnology Progress*, **25**, 1184–1193.

66 Ogawa, H., Fukuhisa, K., Kubo, Y., and Fukumoto, H. (1990) Pressure inactivation of yeasts, molds, and pectinesterase in satsuma mandarin juice: effects of juice concentration, pH, and organic acids, and comparison with heat sanitation. *Agricultural and Biological Chemistry*, **54**, 1219–1225.

67 Yen, G.-C. and Lin, H.-T. (1996) Comparison of high pressure treatment and thermal pasteurisation effects on the quality and shelf life of guava puree. *International Journal of Food Science and Technology*, **31**, 205–213.

68 Parish, M.E. (1998) High pressure inactivation of *Saccharomyces cerevisiae*, endogenous microflora and pectinmethylesterase in orange juice. *Journal of Food Safety*, **18**, 57–65.

69 Cano, M.P., Hernandez, A., and De Ancos, B. (1997) High pressure and temperature effects on enzyme inactivation in strawberry and orange products. *Journal of Food Science*, **62**, 85–88.

70 Hernández, A. and Cano, M.P. (1998) High-pressure and temperature effects on enzyme inactivation in tomato puree. *Journal of Agricultural and Food Chemistry*, **46**, 266–270.

71. Butz, P., Koller, W.D., Tauscher, B., and Wolf, S. (1994) Ultra-high pressure processing of onions: chemical and sensory changes. *LWT – Food Science and Technology*, **27**, 463–467.
72. Knorr, D., Heinz, V., and Buckow, R. (2006) High pressure application for food biopolymers. *Biochimica et Biophysica Acta*, **1764**, 619–631.
73. Mozhaev, V.V., Lange, R., Kudryashova, E.V., and Balny, C. (1996) Application of high hydrostatic pressure for increasing activity and stability of enzymes. *Biotechnology and Bioengineering*, **52**, 320–331.
74. Masson, P., Bec, N., Froment, M.-T., Nachon, F., Balny, C., Lockridge, O., and Schopfer, L.M. (2004) Rate-determining step of butyrylcholinesterase-catalyzed hydrolysis of benzoylcholine-and benzoylthiocholine: volumetric study of wild-type and D70G mutant behaviour. *European Journal of Biochemistry*, **271**, 1980–1990.
75. Dallet, S. and Legoy, M.-D. (1996) Hydrostatic pressure induces conformational and catalytic changes on two alcohol dehydrogenases but no oligomeric dissociation. *Biochimica et Biophysica Acta*, **1294**, 15–24.
76. Delgado, A. and Hartmann, C. (2002) Pressure treatment of food: instantaneous but not homogeneous effect, in *Advances in High Pressure Bioscience and Biotechnology II* (ed. R. Winter), Springer, Heidelberg, pp. 459–464.
77. Denys, S., van Loey, A.N., Hendrickx, M.E., and Tobback, P.P. (1997) Modeling heat transfer during high-pressure freezing and thawing. *Biotechnology Progress*, **13**, 416–423.
78. Sanz, P.D., Otero, L., de Elvira, C., and Carrasco, J.A. (1997) Freezing process in high-pressure domains. *International Journal of Refrigeration*, **20**, 301–307.
79. Denys, S., van Loey, A.M., and Hendrickx, M.E. (2000) Modeling conductive heat transfer during high pressure thawing processes: determination of latent heat as a function of pressure. *Biotechnology Progress*, **16**, 447–455.
80. Otero, L., Molina-García, A.D., and Sanz, P.D. (2000) Thermal effect in foods during quasi-adiabatic pressure treatments. *Innovative Food Science & Emerging Technologies*, **1**, 119–126.
81. Hartmann, C. (2002) Numerical simulation of thermodynamic and fluid-dynamic processes during the high-pressure treatment of liquid food systems. *Innovative Food Science & Emerging Technologies*, **3**, 11–18.
82. Hartmann, C., Delgado, A., and Szymczyk, J. (2003) Convective and diffusive transport effects in a high pressure induced inactivation process of packed food. *Journal of Food Engineering*, **59**, 33–44.
83. Ghani, A.G.A. and Farid, M.M. (2007) Numerical simulation of solid–liquid food mixture in a high pressure processing unit using computational fluid dynamics. *Journal of Food Engineering*, **80**, 1031–1042.
84. Knoerzer, K., Juliano, P., Gladman, S., Versteeg, C., and Fryer, P.J. (2007) A computational model or temperature and sterility distributions on a pilot-scale high-pressure high-temperature process. *AIChE Journal*, **53**, 2996–3010.
85. Khurana, M. and Karwe, M.V. (2009) Numerical prediction of temperature distribution and measurement of temperature in a high hydrostatic pressure food processor. *Food and Bioprocess Technology*, **2**, 279–290.
86. Pehl, M. and Delgado, A. (1999) An *in-situ* technique to visualize temperature and velocity fields in liquid biotechnical substances at high pressure, in *Advances in High Pressure Bioscience and Biotechnology* (ed. H. Ludwig), Springer, Heidelberg, pp. 519–522.
87. Pehl, M., Werner, F., and Delgado, A. (2000) First visualization of temperature fields in liquids at high pressure using thermochromic liquid crystals. *Experiments in Fluids*, **29**, 302–304.
88. Kowalczyk, W., Hartmann, C., and Delgado, A. (2004) Modelling and

numerical simulation of convection driven high pressure induced phase changes. *International Journal of Heat and Mass Transfer*, **47**, 1079–1089.

89 Otero, L. and Sanz, P.D. (2003) Modelling heat transfer in high pressure food processing: a review. *Innovative Food Science and Emerging Technologies*, **4**, 121–134.

90 Kilimann, K.V., Kitsubun, P., Delgado, A., Gänzle, M.G., Chapleau, N., Le Bail, A., and Hartmann, C. (2005) Experimental and numerical study of heterogeneous pressure–temperature-induced lethal and sublethal injury of *Lactococcus lactis* in a medium scale high-pressure autoclave. *Biotechnology and Bioengineering*, **94**, 655–666.

91 Chen, C.R., Zhu, S.M., Ramaswamy, H.S., Marcotte, M., and Le Bail, A. (2007) Computer simulation of high pressure cooling of pork. *Journal of Food Engineering*, **79**, 401–409.

92 Rauh, C. (2008) Modellierung und simulation von Kurzzeit-ultra-Hochdruckprozessen. PhD thesis. University Erlangen-Nuremberg.

93 Rauh, C., Baars, A., and Delgado, A. (2006) Analysis of inhomogeneous thermofluiddynamical processes in short time high pressure treatment of liquid foods. Proceedings of the 4th International Conference on High Pressure Bioscience and Biotechnology, Tsukuba, Japan, pp. 186–191.

94 Hartmann, C., Schuhholz, J.-P., Kitsubun, P., Chapleau, N., Le Bail, A., and Delgado, A. (2004) Experimental and numerical analysis of the thermofluiddynamics in a high-pressure autoclave. *Innovative Food Science & Emerging Technologies*, **5**, 399–411.

95 Baars, A., Rauh, C., and Delgado, A. (2007) High pressure rheology and the impact on process homogeneity. *High Pressure Research*, **27**, 77–83.

96 de Heij, W.B.C., van Schepdael, L.J.M.M., Moezelaar, R., Hoogland, H., Matser, A.M., and van den Berg, R.W. (2003) High-pressure sterilization: maximizing the benefits of adiabatic heating. *Food Technology*, **57**, 37–41.

97 Hartmann, C. and Delgado, A. (2002) Numerical simulation of convective and diffusive transport effects on a high-pressure-induced inactivation process. *Biotechnology and Bioengineering*, **79**, 94–104.

98 Song, K., Rauh, C., and Delgado, A. (2008) Experimental *in-situ* investigations on fluid flow during high pressure processing by means of LDA and HWA. *Proceedings in Applied Mathematics and Mechanics*, **8**, 10603–10604.

99 Song, K., Regulski, W., Jovanovic, J., Rauh, C., and Delgado, A. (2009) *In-situ* investigation of the turbulent–laminar transition of temperature fluctuations during the pressure building up to 300 MPa. *High Pressure Research*, **29**, 739–745.

100 Song, K., Al-Salaymeha, A., Jovanovic, J., Rauh, C., and Delgado, A. (2009) Experimental in-situ investigations of turbulence under high pressure. *Annals of the New York Academy of Sciences*, **1189**, 24–33.

101 Rauh, C., Baars, A., and Delgado, A. (2009) Uniformity of enzyme inactivation in a short-time high-pressure process. *Journal of Food Engineering*, **91**, 154–163.

102 Pehl, M., Werner, F., and Delgado, A. (2002) Experimental investigation on thermofluiddynamical processes in pressurized substances, in *Trends in High Pressure Bioscience and Biotechnology* (ed. R. Hayashi), Elsevier Science B.V., pp. 429–435.

103 Özmutlu, Ö., Hartmann, C., and Delgado, A. (2006) Momentum and energy transfer during phase change of water under high hydrostatic pressure. *Innovative Food Science & Emerging Technologies*, **7**, 161–168.

104 Denys, S., van Loey, A.M., and Hendrickx, M.E. (2000) A modelling approach for evaluation process uniformity during batch high hydrostatic pressure processing: combination of a numerical heat transfer model and enzyme inactivation kinetics. *Innovative*

Food Science & Emerging Technologies, **1**, 5–19.

105 Kitsubun, P. (2006). Numerical investigation of thermofluiddynamical heterogeneities during high pressure treatment of biotechnological substances. PhD thesis. TU Munich.

106 Delgado, A., Baars, A., Kowalczyk, W., Benning, R., and Kitsubun, P. (2007) Towards adaptive strategies for high pressure bioprocesses. *High Pressure Research,* **27**, 7–16.

107 Otero, L., Molina-García, A.D., Ramos, A.M., and Sanz, P.D. (2002) A model for real thermal control in high-pressure treatment of foods. *Biotechnology Progress,* **18**, 904–908.

108 de Heij, W.B.C., van Schepdael, L.J.M.M., van den Berg, R., and Bartels, P. (2002) Increasing preservation efficiency and product quality through control of temperature distributions in high pressure applications. *High Pressure Research,* **22**, 653–657.

109 Hartmann, C., and Delgado, A. (2003) The influence of transport phenomena during high-pressure processing of packed food on the uniformity of enzyme inactivation. *Biotechnology and Bioengineering,* **82**, 725–735.

110 Hartmann, C. and Delgado, A. (2005) Numerical simulation of thermal and fluiddynamical transport effects on a high pressure induced inactivation. *Simulation Modelling: Practice and Theory,* **13**, 109–118.

111 Denys, S., Ludikhuyze, L., van Loey, A.M., and Hendrickx, M.E. (1999) Modelling conductive heat transfer and process uniformity during batch high-pressure processing of foods, in *Trends in High Pressure Bioscience and Biotechnology* (ed. H. Ludwig), Springer, Berlin, pp. 381–384.

112 Carroll, T., Chen, P., and Fletcher, A. (2003) A method to characterise heat transfer during high pressure processing. *Journal of Food Engineering,* **60**, 131–135.

113 Otero, L., Ramos, A.M., de Elvira, C., and Sanz, P.D. (2007) A model to design high-pressure processes towards an uniform temperature distribution. *Journal of Food Engineering,* **78**, 1463–1470.

114 Minerich, P.L. and Labuza, T.P. (2003) Development of a pressure indicator for high hydrostatic pressure processing of foods. *Innovative Food Science & Emerging Technologies,* **4**, 235–243.

115 Ardia, A., Heinz, V., and Knorr, D. (2004) Very short treatment times for high pressure processing: a new concept. Proceedings of the International Congress on Engineering and Food (ICEF 9), Montpellier, France, pp. 1–5.

116 Ardia, A. (2004) Process considerations on the application of high pressure treatment at elevated temperature levels for food preservation. PhD thesis. TU Berlin.

117 Valdramidis, V.P., Geeraerd, A.H., Poschet, F., Ly-Nguyen, B., van Opstal, I., van Loey, A.M., Michiels, C.W., Hendrickx, M.E., and van Impe, L.F. (2007) Model based process design of the combined high pressure and mild heat treatment ensuring safety and quality of a carrot simulant system. *Journal of Food Engineering,* **78**, 1010–1021.

10
High Pressure in Renewable Energy Processes
Nicolaus Dahmen and Andrea Kruse

10.1
Introduction

Biomass is presently the fourth largest energy source in the world, representing about 12% of the world's consumption of primary energy today, mainly used for district and home heat production. Unlike the other renewable alternatives for heat and electricity production, biomass as the only renewable carbon-containing feedstock may be increasingly used for the production of chemicals and fuels. Biomass comprises not only the primary production generated by photosynthesis, but also the secondary production (zoo mass), products derived from biomass (e.g., wood pellets, biogas, biodiesel, and bioethanol), and the wide range of residues from the different value chains in food, feed, materials, and energy production. In the context of this chapter, biomass refers to the broad variety of plant matter, grown from many different species and a broad variety of biodegradable wastes. The particular material used should not influence the quality of the end products, but does affect the processing of the raw material. To treat the multitude of different biomass feedstocks, conversion technologies are required, which provide a high feed flexibility. This can be achieved by thermochemical processes such as pyrolysis, gasification, and hydrothermal reactions. Due to the complexity of these technologies, economic operation is only reasonable in large-scale facilities. On the other hand, biomass from agriculture and forestry is distributed over large areas and has to be harvested, collected, and transported. Therefore, the logistics of biomass supply chains are of great importance for feasible economical processes.

Supplying the chemical industry with fuels and raw materials, is mainly determined by the availability of sources of raw materials. In the past, coal used to be the source of raw materials. Its disadvantage is that it occurs as a solid and does not have functional groups for use as a chemical raw material. A way out was the hydrogenation of coal. Reaction with hydrogen under pressure gave a liquid fuel, which, being a hydrocarbon, required large quantities of hydrogen for its preparation. It could be provided by producing synthesis gas, followed by reaction with water generating hydrogen using the water gas shift reaction. Basic chemicals could be prepared via the production of calcium carbide: Its reaction with water as the hydrogen source produced acetylene,

Table 10.1 Typical elemental composition of natural gas (type H), crude oil, lignite, black coal, and lignocellulosic biomass (wood) in % (g/g) on a water and ash free basis.

Element	Natural gas	Crude oil	Lignite	Black coal	Biomass
C	75–77	85–90	60–75	80–90	50
H	23–25	10–15	4–8	3–6	6–9
O	<2	0–1.5	17–34	2.5–10	43
S		0.5–6	0.5–3	0.5–3	<1

which is the basic chemical for all coal-based chemical raw materials. The raw material crude oil had two superior advantages: first it is liquid allowing for easier handling and processing, and second it has a more suitable C–H ratio. Increasingly, also natural gas is used e.g. in Malaysia or Qatar to produce fuels and chemicals via reforming and Fischer-Tropsch reaction providing a wide range of hydrocarbon products. The production of basic chemicals requires functionalization. A typical reaction is chlorination, followed by displacement of chlorine, allowing oxygen-containing groups to be introduced.

Compared to fossil carbon sources, biomass is "overfunctionalized." First of all, its high oxygen content prevents its use as a fuel and as a starting material for the preparation of basic chemicals. In addition, biomass is a solid that initially is formed with a high water content and a widely varying composition (Table 10.1). Moreover, when biomass is to be used on a large industrial scale, logistic problems play an important role. This is due to the fact that biomass is usually produced periodically and is distributed over large areas. Its decentralized cultivation structure in rural areas, its sustained cultivation, and long-term assurance of raw material supply have to be taken into account.

10.2
Thermochemical Processes

The technologies suitable for fuel conversion of dry biomass are, in principal, same as those for the conversion of coal, lignite, or peat. This is why the state-of-the-art and the current development of coal conversion technology will also be mentioned, insofar as high-pressure processes are used, mainly referring to the comprehensive work on coal refining edited by Schmalfeld [1]. However, the above-mentioned special features of biomass must be taken into account when biogenic feedstocks are used in developing and adapting these processes. The basically possible conversion processes of these carbon-containing fuels can be generally described by the diagram shown in Figure 10.1. They result in solid, liquid, and gaseous products of different grades, which can be used in different ways.

The different process routes, in particular with reference to pressure applications, will now be described in more detail. Hydrothermal conversion of highly water-containing biomasses will be described in a different section, due to the special features of these processes.

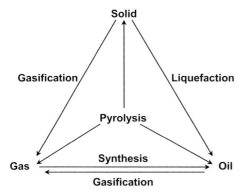

Figure 10.1 Carbon fuel conversion processes for coal and biomass.

10.2.1
Pyrolysis

Pyrolysis, also referred to as dry or destructive distillation, cracking, or coking, means the thermochemical decomposition of condensed matter, leaving a solid residue consisting of char and ashes as well as generating gaseous and tar-like liquid products. Product ratio and composition depend not only on the feed material, but also on temperature (typically between 200 and 500 °C), pressure, and treatment time.

While pyrolysis of coal plays an important role in the production of blast furnace coke on a large industrial scale (approximately 600 Mt of coal/year), it is of minor importance for biomass today. Around 50 Mio.t charcoal are used especially in South America as blast furnace coke. However, modern iron smelting processes would allow for substitution of coke by secondary reduction media and fuels produced from biomass and organic waste. Nonmetallurgical production of carbon is estimated at 18 Mt/year, approximately 600 000 t/year being used as adsorbent with its diverse applications. The latter is prepared mainly from coke, charcoal, and coconut shell coke.

The effect of pressure on the production of coke and char coal is less pronounced in terms of the coke yield, but rather on morphology and activity. Knowledge on pressure effects is not only important in regard to stand-alone pyrolysis technology, but also in regard combustion and gasification processes. Generally, pyrolysis includes two processes: (i) the removal of moisture and volatile substances by breaking weak bonds and (ii) secondary pyrolysis of the condensed carbon matrix. The trends of pressure effects on the char yield observed experimentally suggest a competition between two opposing effects with increasing pressure: (i) pressure inhibits the release of tars and volatiles and (ii) the increase of heat and mass transfer rates accelerates the secondary reactions of large molecules in the coal volatiles, causing an increased weight loss. For the example of Chinese coal, at short residence times (<1 min) the inhibiting effect of the pressure was found to be dominant, resulting in an increased char yield by 5% when the pressure was raised from 0.1 to 0.7 MPa increasingly suppressing the release of tar and volatiles [2]. At longer residence times (5–30 min), the increased rate of the secondary reaction at elevated pressures became dominant, leading to reduced char yields to nearly the same extent. The secondary decomposition is enhanced as the pressure prolongs the residence time of the volatile product inside the particles. Also, at longer residence times, the higher pressures increase the specific density of the residual solid material. Investigating the influence of pyrolysis conditions on morphology using different types of biomass, it was found that increasing the pressure leads to the formation of larger char particles [3]. This could be due to swelling as well as the formation of particle clusters as a result of melting and subsequent coagulation of particles. Pyrolysis pressure influenced the size and the shape of particles through a general increase in void proportion and decrease in cell wall thickness. Swelling occured mainly at low pressures, while higher pressure leads to bubble formation and larger char particle size.

10.2.2
Liquefaction

When producing fuels from coal, its high molecular weight, the low hydrogen content compared to crude oil, the chemically bound heteroatoms (sulfur, nitrogen, and oxygen), and inert material, so called ashes, have to be taken into account. In contrast to the indirect liquefaction via gasification, breaking down the feedstock to CO and H2 molecules, and subsequently following synthesis by e.g. the Fischer–Tropsch (FT) process, the direct hydrogenation of coal does not require complete decomposition of coal. Instead, cracking of the molecular structure occurs with the take-up of hydrogen and parallel reductive removal of heteroatoms. In Germany, the two-stage hydrocracking process developed by IG Farben in Germany during the 1920s basing on the pioneer work of Bergius and Pier, coal is converted into a gasoline type fuel utilizing pressurized hydrogen between 25 and 70 MPa at a temperature of about 480 °C in the presence of a heterogeneous catalyst suspended in a hydroaromatic solvent acting as hydrogen donor and heat carrier medium (e.g., tetralin, naphthalene or phenanthrene). Both, pulverized coal and catalyst, had been suspended in a solvent, acting as additional hydrogen donor and heat carrier medium. As the solvent, usually hydroaromatics such as tetralin, naphtalin or phenantrere obtained from the tar fraction after coal gasification, were used. The middle and light oils produced in the first step of the process are then further were converted in a fixed bed catalyst to benzene and C_1–C_4 gases at pressures of up to 30 MPa. In 12 large facilities, applying the IG Farben process, lignite, black coal, and tars were converted producing about 4 Mt/year of motor fuels in Germany during the 1940s. At the same time, only 15% of that amount was produced by the Fischer–Tropsch process. In 1963, the last coal hydrogenation plants were shut down in Germany. In the 1970s, the project *Kohleöl* was launched in Germany, aiming at improving oil liquefaction technology, in the 200 t/day test facility in Bottrop (DT process). The light and middle distillates were transferred to the mineral oil refinery in Gelsenkirchen for further benzenation. Later, two gas-phase reactors were added to the Bottrop plant within the IGOR integrated refining concept (at 30 MPa, 475 °C and 390–420 °C for liquid-phase and gas-phase reactions, respectively), leading to a highly efficient integrated coal conversion process with an overall C_{5+} yield of >55%, relative to water- and ash-free feed coal. However, no commercialization followed. Based on feasibility studies with

Japan, Germany, and the United States recently the Shenhua Group Corp. in China took the initiative to start the construction of a commercial 5000 t/day plant for mainly diesel fuel production in the Inner Mongolia.

Also, high-pressure liquefaction of biomass was looked at in developments of the 1970s and later. In coal liquefaction, the addition of hydrogen directly leads to an increase of the H:C ratio in the products, while only small amounts of hydrogen used to remove oxygen by water formation. During liquefaction of biomass with its much higher oxygen content, oxygen has to be removed to a larger extent as water, but also by CO_2 formation. Further treatment with hydrogen eventually leads to the uptake of hydrogen into the products. In the PERC (Pittsburgh Energy Research Center, USA) process, cellulosic waste was mashed with recycled pyrolysis oil and treated with synthesis gas in a tubular reactor at 300–340 °C and pressures of up to 20 MPa with sodium carbonate as the catalyst. At reaction times of 10–30 min, around 40–55% (g/g) oil could be recovered along with a solid residue below 1% related to waf biomass. In the LBL (Lawrence Berkeley Laboratory, USA) process, the woody biomass was pretreated by sulfuric acid to increase the degradability of the feed material. The neutralized, aqueous slurry with a dry matter content of 16–20% (g/g) was then converted at 17–24 MPa and 330–360 °C. The CO of the syngas added to the reaction converted to via the water gas shift reaction to active hydrogen *in statu nascendi*, acting as a radical scavenger and thus suppressing the formation of soot and tar. For both processes, semi-technical plants were erected and operated in Albany, OR, USA.

At the same time in Germany, liquefaction was tested based on the Bergius–Pier process for coal conversion by slurry-phase hydrocracking and hydropyrolysis. Here, the BFH (Bundesforschungsanstalt für Forst- und Holzwirtschaft) (today von Thünen Institut - Holzforschung) did pioneering work. In the first process, different types of biomass were converted in numerous screening experiments on a lab scale at typically 10 MPa and 380 °C with a variety of catalysts. Even though Pd on charcoal turned out to be the best choice, Mo, Cr, Ni, and Fe had also been tested successfully. Conversion of lignin led to twice the amount of oil product compared to cellulose and hemicellulose, namely, 60 and 30%, respectively. At best, about 40% yield of light and middle distillate fractions could be obtained, containing about 60% of the biomass input energy. In addition, up to 5% of solids, an aqueous phase containing water-soluble organic material, and a gas consisting of mainly CO_2 and C_1–C_4 compounds were produced. Significant effects of pressure on the product distribution were observed. By increasing the pressure from 3 to 13 MPa, the char formation was reduced from about 40% to below 5% and the total oil yield (including light, middle, and low-boiling components) was increased from about 23 to 60%. Hydropyrolysis without a cosolvent was performed under the same experimental conditions as the hydrocracking experiments described above. Lignin, in particular, resulted in high liquid yields of up to 80% [4].

Alternatively, flash pyrolysis processes were developed for biomass liquefaction as well [5]. On a water- and ash-free basis, from wood typically 75% liquids (including 25% of water), 10% of solid char, and 15% of gases, mainly CO_2 and CO, are formed at 500 °C with gas retention times of only a few seconds. Several reactor concepts such as stationary and fluidized fluidized beds, the mechanically agitated rotating cone and Auger reactors, a well as ablative and vacuum pyrolysis have been carried out and operated on a semi-technical and pilot scale. For fast pyrolytic treatment of

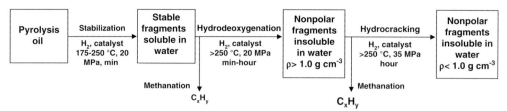

Figure 10.2 Proposed pathways for a mild hydrotreating of pyrolysis oils according to Ref. [6].

lignocellulose at 500 °C the following reaction can be assumed:

$$\underset{\text{Lignocellulose}}{C_6H_9O_4} \rightarrow \underset{\text{Char}}{C_{2.25}H_{2.2}O_{0.35}} + \underset{\text{Condensate}}{C_{2.75}H_{3.2}O_{0.75}} + \underset{\text{Water}}{1.55\,H_2O} + \underset{\text{Gas}}{CH_{0.5}O_{1.35}} \quad (10.1)$$

The dark brown and strongly smoke smelling oils are rich in oxygenated compounds, making them thermally unstable at elevated temperatures, due to condensation reactions of the numerous functional groups in the molecules. In fact, pyrolysis oils comprise of 300–400 different chemical compounds. In addition, they exhibit a low pH value of 2–3, due to the substantial amounts of formic and acetic acids contained in the liquid product. At higher water contents, taking into account the initial water content of the biomass around, and the typically 15% of water formed during the pyrolysis reactions, spontaneous separation of the pyrolysis liquids may occur, forming a tar and an acidic, aqueous phase. When using biomass from fast growing plants, usually containing substantial amounts of salts and minerals, the gas yield is increased at the expense of liquid product, due to the catalytically active alkali components, which increase the degradation of condensable gases. For these reasons, direct substitution of fossil based fuels by pyrolysis oil presents some obstacles. Prior to a subsequent use of fast pyrolysis oils, upgrading is required to decrease the oxygen content mainly by H_2O or CO_2 formation. Hydrodeoxygenation, hydrogenation, and catalytic cracking may be suitable treatment processes to obtain fuels of appropriate quality. The treatment of pyrolysis oils may occur according to Figure 10.2 as proposed by Venderbosch et al. [6]. By mild hydrotreatment, nearly 75% of the oxygen contained in the pyrolysis oil is removed according to the following equation:

$$CH_{1.47}O_{0.56} + 0.40\,H_2O + 0.386\,H_2 \rightarrow 0.742\,CH_{1.47}O_{0.11} + 0.192\,CH_{3.02} + 0.685\,H_2O \quad (10.2)$$

10.2.3
Gasification

Fuel gases play an important role as energy carriers and for synthetic chemistry. They can be generated from carbon-containing solid fuels by gasification on a large scale typically above 700 °C, in which they are converted usually by means of a gasification agent such as steam, oxygen or air, hydrogen, and carbon dioxide. The conversion of liquid and gaseous fuel is referred to as reforming. Solid feed materials can be coal, lignite, peat, biomass, and organic waste; liquid fuels are fractions or residues from crude oil refining, pyrolysis tars, and oils; gaseous fuels are natural gas or gaseous by-products with a high enough calorific value. Oxygen is used in less than

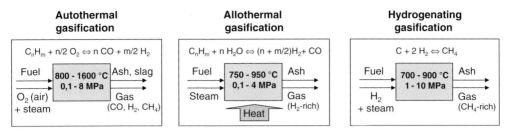

Figure 10.3 Classification of gasification processes according to Ref. [1].

stoichiometric amounts with so-called lambda values between 0.4 and 0.6. The produced gas consists of the following target molecules: carbon monoxide, hydrogen, and/or methane. Synthesis gas for chemicals and liquid fuel production, producer gas for electical power generation, as well as hydrogen or methane (SNG, substitute natural gas) may be the end products. Besides those combustible components, carbon dioxide, water, and nitrogen may be present. Depending on the process applied and the feed material, sulfur-, nitrogen-, and chlorine-containing contaminants, organic compounds, metal vapors, and particles are carried in the gas, which have to be removed to the extent required by the subsequent process.

Gasification is an endothermic process. In Figure 10.3, the different variants concerning heat management at the example of coal gasification are shown. In the most common autothermal processes, an oxygen-containing gasification agent is used, providing internal heat supply by partial oxidation in the reactor. During allothermal gasification, water steam is the gasification agent, while the required heat is supplied externally. Hydrogenating gasification, using hydrogen as the gasification agent, but requiring higher pressures of up to 10 MPa, leads to gases rich in methane.

The amount of ash in different types of feed materials varies widely and has a strong influence on both the type and design of suitable reactors. The melting behavior of ash is determined by its composition and can give rise to liquid slag formation during gasification. Reactors must therefore be designed to accommodate such a behavior. This has been accomplished with slagging gasifiers, in which the ash components melt in the gasifier, leaving the reactor as a liquid slag. In other cases, aggregate materials such as kaolin are added, which are incorporated in the ash matrix and prevented from forming low-melting eutectic mixtures.

Soon after the first application of gasifiers to coal conversion operated at atmospheric pressure, large-scale operation of pressurized gasifiers at pressures between 2.5 and 4.0 MPa became state of the art. The advantages motivating their development were

- increase in the reaction rate;
- higher specific throughput;
- increased methane yield at low-temperature operation (for SNG production);
- reduction of the gas volume to be treated;
- saving the work of compression for the subsequent use of the gas produced (gas turbine, methanol, ammonia, Fischer–Tropsch synthesis).

Fuel supply under elevated pressures is effected for solids by lock hopper systems or pneumatic devices and for liquids and slurries by pumps and screw feeders. The types

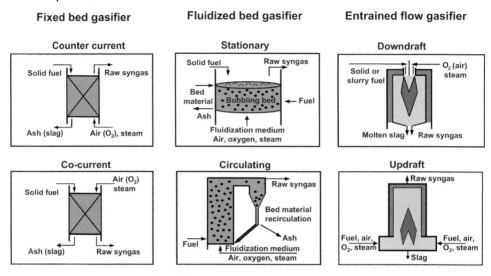

Figure 10.4 Types of gasification reactors.

of gasifiers that are operated at elevated pressures will now be described in more detail referring to the comprehensive compilation in Ref. [1] and are schematically shown in Figure 10.4. Typical process characteristics and gas compositions are given in Table 10.2.

10.2.3.1 Fixed Bed Gasifier

In fixed bed reactors, the feedstock is exposed to the gasifying agent in a packed bed that slowly moves from the top of the gasifier to the bottom, where the ash or slag is discharged. By moving through the reactor, the biomass passes through distinct

Table 10.2 Conversion degrees and product gas composition of coal gasification.

	Fixed bed (countercurrent)	Stationary fluidized bed	Entrained flow
Feed grain size	Coarse (10^{-3} to 10^{-2} m)	Fine (10^{-3} m)	Dust ($<10^{-4}$ m)
Oxygen demand	Low	Medium	High
Carbon conversion	80–90%	85–95%	>95 to >99%
Tar decomposition	Little	Largely	Completely
Examples of syngas composition (lignite, steam/O_2, 2.5 MPa) in % of volume	10	38	59
	38	35	34
	13	5	<0.1
	39	22	7
Example	Lurgi fixed bed	High-temperature Winkler	Siemens fuel gasification

zones of drying, pyrolysis, oxidation, and reduction. Usually, the different types of fixed bed reactors are characterized by the direction of the gas flow through the reactor and consequently are denoted as updraft, downdraft, and horizontal (crossdraft) gasifiers. Depending on fuel and product gas application, a multitude of fixed bed gasifier designs exist. On a small scale, fixed bed reactors are used for district heat and power production up to thermal fuel input capacities of 20 MW$_{th}$. On a large scale, the updraft pressurized type has been successfully used since decades. Among all gasification technologies, the Lurgi pressurized fixed bed gasifier is the economically most successful one. The gas of high calorific value generated by autothermal countercurrent pressurized gasification, with a water steam/technical oxygen mixture being the gasification agent, is used for town gas, SNG, and synthesis gas production for ammonia and FT syntheses as well as being integrated in IGCC power plants. Today, around 80% of the coal gasification capacities are based on that process. To study the influence of the operating pressure on the gas yield and composition, specific performance, and thermal efficiency, a 10 MPa coal-fired pilot plant (Ruhr 100) was constructed and operated between 1979 and 1983. By increasing the pressure from 3 to 9 MPa, the methane yield was improved from 10.3 to 15.5%. At the same time, the oxygen demand in this plant was reduced by about 12%, due to the increased exothermic methane formation. The cold gas efficiency was raised from 70 to 80%. Based on the results, a detailed comparative study for SNG production for a plant design for 3 and 9 MPa was performed. Interestingly enough, it turned out that lower investment costs and thus capital-dependent and maintenance costs are required for the high-pressure alternative.

10.2.3.2 Fluidized Bed Gasifiers

For the production of synthesis gas for chemical syntheses, fluidized bed gasifiers with high carbon conversion efficiencies and low yields of tars have been developed. In fluidized bed gasifiers, fine fuel particles are rapidly mixed and heated, for example, by hot fluidized sand. Due to the intense mixing, the gasification reactions cannot be divided into local zones as in the case of fixed bed reactors, but occur throughout the whole bed, leading to a uniform reaction regime and temperature distribution. The degree of fluidization can be small (bubbling fluidized bed, BFB) or high (circulating fluidized bed, CFB). The former reactors, which have a well-defined interface between the reaction zone of the fluidized bed and the freeboard above the bed surface, are commonly used because of their robust operation. In a CFB gasifier, there is no distinct interface between the fluidized sand bed and the freeboard; the entrained sand and char are recycled back to the gasifier via a cyclone. The carbon conversion is considerably better than that in BFB gasifiers, but operation is more complex and less robust.

High-temperature Winkler (HTW) generators exhibit high carbon conversion efficiencies and a low tar content in the raw syngas. This has been achieved by two reaction zones in the reactor. In improved BFB zone (ca. 1/3 of reactor length), the fuel is contacted with the main quantity of gasification agent, a mixture of steam and oxygen or air. Above the fluidized bed, additional gasification agent is added to increase the temperature in the post-gasification zone (ca. 2/3 of reactor length) for tar decomposition and more complete conversion. HTW syngas generators have been commercially operated between 1956 and 1997 with feedstock capacities of up to 30 t/h and at pressures of up to 2.5 MPa.

Efficient hydrogenating gasification to produce SNG was demonstrated with coal in a stationary fluidized bed gasifier on pilot (7.5 t/h, 8 MPa) and semi-technical scales (300 kg/h of lignite, 10 MPa) within the prototype nuclear process heat project (PNP) in Germany. Helium at high temperatures of about 900 °C, generated in a high-temperature nuclear reactor (HTR), should be used for steam reforming to provide hydrogen in the subsequent coal gasification unit. Different types of coal were converted at around 920 °C to a methane-rich raw gas with a carbon conversion degree of 65%. Within the same project, steam gasification of coal in an allothermal BFB reactor was studied on a semi-technical scale plant (230 kg/h). The required process heat should be generated in the HTR reactor and transferred to the gasification unit by 1000 °C hot helium, leading to an increased carbon efficiency of the overall process. In the test facility, electrically heated helium was supplied to a tube bundle heat exchanger installed in the fluidized bed of the gasification reactor. Gasification temperature and pressure were 800–850 °C and 4 MPa, respectively.

The first pressurized CFB pilot gasifier for biomass has been operated in Värnamo, Sweden. The plant was run between 1996 and 1999 and has been shut down in 2000 after successful operation as a biomass integrated gasification combined cycle (BIGCC) demonstration plant and a gasifier for combined heat and power production (CHP). The feedstocks used in the plant were different wood fuels, bark, straw, and waste-derived fuels. Of the total thermal fuel input of 18 MW, 6 MW_{el} was fed into the public network and 9 MW_{th} was supplied to the district heating network of Värnamo during the EU-funded Chrisgas project. In that process dried, comminuted biomass (e.g., wood chips) was fed by a lock hopper and a screw into the air-blown CFB gasifier. Hot syngas carried the bed material up into a cyclone; solids returned to the bottom of the gasifier. Average gasification temperatures were slightly below 1000 °C at 1.8 MPa of operating pressure. Reorganized to the Växjö Värnamo Biomass Gasification Centre (VVBGC), the plant has been rebuilt to produce clean synthesis gas for chemical syntheses expected to start operation in 2012.

10.2.3.3 Entrained Flow Gasifiers

To provide chemical industries with clean syngas even when using low-rank fuels such as salt-rich lignite or biomass entrained flow-type gasifiers were developed, offering many advantages in synthesis gas generation and downstream synthesis:

- tar-free syngas low CH_4 content in the syngas,
- high carbon conversion degree,
- adaptability to high pressure downstream processing,
- high thermal fuel capacities in excess of 1 GW, and
- high feed and fuel flexibility accommodating gases, liquids, fine powders, and slurry feeds.

At the so-called Gaskombinat "Schwarze Pumpe" (Black Pump) aiming at the utilization and refining of local lignite in East Germany, the GSP-type gasifier was constructed, specially designed to be fed with local salt lignite. This gasification complex with at maximum 36 fixed bed and 2 entrained flow gasifiers ranked second in the world for coal gasification after the Sasol in South Africa, supplying East Germany with town gas, briquettes, and many other products derived from lignite. After German reunification in 1990, lignite for town gas production was substituted stepwise by natural gas. The gasifier park was restructured to SVZ (Sekundär-Rohstoff Verwertungszentrum) within the Sekundärrohstoff-Verwertungszentrum (SVZ) in a way that solid and liquid waste, including solid-containing slurries, could be converted to syngas for methanol and electricity production in a unique combination within the gasifier park comprising of six fixed bed gasifiers of 50 MW_{th} each and a 130 MW_{th} GSP-type entrained flow gasifier, all oxygen blown and run at 2.5 MPa for treatment of a mix of lignite and waste (about 500 000 t/year). Sewage sludge, municipal solid waste, and plastics were fed in addition to lignite used to balance out the heating value variation resulting from the diversity of wastes. Different types of mechanical and thermochemical pretreatment processes were implemented, to make the heterogeneous multitude of feed materials suitable for gasification. In the Noell conversion process, a rotating kiln pyrolysis was combined with the GSP gasifier for waste treatment. The gasification of waste instead of its combustion avoids the undesired formation of dioxins and furans and toxic fly ashes. The fixed bed gasifiers provided with a rotating grate have been operated below the ash melting point (800–1300 °C), while entrained flow gasifiers operate above the ash melting point of 1200–1600 °C. The thereby formed viscous slag is drained down the inner wall of a cylindrical cooling screen, thus protecting the reaction chamber from corrosion and high temperatures. The tar removed from the raw syngas of the fixed bed gasifiers was completely gasified in the entrained flow gasifier. At the end of 2007, the SVZ suspended its gasification activities due to the uneconomic German waste market. On the mid-term, a reactivation of the gasification facilities is planned.

In the GPS-type pilot plant of the former East German Fuel Institute in Freiberg (3–5 MW_{th}), co-gasification of coal powder and biomass was tested. While the coal particles (<0.5 mm) were converted in the expected manner, the straw particles (<10 mm) were blown through the reactor and remained partly or completely unconverted. Thus, pretreatment of biomass prior to entrained flow gasification is required in order to customize the feed material to the process in terms of gasification behavior and heating value. Different options, for example, by applying torrefaction under relatively mild conditions at around 200 °C and leading to a peat-like solid product, liquefaction, or the use of slurries of liquid and solid pyrolysis products, can be considered [7].

Combining Shell's expertise in pressurized oil gasification, developed in the 1950s, and the Koppers know-how on coal gasification led to a joint development resulting in the Shell–Koppers coal gasification (later Shell coal gasification process, SCGP). In this process, powdered coal is cocurrently gasified with water steam and oxygen or air in the updraft water-cooled reactor. In the first commercially operated gasifier of this type, integrated into the IGCC plant in Buggenum, the Netherlands, besides coal also waste-derived fuels and biomass are co-fed up to 30% (g/g) in the fuel mix of 2000 t/day. Krupp-Koppers performed an own development on high-pressure gasification after separating from Shell, resulting in the Prenflo process operated at pressures of up to 3 MPa. The cooled membrane wall inside the pressure vessel is equipped with four radial burners, stabilizing the gasification flame. After pilot plant testing (48 t/day), a first plant on a large scale was erected as part of a coal-fired CHP plant in Puertollano, Spain, making it an IGCC plant of 300 MW of electrical power.

The VEBA VTA process was originally developed for coal gasification, but already in the development has been tested with other feedstocks such as waste and other residues, too. Materials that could not be powdered sufficiently were pretreated by pyrolysis in a rotating kiln reactor. The dry fuel has been fed to the reactor by a twin-screw extruder, compacting the feed material to the desired gasification pressure. This technique was expected to be superior to the common use of coal/water slurries with high water fractions. The fuel paste leaving the screw device was deagglomerated prior to gasification by steam and the product cycle gas or nitrogen. The reactor of the pilot plant (1.5 t/h), erected in Gelsenkirchen, Germany, was designed for pressures and temperatures of 8 MPa and 1600 °C, respectively, and was equipped with a ceramic refractory lining.

The VEW coal conversion process was developed to act as a pretreatment process before coal combustion in a highly flexible power plant concept. In the brick-lined reactor, coal dust, fed by a rotary feeder being less sensitive toward pressure variations, was subjected to high-speed gasification and partly gasified with air as the gasification agent. The fuel gas generated has been converted into electricity b a gas turbine; the residual char was burned in a conventional boiler with the gas turbine flue gas. After testing in the 1 t/h semi-technical plant operated at atmospheric pressure, the 10 t/h pilot plant (Werne, Germany) was designed for higher pressures to reduce the gas volume for gas cleaning, to allow for physical gas cleaning processes, and to avoid gas compression before the gas turbine.

Today, entrained flow gasifiers are a key element of some biomass-to-liquid (BtL) process developments to produce synthetic fuels from biomass. They are particularly suited for the inhomogeneous biogenic materials, particularly in termis of ash-rich feed materials (e.g., wood <1% (g/g), straw 6% (g/g), and rice straw 16% (g/g) on a waf basis) and changing fuel properties. In BtL processes based on entrained flow gasifiers, biomass is pretreated in order to increase the calorific heating value, homogenize the feed material, and convert it into a liquid, slurry, or fine powder easy to feed for efficient atomization in the gasifier with the gasification agent. In that respect, the application of entrained flow gasification to biomass depends more on biomass provision and preparation than on technological issues that of course have also to be adapted to the biomass-specific fuel properties. All BtL concepts outlined below may be backbones of future thermochemical biorefineries producing chemicals and biosynfuels plus energy.

In the "black liquor gasification to automotive fuels" (BLGAF) process of Chemrec AB, Sweden, biomass is added to a pulp mill and converted into methanol and further to dimethyl ether (DME) by black liquor gasification. Currently, the black liquor produced in pulp and paper mills is mainly burned in recovery boilers to recycle the cooking chemicals and to generate energy. In the DP-1 gasifier, concentrated black liquor from the evaporator is fed to a 3 MPa oxygen-blown black liquor gasifier for conversion at about 950 °C to raw syngas and recycling the cooking chemicals resulting from pulping. Raw syngas saturated with moisture at 220 °C is cooled and scrubbed; this generates steam at medium and low pressures. The synthesis gas is purified in activated carbon filters and in an amine wash. The H_2/CO ratio is adjusted to 2 by a water gas shift reactor. The conditioned syngas is compressed from 3 to 6 MPa and routed to the methanol/DME synthesis and purification section. This concept has been realized in 2011 by a consortium around Chemrec within the BioDME project, supported by the EU and the Swedish Energy Agency. Aiming at the production of 4–5 t/day of DME, applying novel synthesis technology from Haldor Topsoe, Denmark. Furthermore, a feasibility study of a 70 000 t/year plant has been conducted. Commercialization of this special technology, is almost certain because of its low feedstock supply cost and because of the high degree of integration into the existing infrastructure of pulp and paper biorefineries.

Choren Company located in Freiberg, Germany, developed the Carbo-V process, which comprises a special entrained flow gasifier operation. The process was first tested in a 1 MW_{th} atmospheric pressure gasifier in the so-called "α-plant" in Freiberg. Several tons of synthetic biofuel had been produced so far in downstream test facilities. A 40 MW_{th} pilot "β-plant" for gasifier operation at 0.4 MPa has been constructed in Freiberg with a design capacity of 15 000 t/year of biosynfuel. The FT synthesis plant, based on the Shell middle distillate synthesis (SMDS) process, was supplied by Shell. Waste heat from the process is first used to dry the biomass to a moisture content of 15–20% (g/g). The comminuted dried biomass is then pyrolyzed by partial combustion at 400–500 °C in a specially designed cylindrical kiln reactor equipped with mixer paddles. In the β-plant, pyrolysis and gasification, permanently connected, are run at 0.4 MPa pressure. This requires tightly locked hoppers. The hot pyrolysis gases and vapors are routed directly into the gasifier, subsequently gasified with pure oxygen above the ash melting point of 1300–1500 °C. The pyrolysis char is cooled down, powdered, and blown into the hot discharge flow from the gasification reactor to cool the gas to about 900 °C by the endothermic carbon gasification reaction. After this chemical quenching step, the remaining char powder is recovered and recycled to the gasifier burner. After a multistage washing and cleaning process, the pure syngas is fed to the synthesis unit. The vitrified solid slag from the gasifier can be granulated for road construction or milled to a fine powder for use as a fertilizer. To facilitate the downstream FT synthesis, it is worth noting that the higher pressure used in this process requires complex intermediate gas compression. After insolvency of Choren Industries in 2011, Linde Engineering, Dresden, will further develop the Carbo-V technology to produce synthesis gas.

Among the current projects on biosyngas generation and utilization within the extensive renewable energy activities in Güssing, Austria, FT synthesis and SNG production are being investigated in side streams in addition to electricity production within a polygeneration concept. The allothermal dual fluidzed bed gasifier generates concentrated syngas containing about 10% by volume of methane in the product gas. After thorough raw syngas cleaning, most of the CO and H_2 can be converted into a biosynfuel mix through FT process (0.5–1 l/h). The unconverted residue, mainly methane, is available for downstream power generation by combustion. Bio-SNG is produced in a 1 MW_{th} semi-technical plant. Polygeneration refers to a variant of syngas use where several products are generated simultaneously with minimum technical effort.

The core element of the bioliq process of the Karlsruhe Institute of Technology (KIT) is the gasification of bioslurries, also referred as to or biosyncrude, in a pressurized entrained flow gasifier operated slightly above the pressure of downstream synthesis (3–10 MPa) avoiding the energy-consuming intermediate syngas compression step prior to chemical syntheses causing high capital expenses and operating costs. Virtually, all chemical reactions with syngas require elevated pressures. The most common variants of FT synthesis typically are run at pressures of 1–4 MPa, while those for methanol or DME synthesis are run at pressures up to 8 MPa. The further conversion of methanol or DME to olefins and gasoline typically occurs at pressures around 2 MPa. Oxosyntheses to produce aldehydes and alcohols are conducted at pressures above 15 MPa. Liquid-like slurries or pastes can easily be fed to a pressurized gasifier. In the bioliq process they are prepared from mixing biocrude oil and char produced in fast pyrolysis plants preserving more than 85% of the energy initially contained in the biomass. In this way, a large and economically operated central gasification and synthesis plant complex can be supplied with biosyncrude from a number of regionally distributed pyrolysis plants overcoming the logistical hurdles of biomass provision arising from its low volumetric energy density and widespread distribution [8]. Presently, a pilot plant is being erected at KIT: the 2 MW$_{th}$ (500 kg/h air-dry biomass) fast pyrolysis plant is based on a twin-screw mixer reactor with a sand heat carrier loop. The 5 MW$_{th}$ high-pressure entrained flow gasifier is based on the Lurgi multipurpose gasification (MPG) technology, but further improved with a water-cooled membrane wall inside the pressure vessel allowing for operation of up to 8 MPa. Both technologies are developed within a cooperation of KIT and Lurgi stream of 700 Nm3 of the biosyngas is cleaned by a hot gas cleaning process with particle filtration and fixed bed sorption for removal of hydrogen chloride and sulfide as well as, alkaline materials. Residual organic material and nitrogen-containing compounds are degraded in a catalytic bed. After CO_2 and water separation, the purified syngas is further converted in a one-step synthesis to DME at around 6 MPa and further to gasoline at a pressure below 2 MPa. Unconverted gas is recycled to the DME reactor. While the pyrolysis plant is already in operation, the other process components have been mechanically completed in 2011 and are expected to be commissioned in 2012. The bioliq project is substantially funded by the german federal Ministry of Food, Agriculture and Consumer Protection.

10.3
Hydrothermal Processes

Virgin biomass as well as many biogenic residues exhibit high water contents typically between 40% and up to 90% (g/g). Besides the option of process internal and external drying down to below 15% (g/g) (for biomass stable on storage) and subsequent use in conventional processes, these types of biomass may be converted by hydrothermal processes. As in the previous chapter, they can be allocated to hydrothermal carbonization, liquefaction, and gasification processes. Hydrothermal in this context refers to conversion in water at elevated temperatures and pressures. In biomass degradation under hydrothermal conditions, water acts as a reactant and solvent at the same time. With increasing temperature, density decreases, and the relative permittivity as a

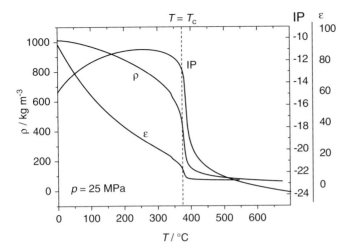

Figure 10.5 Density ϱ, relative permittivity ε, and ion product (IP) of pure water as a function of temperature at a constant pressure of 25 MPa.

measure for the solvent character changes from a highly polar medium under ambient conditions to solvent comparable to a nonpolar organic liquid (see Figure 10.5). At temperatures not too far below the critical point of water, advantage can be taken of the high ion product showing a maximum in this region facilitating, for example, hydrolysis reactions. These special properties of water as a reaction medium under hydrothermal conditions and its participation in the reactions enable fast and complete biomass conversion. Fast hydrolysis of biomass leads to a rapid degradation of the cellulosic and hemicellulosic structure of biomass. Lignin has been found to decompose more slowly due to the more resistant ether and .....bonds. The subsequent reactions also are rather fast, as a result of which the gas is formed at lower temperatures compared to dry gasification processes. Important for gasification, a high solubility of the intermediates formed during biomass degradation significantly inhibits the formation of tar and coke: the reactive species are solubilized in water so that polymerization to give higher molecular products is reduced or even completely suppressed. As a measure for the solubilization ability, the relative dielectric constant ε is shown in Figure 10.5, which at elevated pressure compares to values typical for organic solvents [9]. The pressure and temperature ranges related to the different hydrothermal processes are depicted in Figure 10.6.

In the reactions mentioned, pressure has the following functions:

- Evaporation of water is prevented, maintaining a solvent like density, which is essential to make it a homogeneous reaction medium, reactant, and catalyst precursor. Typically, pressures not to far above the corresponding vapor pressure is adjusted in subcritical applications. In the supercritical state, the density is reduced by a factor of 5-10 compared to water at ambient conditions.
- Pressure provides suitable solvent properties in the near and supercritical state necessary for reducing the coking of catalysts.

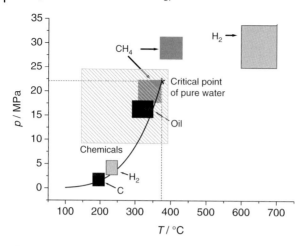

Figure 10.6 Hydrothermal processes mapped into a pressure–temperature diagram (solid curve: vapor pressure curve of water).

- The high pressure makes for particularly good heat transfer. Thus, the use of high-pressure heat exchangers compensates the disadvantage of energy consumption brought about by heating the water.

10.3.1
Hydrothermal Carbonization

The reaction conditions applied to hydrothermal carbonization (HTC) are typically 180-250 °C of temperature and 1-3 MPa of pressure. In this area, already in 1913 Bergius performed systematic studies in the context of coal research, investigating the natural coalification process and describing the detailed biological and chemical reactions. Recently, there has been a growing interest in processes of this type for the production of so-called bio-coals for several reasons. On the one hand, highly water-containing raw materials can be converted to energized coke as CO_2-neutral biogenic energy carrier for energy production.

On the other hand, the porous cokes with their adjustable capacity of nutrient and water storage are discussed for their ability to condition or to recultivate degraded soil. Bringing coke into soils is also discussed as a potential contribution to carbon sequestration in connection with the CO_2 emissions trading system.

The result of a comprehensive literature review by Behrendt consistently confirmed that the hydrothermal treatment of biomass coals by dehydration and decarboxylation yields a product with calorific value of about 30 MJ/kg whose elemental composition is similar to that of lignite [10]. The carbon content and the properties of such cokes strongly depend on the starting materials and the process paramters applied, namely temperature and residence times of 4–14 h. The coke yields can reach values of up to 70% (g/g) and even higher values if catalysts such as citric acid are used. Those products have an increased calorific value of 15–20%, relative to the dry weight. About 80–85% of the carbon remains in the coke, about 10–15% remains in the aqueous phase in the form of organic acids and other

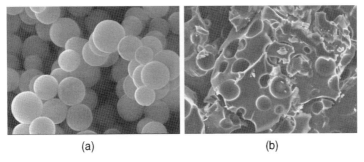

Figure 10.7 Char and coke obtained after hydrothermal carbonization at 200 °C of hydroxymethylfurfural (a) and lignin (b).

oxygen-containing products, and the remainder is in the gas phase. Oxygen is removed from the biomass mainly in the form of water. Newer applications of hydrothermal carbonization aim at the use of secondary cokes formed by repolymerization of hydrolysis intermediates of the biomass that have already gone into solution. In this context, differnce is made between char, formed by solid-solid-reactions, and coke formed by condensation and polymerization of dissolved species. Figure 10.7 shows this clearly for coke formed hydrothermally directly from hydroxymethylfurfural (HMF), a degradation product of cellulose, and char from lignin. Such materials could be used as carbon black substitute or could be designed as nonomaterials with advanced properties by proper process control [11]. In addition to hydrothermal carbonization several related processes utilizing water and steam are under development such as flash carbonization and steam assisted carbonization.

10.3.2
Hydrothermal Liquefaction

Dry biomass can be liquefied by different pyrolysis methods. The resulting pyrolysis oils still have approximately the same oxygen content as the biomass used. Hydrothermal liquefaction (HTL) can be achieved at much lower temperatures of 300–350 °C and the corresponding equilibrium pressures of 12–18 MPa in reaction times of 5–20 min [12]. Hydrothermal liquefaction is often also designated as hydrothermal upgrading (HTU) after the process originally developed by Shell. Compared to pyrolysis oils, HTU oils have substantially reduced oxygen content (10–15%) and a correspondingly high calorific value of 30–35 MJ/kg. However, they are highly viscous and usually do not become liquid until they reach 80 °C. Based on waf biomass, 45% (g/g) of biocrude, 25% (g/g) of gas (>90% of CO_2), 20% (g/g) of water of reaction and, in the aqueous excess phase, soluble organic compounds (acids, alcohols) are produced. Oxygen is removed from the biomass as CO_2 and H_2O. The reason for the high CO_2 content is the water gas shift reaction, in which intermediately formed carbon monoxide reacts with excess water to hydrogen and carbon dioxide. The hydrogen saturates the reactive intermediates, thus suppressing almost completely the formation of coke. The thermal efficiency of the HTU process ranges between 70 and 90%.

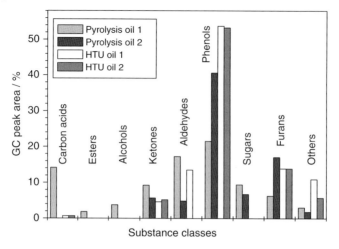

Figure 10.8 Classes of compounds found in pyrolysis oils and HTU oils. Data from: [13].

The content of certain polar compounds, such as acids and sugars, is always lower in HTU biocrude than in pyrolysis oils, vice versa the phenols are more dominant (Figure 10.8). The high phenol content makes HTU biocrude an interesting potential feedstock for the production of resins. The low concentrations of aldehydes and acids lead to a less corrosive behavior of HTU biocrude compared to pyrolysis oil. The difference in composition shown in Figure 10.8 can be attributed perhaps to the special properties of hot compressed water promoting the elimination of water and carbon dioxide. However, in regard to the use of HTL products as liquid fuels, application specific upgrading will be required.

Numerous studies on the use of catalysts are being conducted to solve this problem by increasing the oil yield and decreasing the amount of residual organic content in water. In most cases, alkali metal hydroxides, bicarbonates, or carbonates are utilized. In addition, heterogeneous catalysts, such as Co_3O_4, have been found to increase oil yield.

The HTU process was demonstrated on a bench scale (120 kg/h) successfully operated at TNO-MEP in Apeldoorn, the Netherlands. A similar process, called sludge-to-oil reactor system (STORS), has been used to convert sewage sludge into oil and coke. Typically, some 44% (g/g) of the initial carbon is recovered in the oil, 20% (g/g) in the coke, 16% (g/g) in the gas, and 20% (g/g) in the aqueous solution. Comparable processes utilizing ZrO_2 as a heterogeneous catalyst and potassium salts as a homogeneous water gas shift catalyst have been developed aiming at converting various organic wastewater and industrial sludge to attain a concentrated oil phase of high heating value without the formation of soot and tar. As examples, the Catliq process or Microrefinery carried out at slightly subcritical temperatures between 280 and 350 °C and 18–25 MPa can be mentioned. In the remaining aqueous phase, still considerable amounts of oxygenates and other organics are found. The high amount of effluent water containing residual organic compounds is a challenge to the further development of hydrothermal liquefaction and also carbonization.

10.3.3
Hydrothermal Gasification

Hydrothermal gasification refers to a broad range of operating temperatures and pressures, aiming at generating a product gas rich in either methane or hydrogen. According to Figure 10.6, three areas of gasification may be identified in the pressure–temperature diagram. In the region of 500–700 °C, hydrogen is the preferred product at supercritical temperatures (related to the critical point of pure water of $T_c = 374\,°C$, $p_c = 22.1\,MPa$) in the presence of homogeneous or heterogeneous catalysts to prevent char formation and facilitate the water gas shift reaction. In the near-critical region (374–500 °C), biomass can be hydrolyzed easily and methane formation is preferred. Here, typical hydrogenation catalysts are used. In the subcritical region below the critical point of water, catalysts are required to increase the gas yield considerably because in this process regime mainly liquefaction takes place. Hydrothermal gasification of biomass is typically carried out at reasonable dry matter contents of 10–20% by weight. At lower concentrations and low temperatures, catalytic aqueous phase reforming (APR) is suitable for hydrogen production.

By APR compounds (glucose, sorbitol, glycerol, methanol, and glycol) at low concentrations have been gasified at around 215–265 °C on a laboratory scale to yield mainly hydrogen as burnable gas and carbon dioxide in the presence of a heterogeneous catalyst (Pt, but also Ni, Ru, Rh, Pd, and Ir). The conversion of "real" biomass has not been successful so far, due to catalyst deactivation. High hydrogen yields have only been achieved for hydrogen-rich compounds and at low concentrations of around 1% by weight.

10.3.3.1 Catalytic Hydrothermal Gasification

In his experiments with wood flour, Elliott showed that the gas yield was raised from 42% to 67% (relative to the fed carbon content) in the presence of a nickel catalyst and a sodium carbonate co-catalyst when increasing the temperature from 350 to 400 °C at reactions times of around 1 h [14]. Accordingly, 39% and 34% by volume of hydrogen and 12% and 24% by volume of methane were found together with 49% and 41% by volume of carbon dioxide. Without a catalyst, only 15% and 19% by volume gas yield were obtained, respectively. Nearly complete carbon conversion to gases could be achieved in other experiments with CH_4/H_2 ratios in the product gas of up to 10. Elliott also showed that a range of typical catalyst supports, such as alumina- and silica-based materials, do not have long-term stability in hot compressed water making them unsuitable for use in this environment. Hydrolysis, phase transitions, and sintering may occur. As useful solid supports, carbon-, zirconia-, or titania-based materials and α-alumina could be identified. For the development of suitable catalysts, model compounds for biomass, mainly cellulose and glucose as its hydrolyzed consecutive product, have been used, as well as a variety of biomass feedstocks with dry matter contents in the feed solutions of typically between 10 and 20% (g/g). After most of the experiments have been carried out in small batch devices so far, increasingly continuously operated systems contacting the biomass slurries with catalyst beds are in use for process development.

10.3.3.2 Supercritical Hydrothermal Gasification

During biomass conversion under supercritical conditions far beyond the critical temperature of water, typically between 550 and 700 °C, and at pressures around 30 MPa, the water gas shift reaction dominates during gasification providing a product gas with high hydrogen and low methane yields. Antal in 1978 already predicted a temperature around 600 °C for complete gasification yielding a hydrogen-rich gas for steam reforming of cellulose [15]. The predominating chemical reactions for the example of the conversion of glucose are given in Eqs. (10.3)–(10.5). At 600 °C, a hydrogen yield of around 45% by volume of lignocellulosic material can be expected from thermodynamic predictions. The methane content should be around 15% by volume. In practice, lower values have been found experimentally most probably due to kinetic hindrance of methane formation. In the process, a catalyst is required to promote the water gas shift reaction, usually provided by alkali metal salts that are natural constituents of biomass.

$$\text{Hydrogen formation}: \quad C_6H_{12}O_6 + 6\,H_2O \rightarrow 14\,H_2 + 6\,CO_2 \tag{10.3}$$

$$\text{Methane formation}: \quad C_6H_{12}O_6 \rightarrow 3\,CH_4 + 3\,CO_2 \tag{10.4}$$

$$\text{Water gas shift reaction}: \quad H_2O + CO \rightarrow H_2 + CO_2 \tag{10.5}$$

Supercritical hydrothermal gasification has been relatively well investigated [16] and the influence of process parameters such as temperature, pressure, concentration, and heating rate, as well as of catalysts (in particular of salts) and biomass composition, has been studied extensively by using different types of biomass and model compounds occurring as key compounds during biomass degradation. This has made it possible to better understand the complex reaction network involving hydrolysis reactions, water elimination leading to formation of double bonds or ethers, aldol splitting or condensation, and rearrangement reactions. A simplified reaction scheme is depicted in Figure 10.9. Differences in experimental results obtained from batch or tubular reactors most often are caused by different heating rates; preheating or pretreatment also being an important issue in this context. Differences could also be observed between tubular reactors and continuous stirred

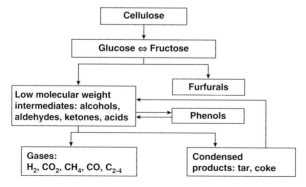

Figure 10.9 Reaction scheme of hydrothermal decomposition of cellulose.

tank reactors (CSTR) with respect to concentration. While in the first reactor type a decrease in the gas yield with increasing initial biomass concentration is observed, the opposite is true in the second reactor type. This can be explained by the assumption that the degradation of the intermediates to gases exhibits a lower reaction order compared to the competing polymerization reactions to form tars and coke (see Figure 10.9). By backmixing as provided in the CSTR, hydrogen comes into contact with reactive intermediates formed in the early state of biomass degradation, stabilizing and preventing them from polymerization reactions.

The process has been tested on laboratory- and bench-scale equipment and was demonstrated in a pilot plant at the KIT, Karlsruhe, Germany, with a throughput of 100 kg/h. The biomass suspension (aqueous slurries with an organic dry matter content of 10–20% (g/g)) is prepared and conditioned in the feed section. It is then delivered by a high-pressure membrane pump of up to 30 MPa to the 30 l reactor passing the heat exchanger and a preheater. The preheater and reactor are heated by hot flue gas up to 650 °C. Heat exchange is very efficient under the prevailing conditions (efficiency up to 80% in the KIT pilot plant) because pinching during heat transfer is avoided in water at elevated pressures. Salts that are insoluble in water under hydrothermal conditions and other solids contained in the biomass can be separated upstream of the preheater by means of a separator. In addition, solids can be removed from the reactor bottom. The retention time in the reactor is on the order of a few minutes. Afterward, the hot, homogeneous product mixture leaves the reactor and heats up the incoming biomass. After cooling, the mixture is separated into the raw product gas and a liquid phase consisting of excess water in a first separator. A second separator is equipped with a carbon dioxide scrubber, in which this gas is efficiently removed by pressurized water due to the much higher solubility of carbon dioxide under these conditions compared to that of hydrogen and methane. The remaining fuel gas can be used directly, for example, in gas engines, or can be further treated for other energetic use. Energy efficiencies up to 60% have been demonstrated in that plant. Salts and minerals remaining in the excess water are collected in the effluent tank. Tests have been conducted on maize silage, greens, residues from wineries and breweries, and those from bioethanol and biogas production. Biocrude oil from pyrolysis and glycerol from biodiesel production have also been used. In Table 10.3, selected results from the pilot-scale

Table 10.3 Fuel gas composition (on a carbon dioxide-free basis) as resulting from supercritical gasification of different types of biomass in the VERENA pilot plant.

	Ethanol	Pyroligneous acid	Glycerol, pure	Maize silage	Herbage
H_2	57.8	50.3	77.3	43.9	35.8
CH_4	31.0	42.8	14.7	38.9	46.4
CO	1.0	0.7	1.3	0.7	1.8
C_{2+}	7.6	4.4	5.3	13.7	16.0
Conversion	99%	97%	—	90%	82%

For experimental conditions, see the text. Data from Ref. [17].

experiments are compiled, which are performed at an operation pressure of 28–30 MPa, reaction temperatures of 590–620 °C, retention times of some minutes, dry matter feed contents of 10–15% in weight, and at a total throughput of 60–100 kg/h.

References

1 Schmalfeld, J. (ed) (2008) Die Veredlung und Umwandlung von Kohle, DGMK, Hamburg.
2 Li, C., Zhao, J., Fang, Y., and Wang, Y. (2009) Pressurized fast-pyrolysis characteristics of typical Chinese coals with different ranks. *Energy Fuels*, **23**, 5099–5105.
3 Cetin, E., Moghtaderi, B., Gupta, R., and Wall, T.F. (2004) Influence of pyrolysis conditions on the structure and gasification reactivity of biomass chars. *Fuel*, **83**, 2139–2150.
4 Meier, D., Bems, J., and Faix, O. (1995) Pyrolysis and hydropyrolysis of biomass and lignins – activities at the Institute of Wood Chemistry in Hamburg, Germany. *ACS Fuel*, **1995**, 298–303.
5 Venderbosch, R.H., and Prins, W. (2010) Fast pyrolysis technology development. *Biofuels Bioproducts Biorefining*, **4**, 178–208.
6 Venderbosch, R., Ardiyanti, A., Wildschut, J., Oasmaa, A., and Heeres, H. (2010) Stabilization of biomass-derived pyrolysis oils. *J. Chem. Technol. Biotechnol.*, **85**, 674–686.
7 Svoboda, K., Pohořelý, M., Hartman, M., and Martinec, J. (2009) Pretreatment and feeding of biomass for pressurized entrained flow gasification. *Fuel Process. Technol.*, **90**, 629–635.
8 Dahmen, N., Henrich, E., and Dinjus, E. (2009) Cost estimate for biosynfuel production via biosyncrude gasification. *Biofuels Bioproducts Biorefining*, **3**, 28–41.
9 Kruse, A., Vogel, G.H. (2010) Chemistry in near supercritical water, in *Handbook of green chemistry, Vol.4 Supercritical solvents*, Leitner, W., Jessop, P.G. (eds.), Wiley VCH, Weinheim, 457–475.
10 Belusa, T., Funke, A., Behrendt, F., and Ziegler, F. (2010) Hydrothermale Karbonisierung und energetische Nutzung von Biomasse – Möglichkeiten und Grenzen. *Gülzower Fachgespräche "Hydrothermale Carbonisierung"*, **33**, 42–54.
11 Hu, B., Wang, K., Wu, L., Yu, S.H., Antonietti, M. and Titirici, M.M. (2010) Engineering carbon materials from the hydrothermal carbonization process of biomass, *Advanced Materials* **22**, 813–828.
12 Peterson, A.A., Vogel, F., Lachance, R.P., Froling, M., Antal, M.J., and Tester, J.W. (2008) Thermochemical biofuel production in hydrothermal media: A review of sub- and supercritical water technologies, *Energy & Environmental Science* **1**, 32–65.
13 M. Kröger (2007), Untersuchung zur Einbindung der hydrothermalen Verflüssigung von nasser Biomasse in das BioLiq-Verfahren, diploma thesis, Hochschule für Technik, Wirtschaft und Kultur Leipzig
14 Elliott, D. (2008) Catalytic hydrothermal gasification of biomass. *Biofuels Bioproducts Biorefining*, **2**, 254–265.
15 Antal, M. (1978) *Energy from Biomass and Wastes* (ed. D. Klass), IGT, Chicago, IL, p. 495–524.
16 Kruse, A. (2008) Supercritical water gasification. *Biofuels Bioproducts Biorefining*, **2**, 415–437.
17 Boukis, N. et al. (2006) Proceedings of the 14th European Biomass Conference; Boukis, N. et al. (2008) Proceedings of the 16th European Biomass Conference; Boukis, N. et al. (2010) Proceedings of the 18th European Biomass Conference.

11
Manufacturing Processes
Andrzej Karpinski and Rolf Wink

High pressure is a proven tool for a number of industrial processes and promising ones in the future. The next chapters give a summary about commonly used high-pressure processes and their industrial applications. The survey is not complete, as the development is changing and progressing continuously. It should be pointed out at this stage that the application of high pressure as a beneficial tool for production procedures is continuously increasing.

This chapter will give general information about hydroforming and isostatic pressing and more detailed information about autofrettage process (Section 11.1) and waterjet cutting technology (Section 11.2).

Hydroforming Hydroforming is one of the new technologies in the manufacturing processes that has become popular in recent years due to the increasing demands for lightweight parts in various fields, such as bicycle, automotive, aircraft, and aerospace industries [1]. In hydroforming process, workpieces are uniformly plastically formed (stretched) in every direction under hydrostatic pressure up to 6000 bar (mostly up to 2500–3000 bar) of a fluid (water or oil) in a controlled manner. The final shape of the hydroformed piece results from the contact with the process fluid from one side and with a male or female die from the other side.

Today, the most known and used types of hydroforming processes are tube (THF) and sheet (SHF) hydroforming. Compared to conventional manufacturing such as stamping, rolling, forging, and welding, THF and SHF offer several advantages, such as decrease in workpiece cost, tool cost, and product weight, improvement in structural stability, increase in the strength and stiffness of the formed parts, more uniform thickness distribution, fewer secondary operations, and so on. Numerous steps of setting the component and polishing the whole manufactured lots are eliminated – the part being in contact with the fluid does not present any scratch.

The hydroforming process is influenced by many factors, among which the matching relation between the internal pressure of the fluid and axial feed (loading paths) is particularly important [2, 3]. Because hydroforming is relatively new compared to conventional manufacturing processes, there is not much knowledge available for the product or process designers [1]. That is, application of pulsating

Industrial High Pressure Applications: Processes, Equipment and Safety, First Edition. Edited by Rudolf Eggers.
© 2012 Wiley-VCH Verlag GmbH & Co. KGaA. Published 2012 by Wiley-VCH Verlag GmbH & Co. KGaA.

pressure is a new and effective method to improve the formability of the THF process. However, the factors that cause this improvement are still unclear [4]. Because of this reason, hydroforming is still under continuous development and will not be further discussed in this chapter.

Isostatic Pressing Pressure applied to a liquid or a gas acts uniformly in every direction at every contacting surface. This basic principle, first postulated by Blaise Pascal, 1623–1662, is commonly called "isostatic pressing" and is widely utilized for compacting powder or densifying porous materials.

Depending on the compressibility and the porosity of the material, the dimensions are reduced proportionally to the applied pressure, but the shape of the product remains unchanged.

According to the applied temperatures, ambient or elevated, the isostatic pressing techniques can be classified into the following main categories: cold isostatic pressing (CIP) and hot isostatic pressing (HIP).

Historically, this method of compacting powder started in the early twentieth century when H.D. Madden patented a method enclosing prepressed "green bodies" made out of refractory metal powder into a rubber mold and putting it under high pressure for compaction [5]. This method was steadily spread toward new applications also in the electroceramic industry. There are a number of applications mentioned in the literature: a comprehensive overview of the current technology can be found in Refs. [5, 6]; the most important are

- the production of refractory liners, insulators, or tool tips in ceramic industry;
- the production of refractory metals, hard metal tools, and dies or wear-resistant parts for valves and pumps in powder metallurgy.

The classical process uses a flexible mold that encloses the powder. This separately handled mold is inserted into the high-pressure vessel and is pressurized thereafter. This method is called "wet bag" or "free mold" method. By the next step of development, the elastomer mold is fixed inside the vessel and is filled directly. The mold and the vessel are closed simultaneously by the closure. This apparatus design and processing technique are called "fixed mold" or "dry bag" method [6].

For high-temperature applications (HIPs), the flexible mold is replaced by a metallic capsule containing the material to be treated. Subsequently, the capsule is evacuated and leak-tight closed. This workpiece is now compressed under isostatic pressure and high temperature in an inert gas atmosphere.

HIP is the state-of-the-art method in the powder-based material industry to obtain products with almost isotropic properties, high density, and low porosity. The method was established in 1970 to produce pore-free tungsten carbide products such as plunger of polyethylene hypercompressors or wear parts such as extruder screws, dies, bushings, and diamond compaction anvils.

The method was steadily spread to ceramic components as well as to sintered metals such as tool steel billets, nickel-based alloys, or soft ferrite for magnetic recording heads [6]. The uniform application of heat and pressure to all surfaces opens a wide field in material technology.

- Posttreatment of material to improve homogeneity and mechanical properties or eliminate inner porosity and imperfections such as pits or flaws. Hard metals, cast parts, or ceramics can also be processed.
- Fabrication of powder metallurgical steels especially hard materials such as tungsten carbide and nickel- or cobalt-based alloys showing maximum abrasion or corrosion resistance. HIP combines sintering and densification simultaneously in one process.
- Fabrication of composites between powder-based and solid materials or composites consisting of weldable and nonweldable solid materials in order to combine different material properties. The materials are bonded by diffusion.

Beyond the applications already mentioned, HIP-treated materials can also be found in components and equipment in centrifuges, subsea and offshore equipment, compressors and turbines, and nuclear energy and space applications [7–9]. Hot isostatic processes are also used to produce open porous material for filter applications or cast iron diamond grinding wheels [10].

Depending on the inner volume, isostatic pressing achieves pressures up to 300 MPa and, for HIP applications, temperatures up to 2500 °C. Some principles about the vessel design will be described in Chapter 12.

Autofrettage The autofrettage treatment is one of the oldest, but very useful methods to create beneficial residual stresses in thick-walled components (e.g., cylinders and valve bodies).

The autofrettage process, which can be described as a self-shrinking process, provides beneficial compressive hoop stresses at the bore of hollow component, for example, a tube.

The residual stresses resulting from autofrettage process reduce the hoop stresses when the normal operation pressure is applied and improve the fatigue lifetime of the component tremendously; for example, lifetime of cross-bored thick cylinders can be improved by three times using autofrettage [11, 12].

Fatigue life refers to the number of the allowed design cycles by the code or the number of cycles to initiate first cracks.

The concerned industries are chemical/petrochemical, power, nuclear, and armament industries, as well as high-pressure food processing. Autofrettage is common practice for equipment used in polyethylene (LDPE) plants, gun barrels, waterjet cutting pumps, and high-pressure pasteurization equipment.

This technology is presented in Section 11.1.

Waterjet Cutting Technology The waterjet cutting technology has been developed from the cutting with plain waterjets in the late 1960s. With the addition of abrasive material (solid particles) to the plain waterjet in the beginning of the 1980s, an abrasive water injection jet was created, which belongs to today's state of the art and is widely applied in industrial manufacturing with pressures up to 6000 bar. The waterjet technology offers its users several important advantages compared to traditional machining methods and other nontraditional machining techniques as well (laser cutting, plasma cutting, and ultrasonic machining). The most important are the following: almost any kind of material can be cut, great material thickness can be cut (up to 500 mm at 6000 bar pressure), no thermal effect on workpiece materials, and many others. These advantages cause the rapid spread of the waterjet technology application fields in the aircraft industry, automotive, military, medicine, job shops, and many others. The number of waterjet cutting systems of different types (2D, 3D, robotized, and medicine systems) quickly and systematically increased and are estimated to be several thousands worldwide. The revenues of the waterjet cutting market worldwide are expected to reach $911 million in 2012 with a compound annual growth rate of approximately 12% [13]. This technology is presented in Section 11.2.

11.1
Autofrettage: A High-Pressure Process to Improve Fatigue Lifetime

Many applications in high-pressure services are characterized by alternating pressures. Such alternating pressures may be caused by the use of piston pumps or compressors or by the process itself. Many of them are batch processes where vessels and components are repeatedly pressurized and depressurized.

The concerned industries are chemical/petrochemical, power, nuclear, and armament industries, as well as high-pressure food processing. Autofrettage is common practice for equipment used in polyethylene (LDPE) plants, gun barrels, waterjet cutting pumps, and high-pressure pasteurization equipment.

Typical apparatuses are cyclically stressed, thick-walled vessels such as tubes and autoclaves and above all components with inherent stress peaks such as threaded fittings and components with cross bores.

The autofrettage process, which can be described as a self-shrinking process, provides beneficial compressive hoop stresses at the bore of hollow component, for example, a tube, and prolongs their fatigue lifetime, where fatigue life refers to the number of the allowed design cycles by the code or the number of cycles to initiate first cracks.

Basically, this process is accomplished by partially overstraining the inner shape of the bore by a hydraulic or mechanical overloading. The applied autofrettage pressure determines the amount of material at the bore that is yielded and thus determines the amount of residual compressive stresses when the pressure is removed. These residual stresses reduce the hoop stresses when the normal operation pressure is applied and improve the fatigue lifetime of the component tremendously.

Li *et al.* [11] as well as Koerner and Wüstenberg [12] reported that the lifetime of cross-bored thick cylinders can be improved by three times using autofrettage. Similar results were observed by Burns *et al.* [14] testing tubes and bends made out of low-alloy steels. It can be summarized that cyclically pressurized equipment achieve sufficient fatigue life by introducing compressive residual stresses using a prestressing technique where autofrettage is the most widely used and most economical process.

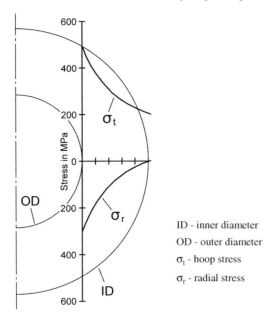

Figure 11.1 Typical stress distribution of a thick-walled cylinder (OD/ID = 2) under internal pressure (300 MPa).

Thick-walled vessels in high-pressure applications are characterized by a very nonuniform stress distribution: the stress peak occurs at the inner wall while descending through the wall with the minimum at the outer wall. The theory of elastically pressured cylinders can be found, among other sources, in Refs. [15–17], and a complete set of formulas has been provided in Ref. [18]. Figure 11.1 shows the stress distribution of a thick-walled tube for a typical diameter ratio of 2 at a pressure of 300 MPa: the hoop stress at the inner surface is 2.5 times the value at the outer surface. While the inner surface achieves the level of the yield strength, the material below outer surface provides reserves that can be utilized for compressing the inner surface.

By further increasing the pressure, the elastic range of engineering design is trespassed: the material at the inner bore gets overloaded and plastically deformed. Thus, the designer has to consider that – in theory – the used pressure may cause full plasticity and therefore incorporates the danger of a final breakdown [17]. To prevent plastic collapse, a safety margin has to be obeyed. For example, the ASME high-pressure vessel code [19] requires a safety margin of $\sqrt{3}$, but it may vary in accordance with the requirements of other pressure vessel codes.

A first engineering approach assumes a nonhardening material with a linear elastic, perfectly plastic stress–strain curve and secondly makes the assumption that dimensional changes are small and can thus be neglected. The analytical solution of elastic–plastic pressurized cylinders is described in Refs. [18, 20].

Figure 11.2a shows the stresses of an autofrettaged cylinder after removing the pressure: the partial plastification causes significant compressive hoop stresses of the inner part of the tube indicated by high negative value of the tangential stress.

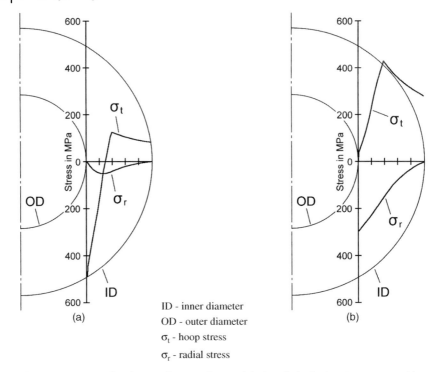

Figure 11.2 Stress distribution of an autofrettaged thick-walled cylinder (OD/ID = 2; yield strength = 965 MPa; autofrettaged at 650 MPa; operating pressure = 300 MPa): (a) residual stresses; (b) superposed stresses.

When operating the tube, the stresses caused by the operating pressure and the autofrettage-caused residual stresses are superposed: since the operating pressure results in a highly positive value and the residual autofrettage stress is highly negative, the final value of the hoop stress inside the bore can be reduced (see Figure 11.2b).

The increased fatigue life of autofrettaged vessels is primarily based on the shift of the mean stress toward the compressive side that allows higher alternating stresses [21]. In addition, the sensitivity against mean stress increases with the strength of the material [22].

By creating compressive hoop stresses through an autofrettage process, it is possible to operate a high-pressure component completely under compressive stresses inside the bore. In terms of a fracture mechanics approach, this gives a negative range of the stress intensity factor: potential surface cracks will not propagate further [19]. Thus, the high-pressure component can be considered not fatigue critical.

As mentioned earlier, the classical solutions assume an isotropic hardening material with fully reversible behavior of the compressive side. These preconditions, however, are not accurately fulfilled for real materials such as the low-alloy,

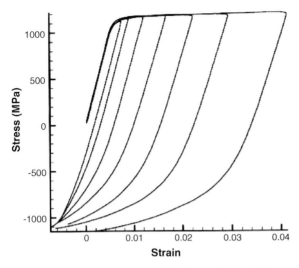

Figure 11.3 Stress–strain tests of 35NiCrMoV12-5 (1.6959) according to Ref. [24].

high-strength materials used for thick-walled high-pressure vessels and tubes. Performing a complete cycle of a tension–compression test, these materials show, on the one hand, a nonlinear hardening behavior and, on the other hand, a loss of yield strength when compressed subsequently. In other words, most of the materials used for high-pressure components have lower compressive yield strength after they were plastically tensioned. This phenomenon is called "Bauschinger effect" and is known since the end of the nineteenth century [23]. Figure 11.3 shows true stress–strain curves of the material 35NiCrMoV12-5 (1.6959) achieved from a number of tension–compression tests. The material is often used for tubular reactor systems in petrochemical industry. Looking at the first loop, it is clear that the yield stress in tension is approximately 1000 MPa while reyielding starts at lower compression stress of approximately −500 MPa. The figure shows, in addition, that this phenomenon depends on the amount of the plastification during the tensioning: increasing the plastic strain, the yield stress achieves approximately 1200 MPa but yielding during compression starts at approximately −100 MPa (see the outer loop).

The influence of the Bauschinger effect on the residual stresses of an autofrettaged cylinder was hardly investigated: improvements of the analytical approach can be found, among others, in Refs. [25–28]. Finite element analysis was performed [29, 30], and experimental verifications are reported by Underwood et al. [31].

The Bauschinger effect results in lower residual compressive stresses than calculated by classical theory assuming an ideal isotropic perfectly plastic material. It cannot be neglected when the portion of the plastically deformed wall exceeds 30%.

Recent investigations focus on a combination of a mechanical swage autofrettage followed by a low-temperature heat treatment and a hydraulic re-autofrettage [32, 33]. Improvements in safety against reyielding and in fatigue lifetime are expected.

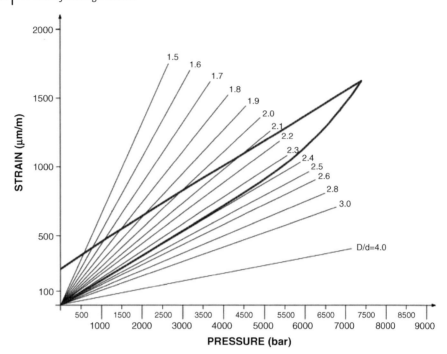

Figure 11.4 Hoop strain at outside surface during autofrettage of a thick-walled cylinder (OD/ID = 2.35; yield strength = 965 MPa).

In mechanical swage autofrettage, the compression of the inner portion of the wall is achieved by a mechanically caused plastic deformation. Therefore, a mandrel of a slightly larger outer diameter than the inner diameter of the high-pressure part is being pressed through the bore.

Because of its benefits to fatigue life, all high-pressure reactor tubes used, for example, in low-density polyethylene plants are autofrettaged. The autofrettage is usually carried out in barricaded areas by means of hydraulic autofrettage. The hydraulic pressure of typically 600–800 MPa is applied steadily, followed by a defined holding time. The process is controlled and documented using strain gauges measuring the hoop strain at the outer surface of the tube. Figure 11.4 shows a typical pressure–strain diagram of an autofrettage process of such a high-pressure tube.

In summary, it can be stated that the use of the autofrettage process has a number of benefits in particular where alternating pressures are to be expected and/or applied. It increases the fatigue life of the respective high-pressure component. It can be considered a smoothing of local peak stresses caused by inner surface injuries or imperfects as well as by engineering-caused discontinuities such as nozzles, corners, or filets. Manning and Labrow [17] mention that autofrettage spreads the stress more evenly over the wall, reducing the danger of brittle failure.

11.2
Waterjet Cutting Technology

Information on the possibility of using a high-pressure waterjet as a professional cutting tool first appeared in the late 1960s – in 1968, the design of the first waterjet cutting system for paper cutting at 700 bar water pressure was patented in the United States [34]. At the end of the 1960s, the first high-pressure intensifier pump was built for a maximum water pressure of 4000 bar [35]. The addition of abrasive material (solid particles) by means of an injection system in the beginning of the 1980s made it possible to achieve higher traverse rates, greater workpiece thickness, and higher cutting quality [36]. Afterward, the number of waterjet cutting systems quickly increased worldwide.

The waterjet technology offers its users several important advantages compared to traditional machining methods and other nontraditional machining techniques as well (laser cutting, plasma cutting, and ultrasonic machining), for example,

- almost any kind of material can be cut (steel, ceramics, composite materials, glass, stone, etc.);
- great material thickness can be cut: up to 300 mm at 4000 bar pressure and up to 500 mm at 6000 bar pressure;
- no thermal effect on workpiece materials;
- no thermal and chemical reaction products;
- no direct contact between the cutting head and the workpiece;
- small cutting width: on the average about 0.5–1.2 mm (depending on the focusing tube size and cutting parameters), contributing to workpiece material savings;
- omnidirectional tool – capability of machining complex geometries;
- low cutting and reaction forces;
- multifunctional tool (cutting, drilling, milling, and cleaning with one tool);
- susceptibility to automation (including robotization).

These advantages caused the rapid spread of the waterjet technology application fields in the aircraft industry, automotive, military, medicine, job shops, and many others. The revenues of the waterjet cutting market worldwide are expected to reach $911 million in 2012 with a compound annual growth rate of approximately 12% [13].

11.2.1
Generation of Waterjets

Today in most industrial applications, two types of the high-pressure waterjets are used – a plain waterjet (WJ) and an abrasive injection waterjet (AIWJ). A schematic drawing of the plain waterjet generation is presented in Figure 11.5a. Compressed in a high-pressure pump, water is conveyed by the high-pressure piping to an orifice with a diameter of 0.08–0.5 mm (Figures 11.5a and 11.6a). Orifice materials are usually sapphire, ruby, or diamond. Sapphire and ruby orifices are cheaper, but prone to wear and crack and due to their lifetime of approximately 50–100 h they must be frequently changed. Diamond orifices offer better reliability and an extended life

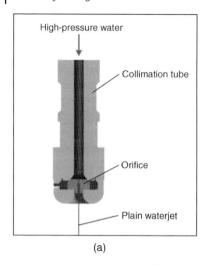 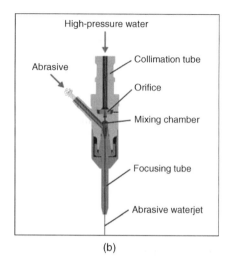

Figure 11.5 (a) Cross section of the plain waterjet cutting head. (b) Cross section of the abrasive waterjet cutting head.

of up to 1000 h, but their high purchase costs have a negative impact on the quick expansion in the waterjet cutting market.

On the orifice, the potential energy of compressed water (today up to 6000 bar) is transformed into high kinetic energy of the waterjet. Its velocity can be estimated according to Bernoulli's law:

$$v_w = \varphi_d \sqrt{\frac{2p}{\varrho_w}}$$

where φ_d is the energy transfer factor (typically 0.9–0.98) [37], ϱ_w is the density of water, and p is the water pressure.

Thus, for example, for the water pressure $p = 4000$ bar, the velocity $v_w = 805 - 875$ m/s is achieved. With such high-energy plain waterjet, soft materials can be cut. Typical applications are seals, foam, rubber, car interior equipment, wood, plastic, paper products, or frozen food products.

A schematic drawing of the abrasive injection waterjet generation is presented in Figure 11.5b. The plain waterjet generated in the orifice flows through a mixing chamber inside the cutting head and generates a vacuum pressure (Venturi effect). This vacuum pressure allows for the pneumatic feeding of dry solid abrasive particles into the mixing chamber of the cutting head. In the focusing tube water, abrasive and air are mixed, accelerated, and focused to the coherent abrasive waterjet. On average, the AIWJ consists, by volume, of approximately 95% air, 4% water, and 1% abrasive particles.

The modern focusing tubes (Figure 11.6a) are made of advanced ceramic materials (carbide alloys without a soft material binder) and their lifetime reaches up to 80 h of precision cutting and up to 140 h of rough cutting [38]. In the abrasive waterjet cutting process, the only function of the waterjet is to accelerate the abrasive particles and to get away the removed material. The abrasive particles carry out a microcutting

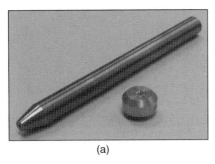 

Figure 11.6 (a) Orifice and focusing tube. (b) Abrasive grains of Australian almandine garnet 80 mesh.

process [37, 39]. The most used abrasives are natural grains such as garnet (Figure 11.6b) and olivine. Their high hardness of 7–9 on the Mohs scale makes it possible to cut any material. Natural abrasives are mined (Australia, United States, India, and Eastern Europe) or either rock forming minerals or heavy minerals from sands. Abrasive concentrates are obtained from the raw material by the technological process consisting mostly of crushing, separation, and sorting. The quality of mineral abrasives is influenced mostly by their geological history [40].

Manufactured abrasives (mostly secondary raw materials or industrial wastes) such as alkaline glass and slag (glassy form of slag, coal slag, Cu slag, and Fe slag) have also been tested [40].

The most widely used particle size for AIWJ cutting is 80 mesh. For rough cutting and cutting of very thick workpieces, 60 mesh abrasives are used. When low roughness or a small diameter of the waterjet is required, the sizes of 120 mesh or 220 mesh are used. Commonly, industrial consumption of the abrasive is between 180 and 600 g/min.

AIWJ is today's state of the art and is worldwide applied in industrial manufacturing with pressures up to 6000 bar. With the AWIJ, all hard materials such as metals, ceramic- and fiber-reinforced composite materials, glass, or stone can be cut. In this field, the abrasive waterjet competes with mechanical sawing, wire sawing, punching, EDM, and laser and thermal cutting techniques.

11.2.2
Cutting Process and Parameters

The waterjet cutting result depends on several process parameters. The most important are

1) for plain waterjet cutting: water pressure, orifice diameter, traverse rate, standoff distance between orifice and workpiece, and type and thickness of the material;
2) for injection abrasive waterjet cutting: same as above for plain waterjet cutting, but also focusing tube diameter and parameters of size, type, and quantity of abrasive material.

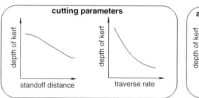

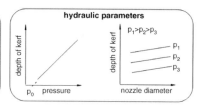

Figure 11.7 Selected general tendencies of abrasive waterjet cutting process parameters influence the cutting result (source: Institute of Materials Science, Leibniz Universität Hannover).

In Figure 11.7, some general relations between the waterjet cutting process parameters and the cutting result are shown.

In AIWJ cutting processes, the stock removal results mainly from microcutting with abrasive particles. Abrasive grains, accelerated in the focusing tube by the high-speed plain waterjet, hit with high velocity against the workpiece material surface inducing in it microscopic decrements (erosion). In abrasive waterjet cutting, four characteristic phases, presented in Figure 11.8, can be distinguished [41]. In the first phase of the cutting process, a curving of jet motion path occurs in the work material (Figure 11.8a). Since the inert mass of moving abrasive grains is big, it leads to a concentration of abrasive grain strokes on the face of the kerf. The consequence of such form of stock removal is the creation of a leap (Figure 11.8b). On this part of the kerf, the concentration of abrasive grains leads to "driving" the leap inside the kerf (Figure 11.8c). As the kinetic energy of the abrasive waterjet significantly decreases, the maximum cutting depth k_{max} (Figure 11.8d) for given machining conditions is obtained.

Generally, the upper part of the cutting kerf, near the AIWJ entrance side of the workpiece, is characterized by a smooth cutting surface, without noticeable striation marks and waviness (h_S – smooth zone, Figure 11.9a and b). This is a result of high energy of the AIWJ in this part of the kerf, where the material is removed by a cutting wear mode [42]. With the decreasing kinetic energy of the AIWJ in deeper parts of the

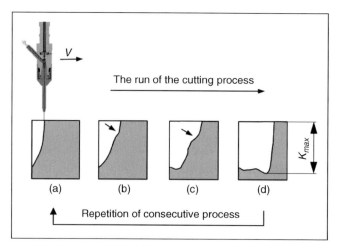

Figure 11.8 Successive phases of stock removal during abrasive waterjet cutting process.

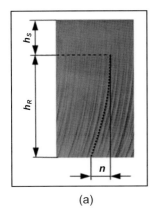 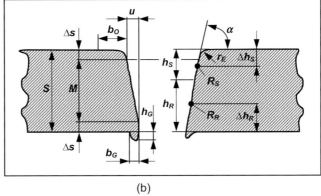

(a) (b)

Figure 11.9 (a) Typical material surface after AIWJ cutting (100 mm thick stainless steel). (b) Quality indicators of the cutting kerf. b_O: width of the jet affected zone; r_E: radius at the jet entrance side; h_S: height of the smooth zone; R_S: roughness of the smooth zone; h_R: height of the rough zone; R_R: roughness of the rough zone; M: measuring distance for u and α; $\Delta_S = 0.1\,S$ if $S < 2$ mm; $\Delta_S = 0.2\,S$ if $S \geq 2$ mm; u: straightness of the kerf; α: kerf taper angle; b_G: width of machining burr; h_G: height of machining burr; S: workpiece thickness; Δh: measuring distance for the roughness. Mostly $\Delta h_S = 0.5 h_S$ and $\Delta h_R = 0.5 h_R$.

cutting kerf (h_R – rough zone, Figure 11.9a and b), the AIWJ becomes increasingly unsteady – the cutting wear mode is systematically replaced with a deformation wear mode. The cutting front AIWJ–workpiece swings sideways and bends in the opposite direction to the moving direction of the cutting head. It results in the appearance of a jet lag (n, Figure 11.9a), visible striation marks, and high waviness on the material surface. Depending on the expected cutting quality, the h_S/h_R ratio can be regulated by changing and adjusting process parameters.

The quantitative evaluation of the technological quality of the cutting kerf machined with a high-pressure abrasive waterjet could be made on the basis of guideline VDI 2906 [43]. The evaluation indicators provided in the guideline are defined in Figure 11.9b.

The erosion process during abrasive waterjet cutting has a very complex character that so far has not been fully understood and theoretically described. In the literature, some abrasive waterjet cutting models can be found to estimate the maximum cutting depth as a function of cutting parameters such as traverse rate of the cutting head, waterjet pressure, and abrasive mass flow rate, but most of the models can be applied to only one type of material [44, 45]. Similarly, some models to describe the roughness of the cutting surface and the quality of the kerf are discussed today [46].

11.2.3
High-Pressure Pumps

In order to obtain water under high pressure suitable for industrial waterjet cutting, the hydraulic intensifier pump or mechanical directly driven (electric or diesel motor) plunger pumps are used.

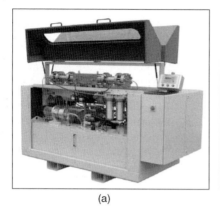
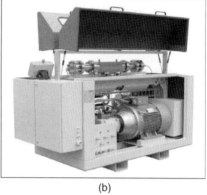

(a)   (b)

Figure 11.10 4000 bar intensifier pump: (a) front view; (b) rear view (*source*: Uhde High Pressure Technologies GmbH).

On the intensifier pump (Figure 11.10), an electric motor drives a hydraulic pump that generates a maximum of approximately 200 bar oil pressure in the hydraulic system. The hydraulic drive is equipped with a dynamic controller to adapt it to the pressure and flow requirements. The hydraulic oil is fed to one side of the hydraulic piston that causes its movement to the other direction (Figure 11.11). On both sides of the hydraulic piston, a plunger is fitted. Because of their higher hardness, better resistance to dirty water, and lower weight than hard metals, today ceramic plungers of partially stabilized zirconia (ZrO_2) are mostly used. This plunger moves into the high-pressure cylinder and compresses the water. Because of the compressibility of water, approximately 12–13% (at 4000 bar) of the stroke of the piston is used to pressurize and compress the water in the

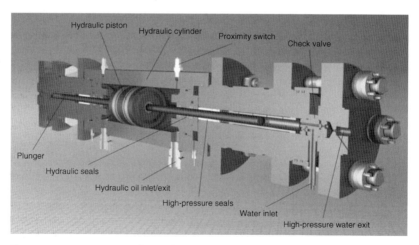

Figure 11.11 Cross section of a 6000 bar intensifier unit (*source*: Uhde High Pressure Technologies GmbH).

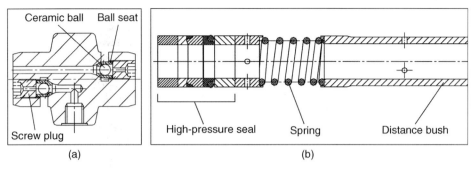

Figure 11.12 (a) 4000 bar check valve. (b) 4000 bar high-pressure packaging (*source*: Uhde High Pressure Technologies GmbH).

cylinder without delivering water to the system. Two check valves (Figure 11.12a) ensure the water flow toward the orifice on one high-pressure cylinder and at the same time close the high-pressure cylinder on the other side of the intensifier. The high-pressure packaging of the intensifier is of particular importance (Figure 11.12b) since its lifetime is a critical factor for the customers. The typical lifetime of this component is 400–600 h on average and mostly depends on the operating pressure.

The end of the piston stroke is detected by a proximity switch installed inside the hydraulic cylinder. The signal from the proximity switch causes a fast 4/3-way valve to redirect the oil flow in a manner that is fast but smooth, in order to avoid damage of the components. In the suction part of the piston movement, the hydraulic oil is discharged to a tank. Cooling of the oil is commonly achieved by either an air cooler or a water cooler, depending on ambient conditions and customer's requirements. The output water pressure of the intensifier pump is determined by the inlet hydraulic oil pressure and the pressure intensification ratio. This ratio is defined as the working area of the hydraulic piston divided by the area of the plunger. Commonly, intensification ratios are between approximately 1:19 for 4000 bar pumps and 1:33 for 6000 bar intensifier pumps.

The interruption of water compression at the reversing point of the piston produces pressure pulsation, which induces a fatigue cycle in the high-pressure system, reducing the lifetime of important system components such as high-pressure seals, check valves, piping, fittings, orifices, and others. Too high level or pressure pulsations have a significant negative influence on the cutting results (e.g., lower cutting depth and lower cutting quality). In order to reduce the pulsation, pulsation dampers (attenuators) are installed in the high-pressure piping (Figure 11.13a). The pressure pulsation level, depending on the specific setup, is typically dampened to ±2–3% of the operating pressure (Figure 11.13b).

In the mechanically driven pumps, several offset oscillating pistons are directly coupled to an eccentric rotating crankshaft (Figure 11.14). The delivered flow of high-pressure water can be changed by changing the speed of the crankshaft using a frequency converter. Due to their mechanical drive, plunger pumps demonstrate an increased (up to 95%) degree of efficiency in comparison to intensifier pumps

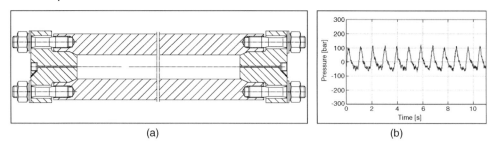

Figure 11.13 (a) Cross section of a 2 l high-pressure pulsation damper (*source*: Uhde High Pressure Technologies GmbH). (b) Pressure fluctuations in a typical intensifier pump at 3000 bar [47].

(approximately 60%). However, today's maximum available operating pressure is 3800 bar and their high stroke rate (up to 900/min) causes higher wear of expensive high-pressure components such as seals, check valves, and piping.

Today, the typical 4000 bar high-pressure pumps allow to achieve a flow rate from 0.8 up to 7.6 l/min, and the 6000 bar intensifier pumps offer the user flow rates between 2.8 and 5.4 l/min.

11.2.4
Waterjet Cutting with 6000 bar

Since waterjet cutting is permanently competing with a lot of other cutting processes, such as laser or plasma cutting, a further development toward higher cutting speeds and better accuracy has to be enforced. Higher operating pressures are the most important way to achieve this goal. In 2001, a first high-pressure 6000 bar intensifier pump as two-stage (cascading intensifier) system was introduced in the European job shop industry [48]. In 2004, a first one-stage 6000 bar pump for industrial use came on the waterjet market. In the 6000 bar waterjet cutting technique, the use of mechanical directly driven plunger pumps is not possible – their high number of strokes causes tribological problems (high-pressure seals) and accelerated fatigue of high-pressure components of the pump and of the cutting machine.

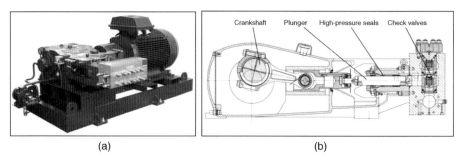

Figure 11.14 Plunger pump: (a) outside view; (b) cross section (*source*: URACA).

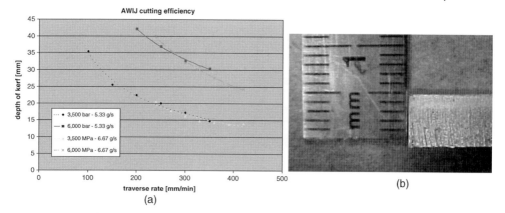

Figure 11.15 (a) Comparison between cutting results of material INOX with 3500 and 6000 bar pressures [49]. (b) 6 mm thick AlCuMg1 cut with 6000 bar plain waterjet. Traverse rate $v = 30\,\text{mm/min}$ [49].

The increase of the operating pressure to 6000 bar improves the working efficiencies and economical benefits [49, 50]:

- increase of the waterjet power density by 83% in comparison to the 4000 bar level;
- up to two times faster cutting speed while the other parameters are kept constant (Figure 11.15a);
- up to 40% reduction in abrasive mass flow;
- up to 25% reduction in operating costs;
- harder materials can be cut effectively;
- deeper cuts of metals and other hard materials are possible without the use of abrasive, for example, for titanium up to 3 mm cutting depth and for aluminum up to 6 mm (Figure 11.15b).

The implementation of the 6000 bar waterjet cutting technique requires at the same time the introduction of new hardware equipment suitable for this pressure level: cutting heads with on/off valves, piping tubes, and fittings. Sometimes a new construction of the equipment is required (e.g., cylinder body can no longer be made from a monoblock material), and sometimes the use of the best engineering materials and advanced manufacturing processes is required. For example, for piping tubes only material HP 160 after autofrettage can ensure a long work with a 6000 bar cutting system. Apart from piping tubes, other high-pressure 6000 bar components such as check valves and pump cylinder bodies have to be autofrettaged (more information about autofrettage process is provided in Section 10.1).

11.2.5
Cutting Devices

In industrial practice, abrasive waterjet cutting is today primarily used to cut 2D workpieces from sheet materials. One or more cutting heads (systems with 15

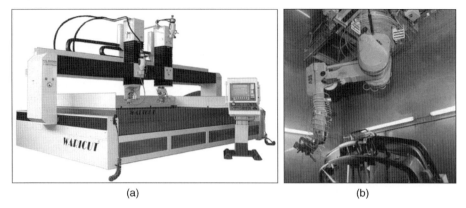

Figure 11.16 (a) Portal cutting machine with a 2D axis and a 3D axis (*source*: H.G. Ridder Automatisierungs-GmbH). (b) Robotized cutting of a car bumper with the plain waterjet in a cutting box (*source*: Uhde High Pressure Technologies GmbH).

cutting heads on one crossbeam are known) are guided along two X–Y CNC-controlled axes of a portal-gantry system. The third axis (Z-axis) allows to adjust the height of the cutting head to the thickness of the workpiece. The remaining energy of the waterjet after cutting is dissipated in a water basin (catcher). Typical industrial sizes of cutting areas are ($X \times Y$) 1500 or 2000 mm × 3000 mm. For large-volume workpieces (e.g., for the aerospace industry), it is possible to build modular cutting tables up to 50 000 mm in length in X-axis direction. The modern precise guide systems allow the cutting machines to achieve a positioning deviation of $< \pm 0.02$ mm and an average positioning spread of $< \pm 0.015$ mm [51]. The motion of the machine is previously programmed using a CAD/CAM software including mostly advanced automatic nesting features and completely controlled by CNC.

New fields of application for abrasive waterjet cutting can be opened up by integration of a 3D (five-axis control) cutting head. Both types of cutting head axes (2D and 3D) can be combined in one machine, too (Figure 11.16a). In applications where the accuracy is more strict (e.g., in the aerospace industry, where large, complex shaped parts of composite materials for airplanes have to be cut), accurate five-axis machining was introduced first.

As a result of hard competition with other nonconventional cutting methods, many new modern equipment components for waterjet cutting machines have been developed during the past few years and put into industrial practice, for example,

- cutting head with tilt function: it allows to correct an inclination angle of the cutting head in relation to the workpiece during cutting. The inclination angle is permanently calculated by a software that allows to achieve virtually zero kerf taper angle (α, Figure 11.9b);
- anticollision protection and height sensor systems for the cutting heads;
- sensors of vacuum level in the cutting head;

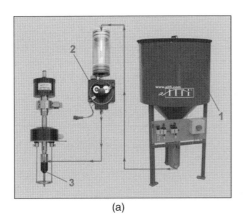 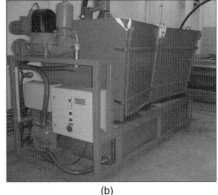

(a) (b)

Figure 11.17 (a) Typical industrial abrasive feeding system (*source*: Allfi AG). (b) Sludge removal system (*source*: Metallbau Müller GmbH).

- remote diagnosis of the cutting machine via Internet;
- high-speed linear driving units with speeds up to 60 000 mm/min [52];
- tube cutting axis, and many others.

For the cutting of products with complex 3D shapes with lower requirements on accuracy, robotized systems are used. A robotized system, with its many degrees of freedom, can in practice cut like a regular five-axis cutting machine. The only limit is its accuracy – today approximately ±0.1–0.3 mm. Therefore, the use of robotized cutting systems is in most cases limited to plain waterjet cutting. Figure 11.16b shows a typical application in the automotive industry – the programmed robot is mounted in an automatically controlled and safe-for-personnel cutting box.

A standard equipment of an abrasive waterjet cutting system is an abrasive feeding system. A typical industrial abrasive feeding system is shown in Figure 11.17a. In an abrasive hopper (position 1), large amounts of abrasive (up to 3000 kg) are stored. Using air pressure or vacuum, abrasive grains are transported to a metering unit (position 2), installed close to the cutting head, which supplies a regular, constant mass flow rate of abrasive to the cutting head (position 3). Electrically controlled metering systems are typically used that allow to change the abrasive quantity during the cutting process and adjust it to the cutting process requirements.

As the sludge in the cutting basin can solidify, this sediment abrasive must be regularly removed from the cutting machine. This occurs, among other things, with scraper conveyors integrated into the machine or with special pump systems with filtration (Figure 11.17b). After separation from water, the collected used abrasive can be prepared for reuse with recycling systems. Feasibility studies have shown that the choice of optimal cutting parameters leads to 70–80% recyclable solid, which normally would be disposed [53]. In recent years, several companies have offered recycling plants for the recycling of abrasives. Because of their high exploitation costs (electricity and water costs), the used abrasive materials are mostly sent to large central recycling locations.

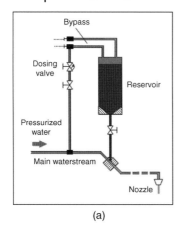 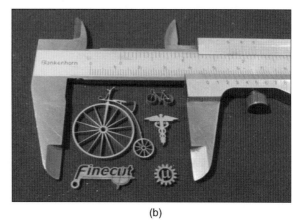

Figure 11.18 (a) Generation of the abrasive water suspension jet – "bypass principle" [39]. (b) Parts cut with the microabrasive waterjet. (*source*: Finecut AB).

11.2.6
New Trends in the Waterjet Cutting

11.2.6.1 Abrasive Water Suspension Jet

The second mixing technique to generate an abrasive waterjet, called abrasive water suspension jet (AWSJ), was developed in the early 1980s and its first industrial application was mentioned in the literature in 1986 [54]. A schematic drawing of the most used method of the AWSJ generation in practice is presented in Figure 11.18a. In this method, widely known as "bypass principle," one part of the pressurized water flow delivered by the pump is used to feed the abrasive particles out of the pressure vessel (reservoir) and to mix it back into the main waterstream. The resulting suspension can be led through elastic high-pressure hoses to the cutting location. The main difference between the abrasive water suspension jet and the injection waterjet is the absence of air. The absence of air gives the AWSJ a higher stability and a larger cutting efficiency than the AIWJ – a cutting thickness of 1000 mm of reinforced concrete is possible [55]. The second important advantage over the AIWJ is that the AWSJ can be transported for long distances up to 1000 m and more on surface and because of this it can be easily used in mobile applications (e.g., for cutting munitions). Underwater distances up to 6000 m are possible [56]. Today's commercial AWSJ systems work with pressures up to 2000 bar. Because of higher wear of the high-pressure components and the complex controlling process of the cutting system process parameters, this mixing technique is still under development [57].

11.2.6.2 Microcutting

Waterjet cutting with its many advantages is an interesting alternative for the fabrication of microcomponents for medicine, biotechnology, electronic, automotive, and fine mechanics applications, as well as the machining of expensive materials

 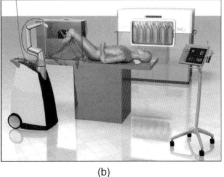

Figure 11.19 (a) A bone cut with AIWJ at 3500 bar with additive of sugar as abrasive material (*source*: Institute of Materials Science, Leibniz Universität Hannover). (b) "Cutting box" for future orthopedic surgery [64].

(gold, platinum, etc.). The market for microelectromechanical systems (MEMS) grows rapidly and its potential worldwide is estimated at $103 billion/year by 2012 [58].

For the long time, the participation in micromachining market has been reserved only for complementary technologies such as laser, EDM, or chemical etching. But in 2000 a miniature abrasive waterjet of 50 μm diameter using AWSJ technique was created [59]. In 2003, the first six-teeth gearwheels with an outside diameter of 780 μm with a plain waterjet (orifice 0.08 mm, pressure of 3500 bar) were manufactured [60]. In 2009, a first professional series-produced waterjet cutting machine, dedicated to microcomponents machining, appeared on the market [61]. It allows to achieve kerf widths of 0.2–0.3 mm and small holes with industrial constant reliability using the AIWJ. In future, reliable cutting with kerf widths of 0.05 mm is expected. Some examples of microcomponents cut with the mini-AIWJ are shown in Figure 11.18b.

11.2.6.3 Medical Applications

First results of the use of waterjets for medical applications have been published in 1982 – a plain waterjet scalpel for the resection of liver tissue has been applied [62]. The use of the plain waterjet offers an important advantage over conventional surgery and ultrasound or laser cutting: athermic and selective cutting – the plain waterjet under low pressure (below 100 bar) cuts the tissue but the blood vessels are not injured, because they are slightly more resistant.

There are several other possibilities to exploit the potential of the plain waterjet in medicine, such as

- removal of dimmed lens capsule of the eyes at the gray star;
- in neurosurgery to cut the cerebral tissue;
- in the operation of tumors in the ear, nose, and throat or in the urology.

However, in medical applications the waterjet technology is actually limited to soft tissue cutting – the plain waterjet has not enough cutting performance to be applied on biological hard tissues such as bones or teeth. Because of this, various projects with the goal to develop the abrasive waterjet to a tool for orthopedic surgery have been undertaken in recent years (Figure 11.19a). This includes, among others, biophysical experiments on the interaction of biocompatible abrasives (e.g., saccharose, sorbitol, and lactose) and somatic cells [63]. For future orthopedic surgery, a special robotized cutting system has been designed (Figure 11.19b) [64]. In March 2003, the worldwide first *in vivo* application of abrasive waterjets for the cutting of bone took place at the University of Veterinary Medicine in Hanover.

References

1 Kocanda, A. and Sadlowska, H. (2008) Automotive component development by means of hydroforming: a review. *Arch. Civil Mech. Eng.*, **8** (3), 55–72.
2 Zhang, Y., Luen, C.C., Wang, C.G., and Wu, P. (2009) Optimization for loading paths of tube hydroforming using a hybrid method. *Mater. Manuf. Process.*, **24** (6), 700–708.
3 Li, S.H., Yang, B., Zhang, W.G., and Lin, Z.Q. (2008) Loading path prediction for tube hydroforming process using a fuzzy control strategy. *Mater. Des.*, **29** (6), 1110–1116.
4 Hama, T., Asakawa, M., Fukiharu, H., and Makinouchi, A. (2004) Simulation of hammering hydroforming by static explicit FEM. *ISIJ Int.*, **44** (1), 123–128.
5 Koizumi, M. and Nishihara, M. (1992) *Isostatic Pressing*, Elsevier Science Publishers, Ltd., Essex, UK.
6 Spain, I.L. and Paauwe, J. (1977) *High Pressure Technology. Vol. II. Applications and Processes*, Marcel Dekker, Inc., New York.
7 Fujikawa, T. et al. (2004) Recent trends of HIP equipment technology in Japan. ASME Pressure Vessel and Piping Division Conference, July 25–29, 2004, San Diego, CA, Paper No. PVP 2004-2274.
8 Richerson, D.W. (2006) *Modern Ceramic Engineering*, 3rd edn, CRC Press/Taylor & Francis Group, Boca Raton, FL.
9 Bengisu, M. (2001) *Engineering Ceramics*, Springer, Berlin.
10 Ishizaki, K. and Nanko, M. (1995) A hot isostatic process for fabricating open porous materials. *J. Porous Mater.*, **1**, 19–27.
11 Li, H., Johnston, R., and Mackenzie, D. (2007) Effect of autofrettage in the thick-walled cylinder with a radial cross-bore. ASME Pressure Vessel and Piping Division Conference, July 22–26, 2007, San Antonio, TX, Paper No. PVP 2007-26319.
12 Koerner, J.-P. and Wüstenberg, D. (1974) Strength of autofrettaged tees for high pressures, in *Chemie-Ingenieur-Technik*, vol. 10, Verlag Chemie GmbH, Weinheim.
13 Frost & Sullivan (2005) World Waterjet Cutting Tools Markets, Market Engineering Research.
14 Burns, D.J., Karl, E., and Ohlsson, L. (2000) Methods for predicting the fatigue performance of non-welded high pressure vessels. Proceedings of International Conference on Pressure Vessel Technology (ICPVT-9), Sydney.
15 Szabó, I. (1985) *Höhere Mechanik*, 5th edn, Springer, Berlin.
16 Schwaigerer, S. and Mühlenbeck, G. (1997) *Festigkeitsberechnung im Dampfkessel-, Behälter- und Rohrleitungsbau*, 5th edn, Springer, Berlin.
17 Manning, W.R.D. and Labrow, S. (1974) *High Pressure Engineering*, Leonardo Hill, London.
18 Timoshenko, S. and Young, D.H. (1956) *Engineering Mechanics*, 4th edn, McGraw-Hill Book Company, Inc., New York.
19 ASME (2009) ASME Boiler and Pressure Vessel Code. Section VIII, Division 3. Rules for Construction of Pressure Vessels, ASME, New York.
20 Hill, R. (1967) *The Mathematical Theory of Plasticity*, Oxford University Press, Oxford, UK.

21. Nicholas, Th. (2006) *High Cycle Fatigue: A Mechanics of Materials Perspective*, 1st edn, Elsevier Ltd., Oxford, UK.
22. Buxbaum, O. (1992) *Betriebsfestigkeit. Sichere und Wirtschaftliche Bemessung Schwingbruchgefährdeter Bauteile*, 2nd edn, Verlag Stahleisen GmbH, Düsseldorf.
23. Bauschinger, J. (1881) Ueber die veraenderung der elastizitaetsgrenze und des elastizitaetsmoduls verschiedener metalle. *Zivilingenieur*, **27**, 289–348.
24. Farrahi, G.H., Hosseinian, E., and Assempour, A. (2008) On the material modelling of the autofrettaged pressure vessel steels. ASME Pressure Vessel and Piping Division Conference, July 27–31, 2008, Chicago, IL, Paper No. PVP 2008-61482.
25. Huang, X. (2005) A general autofrettage model of a thick-walled cylinder based on tensile-compressive stress–strain curve of a material. *J. Strain Anal. Eng. Des.*, **40**, 23–27.
26. Parker, A.P. (2005) Assessment and extension of an analytical formulation for prediction of residual stress in autofrettaged thick cylinders. ASME Pressure Vessel and Piping Division Conference, July 17–21, 2005, Denver, CO, Paper No. PVP 2005-71368.
27. Hosseinian, E., Farrahi, G.H., and Movahhedy, M.R. (2008) An analytical framework for the solution of autofrettaged tubes under constant axial strain condition. ASME Pressure Vessel and Piping Division Conference, July 27–31, 2008, Chicago, IL, Paper No. PVP2008-61573.
28. Perry, J. and Perl, M. (2008) The evaluation of the 3-D residual stress field due to hydraulic autofrettage in a finite length cylinder incorporating the Bauschinger effect factor based on the "zero offset yield stress". ASME Pressure Vessel and Piping Division Conference, July 27–31, 2008, Chicago, IL, Paper No. PVP 2008-61032.
29. Troiano, E., Underwood, J.H., and Parker, A.P. (2006) Finite element investigation of Bauschinger effect in high strength A723 pressure vessel steel. *J. Press. Vessel Technol.*, **128**, 185–189.
30. Gibson, M.C., Hameed, A., Hetherington, J.G., and Parker, A.P. (2007) Custom material modelling within FEA for use in autofrettage simulation. ASME Pressure Vessel and Piping Division Conference, July 22–26, 2007, San Antonio, TX, Paper No. PVP 2007-26341.
31. Underwood, J.H., de Swardt, R.R., Venter, A.M., Troiano, E., Hyland, E.J., and Parker, A.P. (2007) Hill stress calculations for autofrettaged tubes compared with neutron diffraction residual stresses and measured yield pressure and fatigue life. ASME Pressure Vessel and Piping Division Conference, July 22–26, 2007, San Antonio, TX, Paper No. PVP 2007-26617.
32. Parker, A.P., Troiano, E., and Underwood, J.H. (2009) Hydraulic re-autofrettage of a swage autofrettage tube. ASME Pressure Vessel and Piping Division Conference, July 26–30, 2009, Prague, Czech Republic, Paper No. PVP 2009-77213.
33. Perl, M. and Perry, J. (2009) Changes in the Bauschinger effect level post the autofrettage process. ASME Pressure Vessel and Piping Division Conference, July 26–30, 2009, Prague, Czech Republic, Paper No. PVP 2009-77107.
34. Hunziker-Jost, U.W. (1991) *Wasser schneidet scharf. Schweizer Präzisions-Fertigungstechnik*, Carl Hanser Verlag, München, Germany.
35. Mohamed, M.A.K. (2004) Waterjet cutting up to 900MPa. Ph.D. thesis, University of Hanover, Germany.
36. Oweinah, H. (1990) Leistungssteigerung des Hochdruckwasserstrahlscheidens durch Zugabe von Zusatzstoffen. Ph.D. thesis, University of Darmstadt, Carl Hanser Verlag, München, Wien, Germany.
37. Guo, N.-S. (1994) Schneidprozess und Schnittqualität beim Wasserabrasivstrahlschneiden. Ph.D. thesis, University of Hanover, Fortschritt-Berichte VDI, Reihe 2: Fertigungstechnik, Nr. 328, VDI Verlag, Germany.
38. Kennametal (2009) ROCTEC® Composite Carbide – Abrasive Waterjet Nozzles. Product Brochure.
39. Brandt, C., Brandt, S., and Louis, H. (1998) Jet cutting technology – state of the art and

further developments. 1st International Conference on Water Jet Machining (WJM'98), Cracow, Poland.
40 Martinec, P., Foldyna, J., Sitek, L., Ščučka, J., and Vašek, J. (2002) Abrasives for AWJ cutting, Institute of Geonics, Academy of Sciences of the Czech Republic, Ostrava.
41 Blickwedel, H. (1990) Erzeugung und Wirkung von Hochdruck-Abrasivstrahlen. Ph.D. thesis, University of Hanover, Fortschritt-Berichte VDI, Reihe 2: Fertigungstechnik, Nr. 206, VDI Verlag, Germany.
42 Hashish, M. (1989) A model for abrasive-waterjet (AWJ) machining. *J. Eng. Mater. Technol.*, **111**, 154.
43 VDI-Fachbereich Produktionstechnik und Fertigungsverfahren (1994) VDI-Richtlinien – VDI 2906, Blatt 10: Schnittflächenqualität beim Schneiden, Beschneiden und Lochen von Werkstücken aus Metall. Abrasiv-Wasserstrahlschneiden.
44 Hoogstrate, A.M. (2000) Towards high-definition abrasive waterjet cutting. Ph.D. thesis, Delft University of Technology, the Netherlands.
45 Momber, A.W. and Kovacevic, R. (1998) *Principles of Abrasive Water Jet Machining*, Springer, Berlin.
46 Annoni, M., Monno, M., and Vergari, A. (2001) The macrogeometrical quality of the kerf in the AWJ process parameters selection. 2nd International Conference on Water Jet Machining (WJM 2001), Cracow, Poland.
47 Karpinski, A., Louis, H., Monno, M., Peter, D., Ravasio, C., Scheer, C., and Südmersen, U. (2004) Effects of pressure fluctuations and vibration phenomenon on striation formation in AWJ cutting. Proceedings of the 17th International Conference on Water Jetting, Mainz, Germany.
48 Koerner, P., Hiller, W., and Werth, H. (2002) Design of reliable pressure intensifiers for water-jet cutting at 4 to 7 kbar. Proceedings of the 16th International Symposium on Water Jetting Technology, Aix-en-Provence, France.
49 Lefevre, I., Lefevre, R., Stinckens, T., Koerner, P., Luetge, C., and Werth, H. (2004) Experiences of a job shop with 6 kbar abrasive water-jet cutting technology during day to day operation. Proceedings of the 17th International Conference on Water Jetting, Mainz, Germany.
50 Hashish, M. (2002) Observations on cutting with 600-MPa waterjets. *J. Press. Vessel Technol.*, **124**, 229.
51 H.G. Ridder Automatisierungs-GmbH (2009) Simply Cut Better! Product Brochure.
52 Sato Cutting Systems (2009) Satronik _WS, Linear Wasserstrahl-Schneidmaschine. Product Brochure.
53 Ohlsen, J. (1997) Recycling von Feststoffen beim Wasserabrasivinjektorstrahlverfahren. Ph.D. thesis, University of Hanover, Fortschritt-Berichte VDI, Reihe 15: Umwelttechnik, Nr. 175, VDI Verlag, Germany.
54 Fairhurst, R.M., Heron, R.A., and Saunders, D.H. (1986) DIAJET – a new abrasive water jet cutting technique. Proceedings of the 8th International Symposium on Jet Cutting Technology, Cranfield, UK.
55 Brandt, S. and Louis, H. (2000) Controlling of high-pressure abrasive water suspension jets. Proceedings of the 15th International Conference on Jetting Technology, Ronneby, Sweden.
56 Brandt, C., Louis, H., Meier, G., and Tebbing, G. (1995) Abrasive water suspension jet – a multifunctional working tool for underwater applications. Proceedings of the 5th International Offshore and Polar Engineering Conference, Hague, the Netherlands.
57 Brandt, S. (2002) Verschleissorientierte Prozessoptimierung einer Wasserabrasivsuspensionsstrahlanlage. Ph.D. thesis, University of Hanover, Fortschritt-Berichte VDI, Reihe 2: Fertigungstechnik, Nr. 628, VDI Verlag, Germany.
58 Yole Développement (2008) Global MEMS/Microsystems – Markets and Opportunities 2008. Market Research Study (http://www.i-micronews.com/

reports/MEMS-Microsystems-Markets-Opportunities-2008-Yole-SEMI/27/).

59 Miller, D. (2000) Development of micro abrasive water-jets. Proceedings of the 15th International Conference on Jetting Technology, Ronneby, Sweden.

60 Aust, E., Bremer, T., and Hoffmann, Ch. (2003) Possibilities of using the high-pressure water jet for the manufacture of microcomponents. *Weld. Cutting*, **55** (2), 100–104.

61 Finecut AB (2009) Finecut Waterjet Machining Center. Product Brochure.

62 Papachristou, D.N. and Barters, N. (1982) Resection of the liver with a water jet. *Br. J. Surg.*, **69**, 93–94.

63 Pude, F., Schmolke, St., Kirsch, L., Schwieger, K., Honl, M., and Louis, H. (2003) Abrasive waterjets as a new tool for cutting of bone – laboratory tests in the field of knee endoprostheses. Proceedings of the 6th International Conference on Management of Innovative Technologies (MIT'2003), Piran, Slovenia.

64 Pude, F. (2005) Vorklinische Studien zum Einsatz des Wasserabrasivstrahls als Osteotomiewerkzeug. Ph.D. thesis, University of Hanover, Deutsche Veterinärmedizinische Gesellschaft Service GmbH, Giessen, Germany.

Part Three
Process Equipment and Safety

12
High-Pressure Components

Waldemar Hiller and Matthias Zeiger

The realization of high-pressure processes with pressures of over several hundred or thousand bars requires pressure components that are specifically designed for the high forces acting upon the internal surfaces and other load bearing parts. The assembly of all high-pressure components together with the necessary utility connections and a control system forms an engineered technical system capable of being used to realize a high-pressure process. The limits of application of a high-pressure process are frequently determined by technological or economical limitations in the design and fabrication of high-pressure components. It is therefore indispensable to understand the construction of these components and the state-of-the-art technological limits when designing a specific high-pressure process or a production plant.

The purpose of this chapter is to show the variety of nowadays designs for high-pressure components. This can then be considered as the current "toolbox" for realizing a high-pressure process from the mechanical side. The second purpose is to illustrate the mentioned technological limits and to indicate some of the research and development work that is carried out in order to overcome these.

12.1
Materials for High-Pressure Components

The starting point for the understanding and design of high-pressure components is knowledge about the possible materials of construction. Since the load that the materials have to withstand is very high, it is evident that the strength of the materials of construction also must be very high. In fact, some key technological advances in high-pressure components have been derived from higher strength materials being developed and utilized in a safe and economical design. The use of high-strength materials is generally also the environmentally more friendly design since it consumes less raw material and energy. Today still, the vast majority of high-strength materials used for high-pressure components are based on steel. As a replacement of steel, composite-overwrapped vessels have been developed especially for aerospace applications, where reducing weight is paramount [1]. Since for high-pressure

Industrial High Pressure Applications: Processes, Equipment and Safety, First Edition. Edited by Rudolf Eggers.
© 2012 Wiley-VCH Verlag GmbH & Co. KGaA. Published 2012 by Wiley-VCH Verlag GmbH & Co. KGaA.

applications the core of these vessels is frequently made from steel, the further introduction on materials will primarily focus on the various steel alloys used [2].

12.1.1
Steel Selection Criteria

The selection of appropriate steel grades is made aiming for the lowest cost while satisfying all necessary design conditions such as

- static strength to hold the pressure;
- ductility to fulfill safety requirements and avoid brittle fracture at the given operating temperatures;
- fatigue strength for the intended number of load cycles;
- weldability, if welding is foreseen;
- resistance to corrosive attack or radiation;
- long-term stability at elevated temperatures.

In addition to the purely technical considerations of the steel selection, also regulatory compliance must be observed. The requirements depend on the national laws of the country where the high-pressure components are to be used. Only approved steels within the limitations of the pressure vessel code (or piping code) may be used [3, 4].

Since the demands of high-pressure applications are unique, there is usually not a big choice of materials available that offer high strength, meet other boundary conditions, and are commercially available. Generally, they fall into one of the below mentioned categories, in each of which a few grades offer the highest possible strength that makes them interesting for high-pressure application. All mentioned steel grades are summarized in Table 12.1.

Table 12.1 Steel selection guide for high-pressure components [9].

Steel type (typical grades)	Yield strength at 20 °C (MPa)	Steel type (typical grades)	Yield strength at 20 °C (MPa)	
High-alloy (>12% Cr) steel		Low-alloy (≤12% Cr) steel		
Austenitic stainless (1.4429, 316LN)	~300	Fine-grain and high-temperature structural steel (1.6368)	~450	Weldable
Austenitic–ferritic duplex (1.4462)	~450	Cr–Mo H_2-resistant steel (1.7779)	~500	Weldable
Soft martensitic (1.4418)	~800	HSLA (1.6580, A723Gr1)	~850	Nonweldable
Precipitation hardening (1.4545, 15-5PH)	~850	HSLA (1.6957, A723Gr3)	~900	Nonweldable

12.1.2
High-Strength Low-Alloy Steel

The standard material for high-pressure components is a forging made from high-strength low-alloy (HSLA) steel. These steels are based on 0.2–0.4% carbon content, 1–2% chrome, 2–5% nickel, up to 1% molybdenum, and some vanadium. They have been developed and refined to offer the highest possible strength at certain guaranteed ductility levels. Depending on the thickness of the required forging, yield strength values of 800–1100 MPa and tensile strength values of up to 1300 MPa can be achieved at safe ductility levels. The permissible temperature range for most HSLA steels lies approximately between -50 and $+350\,°C$. At the lower temperature end the steels are limited by a reduction in toughness, while at the high temperature end long-term high-temperature embrittlement effects may lead to a reduction in toughness. Because of the high strength and the high carbon equivalent content, these steels are not approved for welding in pressure vessel construction. Another limitation of application is their susceptibility to corrosion. If HSLA steels are to be used with corrosive fluids, then protective liners or coatings should be applied. HSLA steels are used for many high-pressure applications covering a wide range of pressures, from 1000 up to 14 000 bar.

12.1.3
Weldable Fine-Grain and High-Temperature Structural Steels

If the design of the pressure vessel requires welding, then steels with lower carbon equivalent content have to be used. The application of welding for a specific steel grade is permitted by the applicable pressure vessel regulation and requires heat treatment, further testing, and examination. Among the standard materials for welded construction are fine-grain and high-temperature structural steels that can achieve yield strength values of up to 450 MPa and tensile strength of 600–800 MPa. Because of the vast economical importance of weldable pressure vessel steels, research activities are undertaken to use even higher strength steels and advanced design methods [5] in order to minimize the weight and the cost of the pressure vessels. The permissible temperature range for these steels is approximately between -60 and $+400\,°C$, depending on the specific grade. When compared to HSLA steels, it becomes immediately apparent that because of the lower strength of these steels welding can only be used for moderate-pressure vessels of approximately up to 1500 bar. Also, fine-grain and high-temperature structural steels are not resistant to corrosive attack and must be protected from corrosive fluids. Since this steel group is weldable, weld overlay with corrosion-resistant steels is possible in addition to the protection methods mentioned above. Another commonly used method is explosion bonding of stainless material on fine-grain or other carbon steel plates.

12.1.4
High-Strength High-Alloy Steels

If corrosion resistance must be inherent to the material, then it is necessary to use alloys with more than 12% of chromium. Both weldable and nonweldable grades are available for the construction of high-pressure components. The highest strength nonweldable high-alloy steels with good general corrosion resistance are precipitation hardening and soft martensitic steels with strength levels similar to HSLA steels (yield strength over 800 MPa). These steels are widely used for high-pressure pump cylinders and exhibit excellent fatigue characteristics [6]. The permissible temperature range for these steels is approximately between −50 and +250 °C.

12.1.5
Austenitic Stainless Steels

For weldable stainless steel, there are mainly the options of austenitic and duplex steels. To a lesser extent, also ferritic stainless steels are used. Since all austenitic stainless steels have relatively low yield strengths of up to 300 MPa, they are generally not suitable for high-pressure applications above 1000 bar. Their advantage though is that they are suitable for a wide range of temperatures from −273 up to +400 °C. Chemically, they are characterized by a chromium content of above 12% and a nickel content of above 8%. Austenitic stainless steels can be enhanced in strength significantly through cold working, although this method can be applied effectively to relatively small cross sections of up to 50 mm only. When using cold worked stainless steel, the loss of strength through recrystallization even at relatively low temperatures of below 400 °C must be considered. Apart from the low strength, austenitic stainless steels have the disadvantage of being susceptible to stress corrosion cracking, which is propelled by the presence of chlorides and tensile stresses.

12.1.6
Austenitic–Ferritic Duplex Steels

Another option for weldable stainless steels of higher strength is austenitic–ferritic duplex steels that exhibit a mixed austenitic and ferritic structure. Their chemistry is based on chromium contents of typically 21–23% and nickel content of 4.5–6.5%. The yield strength of duplex steels achieves values of up to 450 MPa, which makes them suitable for moderate high-pressure vessels of up to approximately 1500 bar. Austenitic–ferritic duplex steels combine some of the advantages of ferritic steels, such as high strength and resistance to stress corrosion cracking, with advantages of austenitic steel, such as high general corrosion resistance and fairly good levels of toughness. Unlike austenitic steels, however, duplex steels are not generally suitable for low temperatures. The use at higher temperatures is limited by temper embrittlement to appr. 250 °C.

12.1.7
Chromium–Molybdenum Hydrogen-Resistant Steels

Another important group of steels for high-pressure components is low-alloy weldable steels containing up to 4% chromium and up to 1% molybdenum.

These steels are resistant against hydrogen attack at higher temperatures and pressures and also exhibit relatively high strengths even at higher temperatures (see also Ref. [7]). Therefore, they are very important for a large number of hydrogenation processes that are operated at elevated pressures of 100–300 bar [8]. The typical yield strength values vary between 300 and 500 MPa. The safe temperature range is approximately from −50 to +500 °C, depending on the specific grade.

12.1.8
Fatigue and Fracture Properties of High-Strength Steels

Apart from the static strength of a selected steel grade, its resistance to fatigue and fracture are important factors for the design of high-pressure vessels. Fatigue is not a sudden, but rather a progressive failure mode, where the steel develops cracks from starting locations like small defects that progress through the wall of the component and lead to its eventual failure. Fracture in contrast is a sudden failure of the pressure component that leads to immediate destruction and must be avoided through safe design, high-quality materials, diligent manufacturing, and careful inspection. On the one hand, the fatigue and fracture strength of a pressure component depend on the design of the component itself and the operating conditions. On the other hand, they depend on the properties of the utilized steel. In general, the fatigue strength of a steel is positively correlated to the ultimate tensile strength, while for the fracture strength the opposite applies. Factors that positively influence both the fatigue strength and the fracture toughness of a steel are

- low rates of nonmetallic inclusions such as phosphorus and sulfur;
- fine-grain structure;
- small internal defects;
- homogeneous distribution of defects.

The most important metallurgical measures to improve the fatigue strength and fracture toughness of a steel are

- use of high-quality raw materials such as clean metal scrap;
- use of modern melting furnaces with degassing capabilities;
- sufficient hot forming of the steel to break up microimpurities;
- remelting of the steel;
- precise temperature control in all phases of the steel production.

The result of such metallurgical exercises can be seen in the strength–toughness relation of the steel, in the cleanliness levels in microscopic examination, and in the absence of defects in ultrasonic examination. Since many important criteria for the fatigue and fracture strength of a steel are not fully defined in the pressure vessel codes or the steel standards, it is recommended to utilize refined steel specifications in order to obtain the highest performing steels within the limits of production technology.

12.2
Pressure Vessels

Pressure vessels that are intended for use in high-pressure applications, in particular those for pressures over 1000 bar, require specific design features that would normally not be used for lower pressures. On the other hand, high-pressure vessels may utilize any useful design from conventional low-pressure vessels as far as it can be demonstrated to be safe and economic for use under higher pressures. It is therefore not possible to draw a strict border between high-pressure and low-pressure vessels on the basis of pressure only. It is rather a combination of size, pressure, number of load cycles, and other operational requirements that promote the application of special high-pressure designs.

One clear element of distinction for high-pressure vessels is the shape of the vessel. While in low-pressure application many different shapes and large openings are mechanically feasible, this becomes increasingly difficult at higher pressure. Therefore, the predominant shape of high-pressure vessels is a cylinder, avoiding large side ports in the wall, with the end closures being used for openings and connections.

Another distinction of high-pressure vessels is the large wall thickness – usually expressed as high diameter ratio of outside to inside diameter – and the use of prestressing techniques in order to reduce the high tensional stresses at the inner side of the vessel wall. Figure 12.1 illustrates the fundamental difference between the operating stresses in a low-pressure vessel compared to a high-pressure vessel. While in a low-pressure vessel the dominating tangential stress σ is nearly uniform over the wall thickness s, in a thick wall high-pressure vessel the tangential stress is highest at the inner bore and then decays toward the outer diameter.

In order to have a first layout of a cylindrical high-pressure vessel without considering temperature effects or fatigue, the diagram in Figure 12.2 can be used to determine the maximum allowable pressure P at a given ratio of outer diameter d_o to inner diameter d_i and a yield strength of S_y.

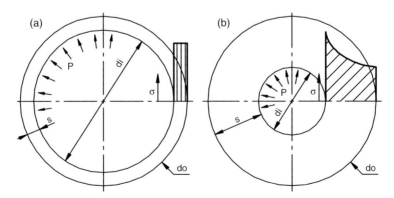

Figure 12.1 Tangential stress of (a) a low-pressure vessel and (b) a high-pressure vessel.

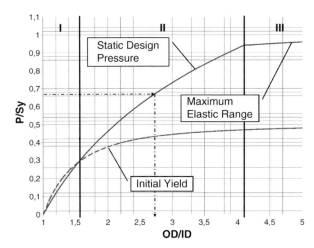

Figure 12.2 Basic high-pressure cylinder design.

The static design pressure of the cylinder is derived based on the plastic limit load of the vessel using the von Mises criterion [10] and a design margin of 1.732 [11]. It is important to acknowledge that this is only one of several possible design pressure definitions and for any specific case the applicable pressure design code must be consulted [10].

$$\frac{P}{S_y} = \frac{2}{3} \ln \frac{d_o}{d_i} \tag{12.1}$$

The initial yield of the pressure vessel occurs at the bore when the following yield condition according to Tresca/Guest is fulfilled [10]:

$$\frac{P}{S_y} = \frac{(d_o/d_i)^2 - 1}{2(d_o/d_i)^2} \tag{12.2}$$

The maximum elastic stress range that can be achieved through prestressing techniques is equivalent to two times the yield condition, since the bore of the vessel can only be prestressed to the compressive yield strength value.

The diagram can then be divided into three zones: Up to a diameter ratio of approximately 1.6, the static design pressure is below the initial yield pressure of the vessel and hence no yielding will occur in operation. Above that diameter ratio, the allowable pressure is higher than the yield pressure and prestressing techniques are required in order to avoid yielding under operating conditions. Above a diameter ratio of 4.1, the stress range at the bore exceeds two times the yield condition and even with prestressing of the inner part of the cylinder up to the compressive yield strength yielding would occur during each load cycle and quickly destroy the vessel. It is therefore not economic to increase the diameter ratio of the vessel further. Since most materials exhibit a reduced yield strength in compression once they are subjected to yield under tension, the maximum elastic range would in many cases be less than two times the yield pressure [12].

In order to illustrate the function of the diagram in Figure 12.2, the following example can be studied: A high-pressure cylinder is to be designed for 6000 bar (equivalent to 600 MPa) for a noncorrosive fluid. Which material should be chosen? What is the required diameter ratio? Does it need to be prestressed?

Answer: The pressure is too high for welded construction and a standard HSLA steel with a yield strength of 900 MPa is best suited for this application. The ratio of P/S_y is 0.67 and from the diagram a diameter ratio of 2.7 is required at the minimum. The operating pressure is above the yield condition so that prestressing techniques have to be used.

12.2.1
Leak Before Burst

When designing a high-pressure vessel, the possible failure mode has to be studied. For this purpose, the designer assumes crack initiation and propagation at different parts of the vessel where they would be likely in operation. He then analyzes the effect of this crack on the structural integrity of the vessel. If propagating cracks lead in all cases to a leakage of the vessel, then the design is considered as leak before burst (LBB). LBB has the advantage that a catastrophic, rapid failure is avoided and only the risk of the released fluid has to be controlled [13, 14]. If LBB cannot be guaranteed by design, there is a potential risk of instant failure of the vessel with a sudden disintegration of the components. This progress can be accelerated by the released compression energy and create massive damage, together with a large amount of expanded process fluid and the associated shock wave. In order to achieve LBB, the fracture situation in a vessel must be controlled in a way that a critical crack depth cannot be reached before the crack creates a leak in the vessel wall. In general, this will be easily achieved in thin wall designs than in heavy wall designs. LBB also depends on the design of the sealing system and the end closure. The higher the pressure of the vessel, the more difficult it will be to design a LBB sealing and closure system, which will be explained in more detail later.

The possible countermeasures to achieve a similar level of safety for a non-LBB vessel are increased number of in-service inspections and safety barriers against the ejection of debris. In order to better understand the amount of mechanical energy stored in a high-pressure system, it is interesting to compare it with the chemical energy of explosives (Figure 12.3).

12.2.2
Welded Pressure Vessels

In order to better understand the need for high-pressure specific designs, first the traditional welded pressure vessel is introduced with the focus on its limitations.

The use of welding for the fabrication of pressure vessels is very common, because it allows the assembly of even complex vessel structures from plate and other

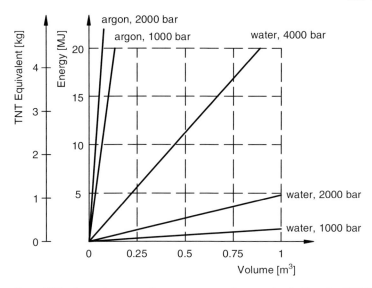

Figure 12.3 Expansion energy in a pressure vessel, pressurized with water (15 °C) or argon [15].

preformed components with a minimum amount of machining work [16]. Heavy plates for shell components are available up to 250 mm wall thickness and more. These plates can be cold rolled to form cylinders of an inner diameter of approximately 20 times the wall thickness or larger. This operation results in a diameter ratio of not more than 1.1. In a purely static design consideration and ignoring further aspects of the weld quality (Figure 12.4a), this would give an upper limit of 6.35% of the yield strength as the maximum allowable design pressure using the assumptions

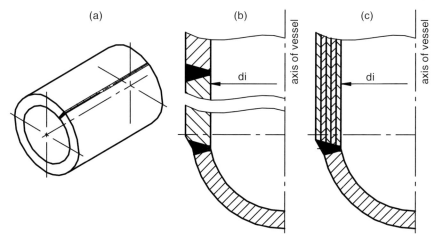

Figure 12.4 Welded pressure vessels: (a) longitudinally welded plate; (b) circumferentially welded cylindrical forgings; (c) layered plate design.

of Figure 12.2. For example, for a fine-grain steel with a yield strength of 450 MPa, it would result in a maximum internal pressure of 286 bar (28.6 MPa). If the pressure vessel is operated at higher temperatures, the allowable pressure reduces accordingly with the reduction in yield strength at elevated temperatures.

In order to raise the allowable pressure, the designer has to increase the diameter ratio. This can be done by hot rolling, by using a hollow forging, or by welding multiple plates around each other to form a larger wall thickness.

Forgings are not limited by diameter ratio. In addition, they offer the advantage of the absence of longitudinal weld, which results in a higher number of admissible load cycles for the forging piece. For larger vessels, the cylindrical parts of forged vessels are divided into several lengths and welded to each other by efficient narrow gap welding techniques. Also, the heads of forged vessels are usually welded to the cylindrical part (Figure 12.4b).

As an alternative to forged vessel, various types of layered plate vessels have been developed. The driving force of this development was mainly commercial in order to create an alternative to expensive heavy wall forgings. In one construction type, the layers consist of segments (see Figure 12.4c) of cylinders being welded with two longitudinal seams. The exact sizing of the weld gap and the plate thickness allows building up well-defined shrink forces as the pressure vessel is being assembled. In this way, the layered vessel not only replaces a thick wall forging, but also employs prestressing in order to improve the stress distribution in the vessel wall. Another advantage is that thin plates can be produced with higher strength than thick forgings, so that the wall thickness of the vessel can also be optimized by this technique.

The disadvantages and limitations of the plate layered vessels lie in reduced fatigue strength at critical welding and discontinuity locations as well as in the difficulty to inspect these locations with the available nondestructive examination methods. If a plate-wound vessel fails due to fatigue cracks or overloading, then the failure mode is relatively safe due to multiple shells and the superior mechanical properties of thin plates [17].

For corrosion protection of welded vessels, the bore of the vessel can be weld cladded with a corrosion-resistant material or fitted with a protective liner. In case of plate layered vessels, the inner core can be made from stainless steel while all other plates may be made from weldable low-alloy steel.

Heating and cooling of welded pressure vessels is frequently done by welding a jacket to the outside of the vessel ducting the tempering fluid. On some occasions, it is preferable to place the tempering jacket to the inside of the vessel in order to have a more direct control in case of temperature changes.

12.2.3
Nonwelded Pressure Vessels

If the pressure vessel can be designed without welded connections, higher strength steels as per Table 12.1 can be used and the design pressure can be elevated accordingly. Also, the fatigue life of the pressure vessel potentially improves since the weld is normally a weak area in the fatigue assessment of a vessel, unless other weak points of the design such as threaded connections exhibit higher stress concentrations.

One characteristic element of a nonwelded vessel is the type of the end closure. Here it can be differentiated between closures attached directly to the pressure vessel and those supported by an external frame. In the first case, the pressure vessel needs to transmit the axial end load. In the other case, the axial load is carried by the frame and the pressure vessel is subjected to circumferential and radial stresses only.

Using the cylindrical part of the pressure vessel itself to transmit the longitudinal load will normally be the more economic solution. The additional axial stress does not affect the pressure limit of the vessel significantly, since the maximum shear stress results from the circumferential and the radial stress [18].

The introduction of the longitudinal force into the cylinder does, however, leads to high localized stresses, which in many cases limits the fatigue life of a pressure vessel. Geometrically, the introduction of the axial force may be realized at the inside, in the middle, or at the outside of the wall at the pressure vessel end. The stress concentration is normally highest and the fatigue life is lowest, when the axial stress is introduced at the inside of the vessel (see Figure 12.5). From a fatigue point of view, it is frequently advantageous to connect the end closure at the outside of the vessel (see Figure 12.6). This does, however, increase the size and the weight. The end closure may also be connected by bolts that are introduced in the middle of the cylindrical wall (see Figure 12.7). Since the bolts are applied with a defined pretension, their cyclic stress is relatively low and will in most cases be below the endurance limit of the component.

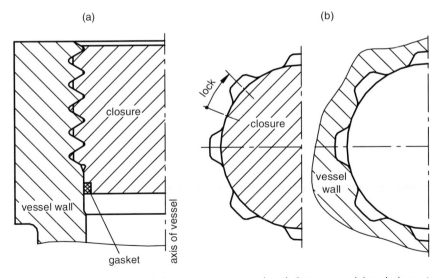

Figure 12.5 Internal thread end closure: (a) continuous thread; (b) interrupted thread: closure is inserted axial into vessel and locked by a small turn.

12 High-Pressure Components

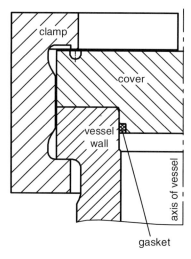

Figure 12.6 Externally clamped end closure. Before opening cover, both clamp halves must be moved to the side.

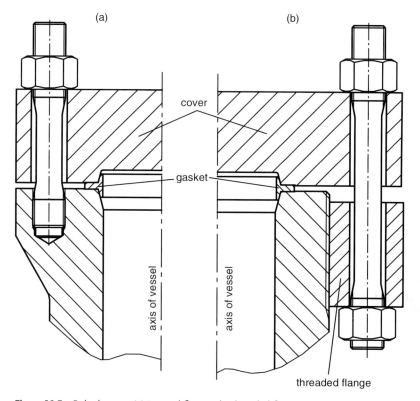

Figure 12.7 Bolted cover: (a) integral flange; (b) threaded flange.

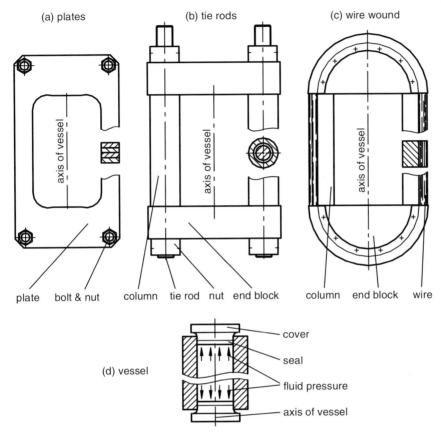

Figure 12.8 External frame-supported end closures: (a) frame design of several plates; (b) frame prestressed by tie rods; (c) frame prestressed by wire winding; (d) appropriate vessel with covers.

For large high-pressure vessels that are subjected to cyclic use, the external frame design offers advantages. The principle is that the cylindrical part of the vessel can be realized as a plain cylinder without any stress raiser. Also, the external frame (see Figure 12.8) can be designed in optimized ways with respect to fatigue life, for instance, by using prestressing techniques (see Figure 12.8b and c). The covers are introduced axially into the cylinder (see Figure 12.8d) and held by the frame. In order to open the vessel, the frame (or the vessel) must be moved to the side before cover can be extracted.

The various designs of end closures can also be differentiated by the required opening and closing times. A bolted connection (Figure 12.7) is very time consuming to open and close since each bolt must be manipulated individually and even bolt load distribution must be achieved. The clamp closure (Figure 12.6) and the interrupted thread closure (Figure 12.5b) can be manipulated very quickly and also automatized easily. The disadvantage is the lack of pretension that leads to a higher cyclic stress compared to the bolted connection and hence a reduced fatigue life.

Where quick opening is less important than fatigue life, the internal thread closure can also be used with a continuous thread (see Figure 12.5a). Also, frame-supported end closures (Figure 12.8) can be operated quickly, although the manipulation of heavy parts is required (vessel or frame).

12.2.4
Prestressing Techniques

In order to use pressure vessels above the initial yield pressure (see Figure 12.2, zones II and III) and generally in order to increase the fatigue life of high-pressure vessels, prestressing techniques are very useful. In a normal pressure vessel, all stresses are equal to zero when no pressure is applied. As the pressure increases, negative radial stress results at the inner surfaces and positive circumferential and axial stresses occur as required to balance the pressure load. In a cyclic operation, the positive circumferential stress will be the main driver of a fatigue crack initiating at the bore and propagating through the wall in the longitudinal–radial plane. If the pressure vessel is prestressed in such a way that the inner bore initially is under compressive stress, the tendency of a crack to initiate and propagate is much reduced. It is therefore a characteristic feature of a high-pressure vessel that prestressing techniques are applied.

The most common prestressing technique is "autofrettage" (Figure 12.9a). It does not require any additional mechanical component to create the prestressing condition, but rather – as the word suggests – the "frettage," which means shrinking, is carried out by its own. This is very simply achieved by pressurizing the vessel (using a hydraulic fluid) over the yield point in a controlled mode – but safely below the burst pressure – and then releasing the pressure. A plastic deformation of the vessel will occur, starting from the inside and proceeding to the outside, which will deform the vessel in an uneven way. The inside will experience a higher degree of deformation than the outside, as it expands plastically in circumference more than the outside does. In effect, the outside of the vessel will be "too small" relative to the inner geometry and will pressurize the inner part from the outside. The compressive stress that can be achieved by autofrettage would in ideal theory reach the compressive yield strength of the material, which for many materials is similar to the yield strength in tension. Unfortunately, the compressive yield strength is reduced by the previous yield in tension, a phenomenon that is known by the name of "Bauschinger effect" [12]. Another point to observe is the increase of the inner diameter that may require additional machining to match tolerances at sealing or other functional areas. Machining of these areas reduces the residual stress. Because of its cost effectiveness and simplicity, it is widely applied.

A method that is capable of achieving a compressive residual stress without the negative Bauschinger effect is shrinking of cylinders (Figure 12.9b). Two or more cylinders are manufactured with a geometric interference and then fitted by means of a temperature difference or hydraulic pressure. Another advantage is that dissimilar materials can be utilized for the different layers, such as corrosion-resistant or very hard inner layers. The disadvantage of the method is the requirement for fabricating

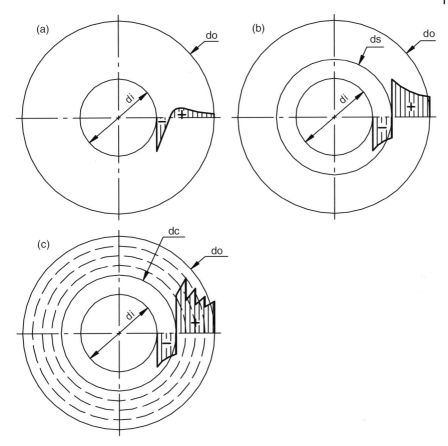

Figure 12.9 Prestressing techniques. Residual tangential stresses in vessel wall, induced by different manufacture processes: (a) autofrettage (monoblock); (b) shrink design, two layers; (c) multilayer, wire/plate winding (ds: shrink diameter; dc: core diameter).

several cylinders with precise machining in order to achieve a tight tolerance on the interference.

A third method is to wind flat wire or plates around a hollow cylindrical inner core in order to compress it sufficiently to achieve the necessary fatigue life (Figure 12.9c) [19]. The advantage of this method is that only one cylinder must be machined precisely and the wire or plate can be used as drawn. While applying the wire to the inner cylinder, the pretension can be precisely controlled by measuring the tension on the wire and the reduction in the inner diameter of the cylinder. A disadvantage is the time required to apply the wire to larger cylinders and the difficulty to achieve a 100% volumetric filling by the wire.

A similar construction type is winding of a corrugated band in a helix on the core shell to form a high-pressure vessel. The additional advantage of the corrugation is the capability to transmit longitudinal forces from one layer to another, which allows further degrees of freedom when designing the heads and end closures. Nevertheless, it has been reported that plate-wound vessels have been utilized for pressures of up to 2700 bar successfully [20].

12.2.5
Sealing Systems

Low-pressure vessels are frequently sealed with a flat seal ring that is pressurized by bolting the face of the vessel against the cover. The clamping force must be high enough to maintain the contact pressure at the sealing as the forces tend to lift the cover. For high pressures, this principle is limited by the required closing force and the deformations that occur in the gasket area. Sealing systems for high pressure therefore frequently have to be designed in a way that the pressure itself supports the function of the sealing element. This principle has been discovered early in twentieth century by Bridgeman [21]. Today, several high-pressure seal designs use the same pressure supporting mechanism in various forms.

The lens ring seal consists of a metallic ring with spherical faces that is clamped between conical seal faces (Figure 12.10). Although the seal ring requires the full pretension in order to seal, it is nevertheless supported by the pressure, as higher pressures push the lens ring stronger into the conical seats and support the sealing effect. On the other hand, the sealing design can be considered LBB by experience, as the connection starts leaking when some of the bolts start failing.

For larger sizes the required clamping forces are so high that the lens ring design becomes very heavy and sealing systems with reduced or without pretension forces offer advantages. For many applications with limited temperature range, soft materials such as elastomer O-rings can be used even for very high pressures up to 10 000 bar to effectively seal a high-pressure connection. From a safety point of view, it is preferable to use sealings on the face of the vessel (see Figure 12.11a). In case of a progressive failure of the closing elements, this system would eventually lead to a leakage and release the pressure. For the other designs (Figure 12.11b–d), LBB

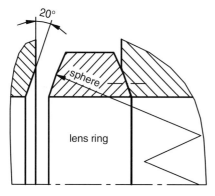

Figure 12.10 Lens ring gasket, sealing element, and adjacent parts.

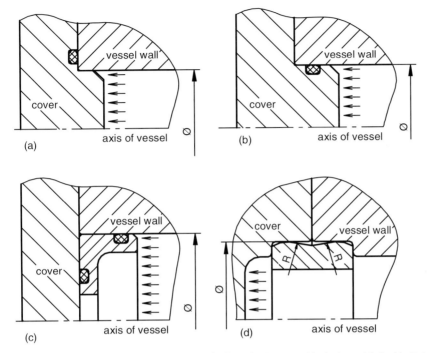

Figure 12.11 Sealing systems: (a) O-ring at the face; (b) O-ring inside the bore; (c) double O-ring carrier; (d) metallic wave ring.

can normally not be assumed for the end closure and special measures have to be taken to ensure the integrity of the vessel for a given number of load cycles.

12.3
Heat Exchangers

Heat exchangers are vessels or piping elements whose primary function is to transfer heat energy from one fluid through the wall to another. In order to best fulfill this function, the resistance to heat transfer and flow restriction should be minimal. As an additional requirement, the design should be compact in order to minimize weight and cost and allow quick temperature changes. These criteria are not easily fulfilled, when the operational pressure of the fluid is high, especially in the range of several thousand bars.

For low- and medium-pressure applications, bundle-type heat exchangers are the dominant type of construction. The process fluid flows through a multitude of small tubes in parallel, while the tempering fluid exchanges energy with the outside of the tubes. This design is very effective but requires welding of the tubes to a tube sheet and is hence limited in pressure by the selection of materials and the stresses and deflections in the tube sheet. When the pressures increase, it becomes necessary to

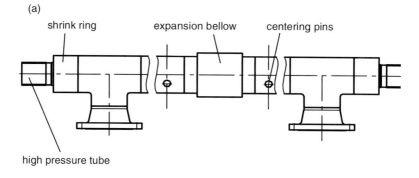

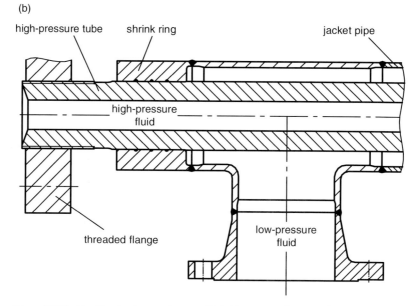

Figure 12.12 Double pipe heat exchanger, PN2500: (a) complete straight pipe; (b) details.

apply a simple design such as the cylindrical shape to the high-pressure part and to avoid welding.

A high-pressure heat exchanger is thus characterized by a tube in tube design with the high pressures acting in the inner tube (Figure 12.12). The large required wall thickness reduces the efficiency of the heat transfer and makes it very inert against sudden changes in operating conditions. Since welding on the high-pressure tube is not permitted, the outer tube is sealed by shrink elements or soft sealings against the inner tube. Another option is to use clamp-on outer half pipes, which are closed containers themselves and do not need to be sealed against the high-pressure core. For smaller applications, electric resistance heating from the outside may be an option.

When designing a high-pressure heat exchanger, the effect of thermal stress must be analyzed. For the axial direction, differential temperatures may require a flexible element in order to limit the resulting stresses and avoid leakage of the connections. For the radial direction, the temperature difference over a thick cylindrical wall causes an increase in tangential stress in the bore when the cylinder is heated and the opposite when it is cooled. Since these thermal stresses are only "secondary," which means that in case of reaching yield level they would limit themselves, they do not affect the static strength of the cylinder. Their main impact is on fatigue life of the heat exchanger since thermal stresses create additional fatigue load cycles and since they may affect the residual stresses that were intentionally introduced with prestressing techniques.

12.4
Valves

Valves have the function to manipulate the fluid flow in a high-pressure plant; they open/close, control, or safeguard the fluid flow against overpressure. High pressures affect the optimum design of valves in comparison to the multitude of low-pressure valves that are built for the various applications and operating conditions. When a valve is designed to close or control a fluid flow, it is regularly of advantage to achieve a minimum flow restriction in the open position. For low-pressure applications, this can be realized by straight valve designs with full bore openings that are closed or controlled by butterfly plates, balls, or sliders. These valve internals are limited in use by their deformation under pressure, which makes it difficult to operate the valves at higher pressures. In the case of closing valves, the deformation of the closing element under pressure will also lead to early leakage. Another way to minimize the pressure drop in a valve is to use flow optimized shapes as they can be realized economically with casting materials. Since casting materials are limited in the strength–ductility performance and homogeneity when compared with forging materials, they can normally not be used for high-pressure applications. The typical shape of a high-pressure valve is thus an angle valve made from a forged, nonweldable body material (see Figure 12.13). The control or closing element is a stem that is pressed into an orifice for closing the valve and manipulated axially inside this orifice to control the flow rate or pressure drop. The stem is sealed against the valve body by flexible packing rings made of graphite, reinforced PTFE, or other high-performance plastics [8]. In order to relief the packing from the high pressure, it may be equipped with a back seat that is engaged in the open position. The increased pressure drop over an angle valve is normally tolerated for high-pressure applications when weighed against the benefits in mechanical stability and pressure resistance. On the other hand, the angle shape always requires a change in piping orientation of 90° that frequently can be utilized in the plant layout. When the angle shape is not acceptable or when the pressure drop should be minimized, a straight valve with a 45° inclined seat can be an interesting option.

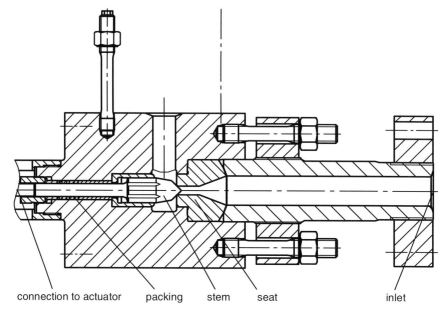

Figure 12.13 Angle control valve, DN70, PN3600.

Valves that safeguard the pressurized systems against overpressures are typically referred to as "safety valves." For low-pressure systems, these are normally spring-loaded valves, where the spring force balances the allowed pressure force on the closing element. For high pressures and larger cross sections, it is extremely difficult to design springs that safely achieve this task. It is therefore common to utilize actively controlled hydraulic valves to depressurize the system (see Figure 12.14). Since an active system always bears some risk of malfunction, this should be then combined with a passive rupture element (see Figure 12.15) as the safety element of last resort.

12.5
Piping

All pressure components in a high-pressure plant are connected by a piping system. The starting point for the design of such a system is again the question of the pressure level and the material choice reflecting all relevant process impacts. If for moderate pressures a weldable material can be used, elements such as straight pipes, bends, and tees can be connected by welding, leaving only few joints with the need of a mechanical connection. The most common type is the bolted lens ring connection (Figure 12.16). It has the advantages to be very robust and easy to install, even with some angular deviation and misalignment at the connection [23]. The disadvantage is the need of an axial pretension force of at least 30% above the pressure load in order to

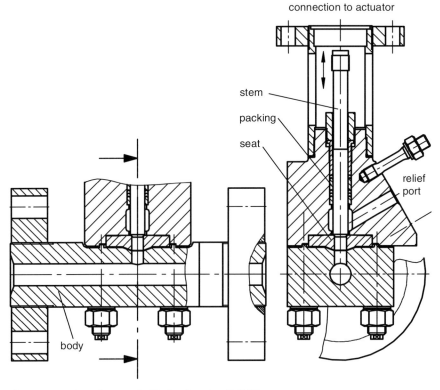

Figure 12.14 Actively controlled hydraulic valve, PN3200.

realize the sealing effect. For very high pressures this will be difficult to achieve, since the sealing element will be deformed excessively. Also, the high pretension force results in additional weight and cost, which is disadvantageous for very large connections.

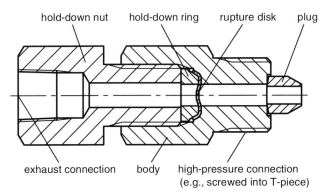

Figure 12.15 Rupture disk assembly: principle sketch for pressure up to approximately 4137 bar; available orifice: 3.15/6.35/9.53 mm [22].

12 High-Pressure Components

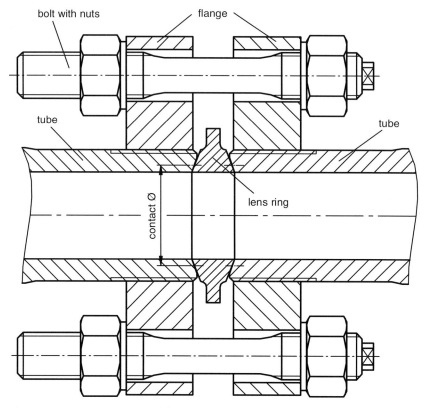

Figure 12.16 Pipe connection with lens ring gasket, DN45, PN500.

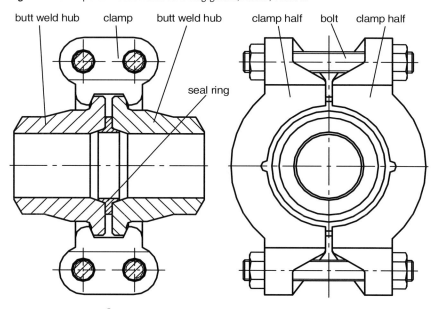

Figure 12.17 Grayloc® clamped pipe connection [24].

As a common alternative, metallic pressure energized seal systems such as the Grayloc® connection are used. Since it only requires a relatively small axial pretension to realize the initial seal effect, the size of the mechanical holding elements can be reduced and the handling of the closure can be simplified by using a clamp mechanism in lieu of a bolted connection (Figure 12.17).

For very small connections, a coupling system based on a conical seal geometry has proven to be very practical and safe even for very high pressure of over 10 000 bar (Figure 12.18). It does not require a seal element but the conical end of the tube rather forms a seal element itself. As the cone angle of the tube end is slightly steeper than the mating part in the connection fitting, it can be introduced easily and with some axial pretension it will form an initial seal at the line of contact. As the pressure increases, the tip of the tube will be widened and pressed even more into the bore of the fitting and the sealing effect is self-supporting [8].

For the fabrication of high-pressure bends, the same principles as for low-pressure bends are available; however, they must be applied with special precautions and adjustments in order to obtain safe components for high-pressure applications. The cold bending process can be used for large radius bends with

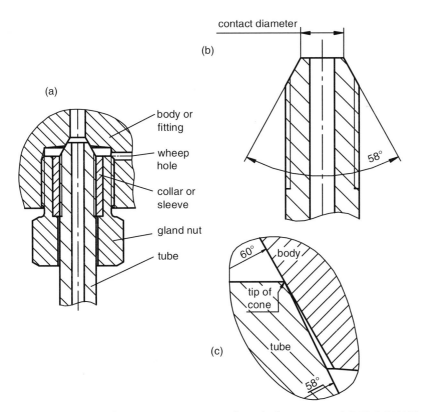

Figure 12.18 Conical pipe connection: (a) typical standard cone-type seal (DN1.6–DN4.76, PN4137); (b) details of tube end (tube outside diameter = 6.35–14.3 mm); (c) details of sealing [22].

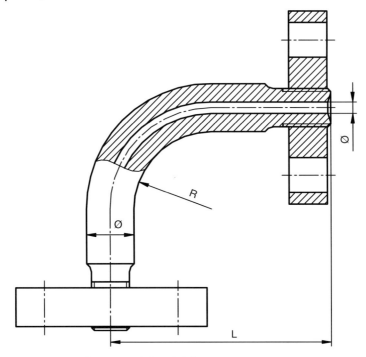

Figure 12.19 Elbow 90° with threaded flanges, DN6-DN60, PN3200 [25].

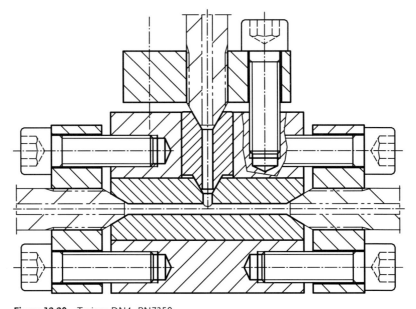

Figure 12.20 T-piece DN4, PN7350.

bending radii of 10 times the outside diameter of the tube, resulting in a maximum plastic deformation of 5% of the tube material. For high-pressure application, it is especially important that no sharp knicks are introduced that can potentially serve as fatigue crack starters. For smaller radii, various hot bending processes can be applied that require a full heat treatment cycle after bending (Figure 12.19). Also, the deformation of the thick walls of high-pressure components needs to be controlled very well in order to avoid excessive undesired deformation such as ovalization and forming of compression waves on the material surface. In case the additional pressure drop is acceptable, 90° block fittings can be used in lieu of a bend. These block fittings are in most cases the preferred type of fluid distributors like T-pieces (Figure 12.20) or headers.

References

1 NASA (2009) Composite Overwrapped Pressure Vessels, http://www.nasa.gov/centers/wstf/laboratories/composite/index.html.
2 Bargel, H.-J. and Schulze, G. (1994) *Werkstoffkunde*, VDI Verlag, Germany.
3 Verbund der Technischen Überwachungs-Vereine e.V . (2009) *AD2000 Regelwerk*, Beuth Verlag GmbH, Berlin.
4 ASME (2009) 2007 ASME BPVC, Section VIII, Division 3, Addition 2007, Addenda 2009: Alternative Rules for Construction of High Pressure Vessels, The American Society of Mechanical Engineers, New York.
5 Langenberg, P. (2003) ECOPRESS: economical and safe application of modern high strength steels for pressure vessels. Proceedings of the 3rd Dillingen Pressure Vessel Colloquium, September 24–25, 2003, Dillingen, Germany.
6 Vetter, G., Lambrecht, D., and Mischorr, G. (1990) The fatigue of thick-walled components with soft martensitic and semi-austenitic chrome–nickel steels under pulsating pressure. Abstract Handbook: 2nd International Symposium on High Pressure Chemical Engineering, September 24–26, 1990, Erlangen, Germany, pp. 499–506.
7 API (2008) *Steels for Hydrogen Service at Elevated Temperatures and Pressures in Petroleum Refineries and Petrochemical Plants*, API RP 94, 7th edn, American Petroleum Institute, Washington, DC.
8 Bertucco, A. and Vetter, G. (eds) (2001) *Industrial Chemistry Library. Volume 9. High Pressure Process Technology: Fundamentals and Applications*, Elsevier, Amsterdam.
9 Verlag Stahlschluessel Wegst GmbH (2007) Stahlschlüssel 2007, Version 5.00.0019, Verlag Stahlschluessel Wegst GmbH, Marbach, Germany.
10 Buchter, H.H. (1967) *Apparate und Armaturen der chemischen Hochdrucktechnik*, Springer, Berlin.
11 Hiller, W., Koerner, P., Wink, R., Keltjens, J., and Cornelissen, P. (2005) Square root 3: background to the new design margin for high pressure vessels. ASME 2005 Pressure Vessels and Piping Conference (PVP2005), July 17–21, 2005, Denver, CO, pp. 55–59.
12 Parker, A.P. (2001) Autofrettage of open-end tubes: pressure, stresses, and code comparisons. *J. Press. Vessel Technol.*, **123**, 271–281.
13 Koerner, P., Hiller, W., and Wink, R., (2002) Leak-before-burst testing of a high pressure tube. ASME 2002 Pressure Vessels and Piping Conference (PVP2002), August 5–9, 2002, Vancouver, BC, Canada, pp. 47–50.
14 Koerner, P. and Hiller, W. (2000) The new trend in the LDPE industry towards large

14. tubular reactors – and how the manufacturer responds. ASME 2000 Pressure Vessels and Piping Conference (PVP2000), July 23–27, 2000, Seattle, WA, pp. 33–36.
15. Radomski, M. and Ros, Z. (1992) Design of HP vessels used in HIP and CIP technologies, in *High Pressure and Biotechnology*, vol. 224 (eds C. Balny, R. Hayashi, K. Heremans, and P. Masson), Colloque INSERM/John Libbey Eurotext Ltd., pp. 541–543.
16. Witschakowski, W. (1974) *Hochdrucktechnik, Schriftreihe des Unternehmensarchievs der BASF Aktiengesellschaft*, Ludwigshafen, Germany.
17. Fryer, D.M. and Harvey, J.F. (eds) (1998) *High Pressure Vessels*, Chapman & Hall, New York.
18. Mraz, G.J. and Kendall, D.P. (eds) (2000) Criteria of the ASME Boiler and Pressure Vessel Code, Section VIII, Division 3: Alternative Rules for Construction of High Pressure Vessels, The American Society of Mechanical Engineers, New York.
19. Tschiersch, R. (1984) *Design Criteria for Layered Vessels*, Thyssen Henrichshütte AG, Hattingen, Germany.
20. Karl, E. (1973) Strip-wound pressure vessels, in *Safety in High Pressure Polyethylene Plants* (prepared by Editors of Chemical Engineering Progress), American Institute of Chemical Engineers, New York, pp. 26–30.
21. Bridgeman, P.W. (1931) *The Physics of High Pressure*, G. Bell and Sons, London.
22. Autoclave Engineers Fluid Components Product Catalog (2011), Autoclave Engineers, Fluid Components Division of Snap-tite, Inc., Erie, PA.
23. Wink, R. (2008) HP-flange connections in LDPE-service. Proceedings of the 3rd Global LDPE Maintenance Workshop, June 9–11, 2008, Hagen, Germany.
24. Grayloc Products LLC (2002) Grayloc® Product Catalog, Grayloc Products LLC, Houston, TX.
25. Uhde High Pressure Technologies GmbH (1989) Catalogue: High-Pressure Valves and Fittings, Pressure Rating 3200, Uhde High Pressure Technologies GmbH, Hagen, Germany.

Further Reading

1. Buchter, H.H. (1979) *Industrial Sealing Technology*, John Wiley & Sons, Inc., New York.
2. Comings, E.W. (1956) *High Pressure Technology*, McGraw-Hill Book Company, Inc., New York.
3. Gleich, D. and Weyl, R. (eds) (2006) *Apparateelemente: Praxis der sicheren Auslegung*, Springer, Berlin.
4. Kecke, H.J. and Kleinschmidt, P. (eds) (1994) *Industrie-Rohrleitungsarmaturen*, VDI Verlag, Germany.
5. Spain, I.L. and Paauwe, J. (eds) (1977) *High Pressure Technology. Volume I. Equipment Design, Materials, and Properties*, Marcel Dekker, Inc., New York.

13
High-Pressure Pumps and Compressors

Eberhard Schluecker

13.1
Selection of Machinery

High-pressure pumps and compressors can be grouped into machines used directly for the generation of high pressures and/or circulation machines in high-pressure processes. The reasons for the respective application in this case – unlike in usual process engineering – are mostly based on the efficiency of the machines at the required pressure levels. Only efficient machinery is suited for high pressure and usually shows pressure-stiff conveying behavior, which can be deduced from the typical throttle curves (Figure 13.1).

Using the example of pumps and, with regard to quality, also compressors, Figure 13.1 shows the typical throttle curves for the three main groups of machinery (turbomachines as well as reciprocating and rotating positive displacement pumps).

The turbomachines (centrifugal pumps and turbocompressors) take the energy to generate pressure from the centrifugal force that is created inside the machine ($F_z = m\omega^2 r$). The bigger the impeller diameter, the higher the drive, and the denser the fluid, the higher the possible pressure will be. Therefore, the following approximation applies:

$$\Delta p = \psi u^2 \varrho / 2 \tag{13.1}$$

In a further approximation, $\psi \sim 1.7$ can be used. This value is typical of radial acting turbomachines, if optimally used and designed.

If such a machine is installed into a system with a throttle valve at the discharge side and this valve is being closed slowly (reducing the flow) while the machine is running, then the flow as well as the achievable pressure will change: starting from the maximum possible flow and a negligible discharge pressure (right), the flow on the left side of the diagram is throttled to zero and the discharge pressure reaches its peak. The entire energy of the impeller is being converted into pressure. Such a working condition is acceptable as long as the supplied energy does not lead to undue heating. The energetically optimal operating zone of the characteristic curve,

Industrial High Pressure Applications: Processes, Equipment and Safety, First Edition. Edited by Rudolf Eggers.
© 2012 Wiley-VCH Verlag GmbH & Co. KGaA. Published 2012 by Wiley-VCH Verlag GmbH & Co. KGaA.

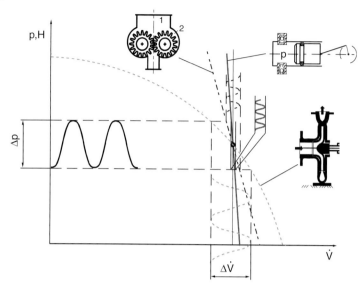

Figure 13.1 Qualitative comparison of characteristic curves for reciprocating and rotating positive displacement pumps and turbomachines and their effects with varying discharge pressures. H: head ($H = p/\varrho g$); $\dot V$: flow (*source*: Institute of Process Machinery and Systems Engineering, University of Erlangen-Nuremberg, Germany).

however, lies at the beginning of the last third of the right part of the curve (crossing point in Figure 13.1) with a usable flow range of about ±15%.

In case such a machine is worked with varying pressure, it is easy to imagine that this will result in a varying flow ($\Delta \dot V$) as well: the machine reacts in a pressure-soft way. For this reason, a low range of flow variation required for dosing purposes cannot be achieved with pumps like these.

The maximum pressures that can be reached by radial acting centrifugal pumps with water (not axial types) are about 2 MPa per impeller stage (i.e., a pump for 150 m³/h, $D_{2,A} = 150$ mm, $n = 50$ Hz, and $p_{max} = {\sim}0.28$ MPa or a pump with $D_{2,B} = 300$ mm and $p_{max} = 1.1$ MPa) depending on the outer diameter (D_2) of the impeller.

If gas is conveyed with the same machines instead, the curve is diminished by the proportion of the fluid densities (i.e., air: 1.22 kg/m³). The resulting pressure therefore would be about 1/1000 of the pressure generated with water. The only chance to increase the discharge pressure therefore is to increase the speed of the impeller or/and the number of stages up to 24 and more. If impellers are placed in series, the achievable pressure will increase at constant levels. Nevertheless, the reachable pressure with such machines in a common size cannot be called high pressure (exception: very big machines). Positive displacement machines, in contrast, suck the flow into the so-called operating chambers and then separate this fluid content from the suction side of the pump by valves or tight fits, open up the pressure side, and displace the fluid toward the pressure in this area. This is also called a volumetric pumping behavior. If such machines are throttled (discharge pressure increased), then the discharge abates much less compared to the turbomachines and as a consequence a pressure variation leads to distinctly lower flow variations. The group of reciprocating displacement pumps has the highest pressure-stiff characteristic because the volume sucked in can only be reduced by leakages through the fluid-controlled valves and the piston seal. If the valves and seals are absolutely leakproof, then the conveyed mass flow is perfectly steady and independent of pressure. In rotating positive displacement machines, however, there are usually gaps between the displacement rotor and the housing, which are supposed to seal off from pressure. Seals like these are never absolutely leakproof! Therefore, these machines react in a more pressure-soft way than the reciprocating types and then again respond to varying pressures with bigger flow variations. Hence, the highest maximum pressure is reached by reciprocating positive displacement machines, the second highest maximum pressure is reached by rotating displacement machines, and the lowest maximum pressure is reached by turbomachines. That is why reciprocating displacement pumps are the typical dosing machines, while reciprocating compressors (piston compressors), due to the additional thermal effects, are not very precise in dosing applications.

Turbomachines in a multistage design are primarily machines for circulating fluids in high-pressure systems with maximum differential pressures of about 20 MPa (exception: very big machines), whereas rotating positive displacement machines can manage pressures of up to 3 MPa with gas (screw compressor) and about 40 MPa with liquid (gear pump with oil) and reciprocating positive displacement machines are suitable for the generation of high and even highest pressures up to 2000 MPa with liquid as well as gas.

13.2
Influence of the Fluid on Selection and Design of the Machinery

Turbomachines can only convey fluids of low viscosity. Fluids of high viscosity consume too much energy from the impeller, so the machine's degree of efficiency distinctly gets lower with increasing viscosity. Positive displacement machines, in contrast, are suitable for a wide range of viscosities. Reciprocating pumps and also a few types of rotating positive displacement pumps are even able to convey fluids of about 2 Pa s viscosity or more without any problem. This, however, no longer works without design modifications. The machines either have to be equipped with charging pumps or spacious inlet channels or valves or have to be supplied with high suction pressures.

The higher the compressibility, the bigger the temperature rise usually is in the fluid when the pressure has increased. Whereas this warming can be neglected in fluids (compressibility of 0.05% (water) up to 0.15% (oils) per MPa), in liquid gases (e.g., supercritical CO_2, compressibility up to 0.7%/MPa) a significant heating can already be noticed. This also leads to heating of the walls in the working chamber so that after a short time of operation the sucked-in fluid hits the heated walls of the working area and gasifies there. This distinctly lowers the flow in the machine and the machine may no longer be able to reach the required pressure due to the high compressibility of the gas. Therefore, cooling measures applied to the suction side (supercooling) and/or the working chamber are unavoidable.

$$T_d = T_s(p_d/p_s)^{(n-1)/n} \quad \text{or} \quad T_d = T_s(p_d/p_s)^{(\kappa-1)/\kappa} \tag{13.2}$$

where p is the pressure, T is the temperature, n is the polytropic exponent, κ is the adiabatic exponent, "d" is the pressure side, and "s" is the suction side.

When pure gas is conveyed, its compressibility is even considerably higher and the heating from suction temperature T_s to conveying temperature T_d can turn out to be so dramatic (Eq. (13.2)) that there is imminent danger of lubrication oil fire. In this case, the pressure increase cannot be executed in one step but has to be achieved in two or more steps with intermediate cooling. Therefore, many high-pressure compressors, often even low-pressure and medium-pressure compressors (turbo or positive displacement types), are designed with multistep and intermediate cooling. If, in addition, aggressive chemicals have to be conveyed, an assured and qualified selection of materials is essential.

13.3
Design Standards for High-Pressure Machines

As long as no noticeable pressure variation is taking effect, the housings of high-pressure machines may be compared to high-pressure vessels (see also Chapter 12). The same regulations apply and the technique of autofrettage and shrinked joints is used here as well. It is a different case, however, if the pressure is created by reciprocating positive displacement machines. In this case, all parts contacting the fluid are exposed to varying or pulsating pressures. These pressures mean a risk of fatigue to components that are dimensioned too weakly, especially those with notches, bores, and cross-section steps. Basically, there are admissible tension values that are valid for dynamically worn components, which could be confirmed by fatigue life tests. Notches even weaken this admissible value by the factor of 2–3. If, in addition, aggressive chemicals are employed, then those components without an absolutely secure durability will be further weakened during operating time. The correct material selection therefore is one of the major requirements for safe machine technology.

From the above facts, it may therefore be concluded that notches and other irregularities in the cross-section areas as well as small and smallest dents in the surface (roughnesses) should be avoided. This is why polished surfaces and cross sections as smooth as possible are a basic requirement for construction as well. The easiest way to realize this requirement is to keep the space that is under pressure as small as possible ("small is beautiful") and to lead the pressure-generated forces to the energy consumption point not circuitously but as directly as possible. Furthermore, shrinked designs as well as the use of autofrettage decisively increase fatigue life (see Chapter 12).

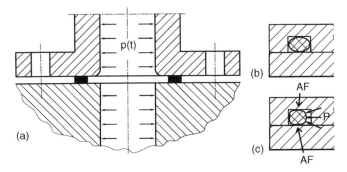

Figure 13.2 Sealing design elements in pulsating high-pressure applications: (a) seal between two parts with different radial stiffness; (b) O-ring unloaded; (c) O-ring pressure loaded; AF: area of friction (*source*: Institute of Process Machinery and Systems Engineering, University of Erlangen-Nuremberg, Germany).

The above statements on design are also valid for all other pressure-loaded assemblies. Figure 13.2 shows some typical design elements. If, for example, a flange is screwed on a massive part of pump housing (below) with a sealing washer in-between (Figure 13.2a), it is absolutely correct to have the same bore diameter for both components. It has also to be kept in mind, however, that both parts will expand in the radial direction differently under internal pressure. The flange side is distinctly less stiff. The seal sitting between the two components has to counterbalance this different expansion under pressure. This will work out only by strong deformation or by shearing. Therefore, the seal will probably show wear and start leaking after only a few pressure load cycles. If such geometry is used for a pulsating pump, the leaking is inevitable. Even if it is made of metal, the seal will wear very quickly. A flange-like geometry on the casing part or a considerably thicker flange could be used as corrective action. Therefore, all sealing parts have to be braced to components that have about the same radial expansion under pressure when dynamic loads are applied.

The well-known option of a lens ring seal has therefore either bigger outer dimensions than the tube or a protrusion of the ring (Figure 12.16, left). Thus, these seals have almost the same radial expansion as the tubes around them. It is also essential to put symmetrical load on both sides of the lens seals because otherwise there is a danger of tilting and wearing as a consequence.

A stationary seal made of plastic of course also suffers from the same phenomenon. If the seal material is elastic (i.e., elastomers), varying loads mean elastic deformations. And if such deformations lead to sliding effects on the seal faces (AF), it is to be assumed that also in this case a noticeable wear of the seals will occur bringing up the risk of leakages. This problem becomes obvious with the example of O-rings (Figure 13.2b). Following the O-ring manufacturers' classical standards of assembly, the O-ring groove is mostly rectangular with a cross section considerably bigger than the O-ring diameter. With static low-pressure conditions, this design works without any problem. With high pressures, however, the O-ring is – depending on the pressure – pressed more or less into the edges (2c) of the groove and is offset to

13.4
Materials and Materials Testing

As long as constant pressures act upon the components, the materials as per Table 12.1 are also suitable for high-pressure engineering and the pressures used there. With pulsating pressures, however, the fatigue properties of the materials have to be considered in addition. Literature provides reliable results on this issue that are always based on fatigue tests [1, 2]. Values such as these are usually determined by extensive testing procedures on up to 30 samples and load cycles up to 10^7 or even more. Figure 13.3a shows such a result with a duplex steel (1.4462) exposed to acid-free oil, artificial seawater, 10% nitric acid, and 10% sodium bicarbonate. As can be seen from the figure, each chemical has a noticeable influence on the fatigue strength. This influence suggests that it is a corrosive attack that works during the whole operating time of the machine and that will further weaken the fatigue strength over time. It therefore does not suffice to know the tested fatigue value, but it is also necessary to know the chemical resistance to dynamic stress. Unfortunately, such figures are scarcely obtained because component testing with chemicals as a hydraulic fluid is quite sophisticated. Some business companies have therefore

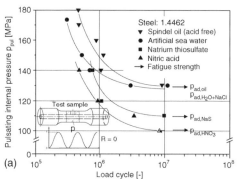
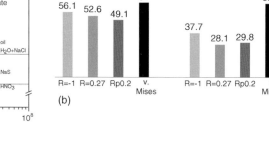

Figure 13.3 Selection of materials for high-pressure machines: (a) Woehler curves for a duplex steel with different test media. (b) von Mises equivalent stress in notched fatigue samples and other specific values: $R = S_b/S_m$ (S_b: minimum of amplitude; S_m: maximum of amplitude). $R = -1$: rotating bending; $R = -0.2$: pulsation tensile with preload; v. Mises (equivalent stress) at $R = 0$ within a cross-bored sample (pulsating internal pressure) (source: Institute of Process Machinery and Systems Engineering, University of Erlangen-Nuremberg, Germany).

worked out admissible tension values based on experiences. Furthermore, tests are done with load cycles of $>10^7$. These investigations have shown that even with very high cycles to failure fractures still occur. The reason is that also in the strength values defined by classic testing procedures tiny cracks can develop that grow very slowly and can lead to fractures even after a very long time [3]. A reliable corrective action therefore is to add an additional safety factor of 1.5 to the fatigue limit value.

As high-pressure pumps for higher and highest pressures are piston pumps, it is necessary to consider the construction of such machines as well when it comes to dynamic testing. Because piston pumps always have channels to the valves (see below), also notched samples (with cross bore) are tested in order to register the effect caused by the notch.

But which material is the ideal one for pulsating pressure stress? Of course, the following rule applies: the higher the pressure, the greater the stability. Nevertheless, the materials have to remain ductile at any pressure. A rupture elongation of $>10\%$ therefore has to be observed implicitly. Thinking of the notches, however, it can easily be imagined that a softer material is able to react on too much tension with plastic strain, and thus reduce the notch factor.

Using the example of materials 1.4462 (duplex) and 1.4405 (austenite), Figure 13.3b shows that the austenite in spite of its much lower yield stress value (Rp0.2: 29.8) has a much bigger rupture elongation and shows almost the same von Mises equivalent fatigue stress in a cross-bored high-pressure sample from duplex steel 1.4462 with a yield stress almost twice as high. This is due to the local autofrettage on the notch by the testing pressure [1, 2]. That is why also the term automatic autofrettage is often used. Selecting the proper material therefore is not an easy task. As a guidance, it is recommended to add a safety factor of 1.5 to the fatigue strength out of the Woehler curves and to implement a rupture elongation of at least 15%.

13.5
High-Pressure Centrifugal Pumps and High-Pressure Turbocompressors

A pressure increase of 2 MPa ($\varrho = 1$ kg/l) per impeller is the attainable maximum for the usual range of applications. If higher pressures are to be realized, this will only be possible by increasing the number of stages or the rotation speed. Increasing the diameter will according to the rules of similarity, $P \approx n^3 D_2^5$, $H \approx n^2 D_2^2$, and $\dot{V} \approx n D_2^3$, cause an increase in the flow in the third power and therefore is only scarcely applied.

Numbers of stages with up to 10 impellers and more in a row are actually realized and these can often be found in high-pressure pumps for chemical applications. The multistage design often leads to quite good energetic efficiencies. Indeed, centrifugal pumps are often used as circulating pumps in high-pressure processes due to their relatively low maximum pressures. Figure 13.4 shows such a circulating pump for a maximum system pressure of 30 MPa and a possible pressure difference of about 2 MPa. The machine is designed in the can and barrel design as usually employed in high-pressure technology. It has four impeller stages (3) and is driven by a canned motor (6/7). The construction is all over suitable for the use in high-pressure

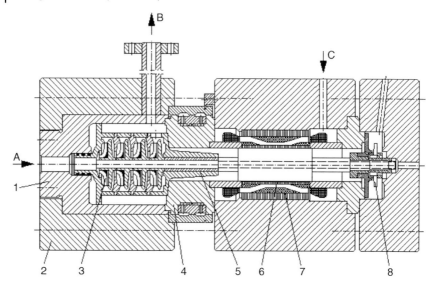

Figure 13.4 High-pressure centrifugal pump for a differential pressure of about 2 MPa and a system pressure of about 30 MPa. A: inlet; B: outlet; C: pressure loading channel for the canned motor (source: Institute of Process Machinery and Systems Engineering, University of Erlangen-Nuremberg, Germany).

engineering. So the engine as well as the pump is surrounded by a thick-walled housing (the barrel), the impellers are placed in a pressure-balanced inner housing, and the bearing can be leak clearance optimized by the surrounding pressure of the cone (5). Moreover, the pump housing (1) is made in shrinked design.

Another way to increase the pressure is to boost the number of revolutions up to 14 000 min^{-1}. It thus enables one impeller to deliver up to 14 MPa. The driving speed for such machines is mostly being reached through transmission gearings (Figure 13.5).

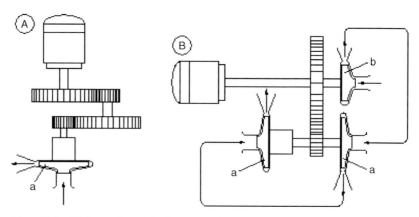

Figure 13.5 High-speed centrifugal pumps with transmission gearing. A: one-stage pump; B: three-stage pump (source: Institute of Process Machinery and Systems Engineering, University of Erlangen-Nuremberg, Germany).

It has to be mentioned, however, that the impellers in such machines are comparatively small and the hydraulic flow behavior in the impellers cannot be adjusted optimally. That is why these machines scarcely exceed an efficiency of 40% and often reach an efficiency far less than this. Above all, there is a high sensibility toward particles in the fluid, which with high speeds can turn into bullets.

Turbocompressors for higher pressures are realized exclusively as multistage designs with radial as well as axial impellers due to the low centrifugal force (low gas density). Speed levels up to 24000 min^{-1} are quite common. Pressures generated with radial turbomachines come up to <10 MPa for medium-sized machines and reach up to almost 100 MPa (up to 250 000 m^3/h) with very big machines. These sizes also come as isothermal versions with interim cooling between the impeller stages. If distinctly smaller flows are to be achieved, the pressure goes down to very small values. High pressure in general can therefore only be achieved with big machines. The flows in high pressures are usually small. Hence, turbomachines play a role in high-pressure use only as circulating machines. There are only a few exceptions (e.g., in long-distance transport of gas).

13.6
Rotating Positive Displacement Machines

The group of rotating positive displacement pumps comprises a lot of different versions. When setting a borderline for high-pressure usability at 10 MPa, only the gear pumps (70 MPa), the screw pumps (32 MPa), and the progressing cavity pumps (15 MPa) remain. The basic conveying characteristics of these pumps can be described by Eq. (13.3):

$$\dot{V} = V_U n \eta_V \quad \text{or} \quad \dot{m} = V_U \varrho n \eta_V \quad \text{with} \quad \eta_V = \dot{V}/\dot{V}_{th} \tag{13.3}$$

where V_U is the theoretic flow per pump revolution, n is the revolution speed, and η_V is the volumetric discharge rate.

As high-pressure compressors are based on the rotating positive displacement principle, preferably screw compressors are used.

13.6.1
Discharge Rate

Positive displacement machines are confined by their displacement geometries to given, exactly definable displacement volumes per revolution (V_U) or stroke (V_S). These theoretical volumes, however, will never really be reached due to inevitable inner leakages or other influences (i.e., cavitation, or with compressors the thermal effects). But the higher the volumetric efficiency of a machine, the better it is suitable for high pressure or the better the quality of the machine. *Remark*: The volumetric efficiency is a part of the total energetic efficiency of a machine.

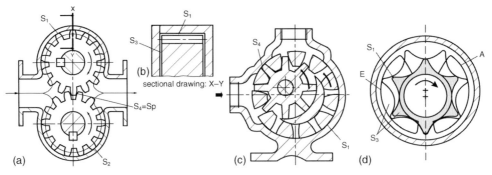

Figure 13.6 Gear pump versions: (a) external gear pump; (b) clearance situation; (c) internal gear pump; (d) trochoidal pump; E: inlet; A: outlet channel; S: clearance channels (*source*: Institute of Process Machinery and Systems Engineering, University of Erlangen-Nuremberg, Germany).

13.6.2
Gear Pumps

Gear pumps exist as externally toothed and internally toothed versions (70 and 10 MPa, respectively; Figure 13.6a and c), as well as trochoidal pumps (17 MPa, Figure 13.6d). The highest pressures are achieved by the externally toothed version. The reason for this is the favorable power transmission. However, the operating principle for gear pumps usually involves big clearance cross sections (S_{1-3}), which cause high inner leakage flow rates. As leakage flows in pumps usually show a laminar characteristic, the Hagen – Poiseuille equation is in good approximation valid for the circular ring gap (Eq. (13.4)):

$$\Delta \dot{V}_L = \frac{\pi d s^3 \Delta p}{12 L \eta} \tag{13.4}$$

where d is the mean circular ring gap diameter, L is the gap length, η is the dynamic viscosity, Δp is the pressure difference to be sealed, and s is the gap height.

The leakage current for oil with 25 mPa s and $p=$ constant therefore comes up to only 1/25 of the leakage current for water. Therefore, gear pumps are preferably used for fluids with higher viscosity. For efficient conveying of fluids with lower viscosities, the gap widths have to be reduced ($f(s^3)$). As a consequence, higher pressures are possible as well. It also has to be observed, however, that pressures cause a widening (strain) of the pump housings and thus always undo in part the reduction of the gap width. One manufacturer therefore carried the gap reduction to the extent that the gears jam within the housing without pressure. Only when a certain static pressure is applied, the gears can move without resistance. Due to this measure, 70 MPa could be reached.

Yet there is a limit to the reduction of viscosity. As the gears work in driving mesh, the tooth flanks roll off over each other. During this process, micromovements occur so that a certain lubrication effect on the fluid cannot be avoided. For this reason, the fluids used in gear pumps should always have a certain lubricating effect and because of the friction of the tooth flanks they should possibly be free of particles.

Nonetheless, gear pumps for fluids with a high content of particles (e.g., adhesives) do exist. In this case, hard materials such as hard metals are used for the gears as well as for the housings. For gases, however, a transferable operating principle does not exist because of the inevitable leak clearance.

13.6.3
Screw Pumps

Screw pumps suck the fluid into a conveying chamber on the suction side and then convey it over the entire length of the screw rotors to the discharge side (Figure 13.7).

There are driving versions and versions synchronized by gearboxes installed outside the discharge chamber, the screws of which do not touch each other. Of course, there is a clearance between the screws and the housing. But here the screw diameter is – with the same flow rate as in a gear pump – distinctly smaller than the gear wheel diameter and there is no end face clearance (Figure 13.6, S_1). As it is possible to manufacture the screws and housings with higher precision, the widening under pressure is reduced and the cross section of the leakage clearance is notably smaller. All these issues lead to a significant reduction of the leak flow. Higher pressures can be achieved and low-viscosity fluids can be conveyed against pressure. The only negative effect is that pressure is built up between the screws by the spindles rolling toward each other, so the screws are bent apart. In order to avoid the screws grinding on the housing wall, this deflection has to be taken into account when it comes to the gap width. This means that certain gap widths are necessary for

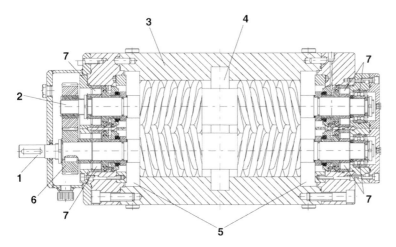

Figure 13.7 Multiphase screw pump for 25 MPa synchronized with external gearbox (2). 1: Driving shaft; 3: housing; 4: discharge channel; 5: suction channels; 6: external gear; 7: bearings (*source*: Institute of Process Machinery and Systems Engineering, University of Erlangen-Nuremberg, Germany).

the machine to work reliably. Considering all these aspects, the machine is even capable of conveying low-viscosity fluids against high pressure, but with high-viscosity fluids it will not reach more than 40 MPa, because of an even higher deflection. This kind of machine does not depend much on the lubrication effect of the fluids as the synchronized version does not need it and, on the other hand, the driving version can more easily build up fluid cushions in the driving area. This is due to the interaction of the screws, which is typical of screw pumps, and there are several meshing points along the screw at the same time. As a result of this, the force necessary for moving the screws is distributed over the whole length of the tube. Of course, this pump version, too, is sensible to particles in the fluid. Yet screw pumps have become common in the oil industry for conveying crude oil against 25 MPa even if there is a good possibility that it contains abrasive sand. However, acceptable lifetimes in such applications can only be reached by plating the screws with hard materials.

There are designs with two to five screws. The two- and three-screw versions are preferred for high-pressure use because of the favorable leak clearances.

The screw pump's counterpart for gas conveyance is the two-spindle screw compressor. Whereas the screw pumps present absolutely symmetric screws and the conveying chambers are steady over the entire length of the screws, the screws of screw compressors are completely unsymmetrical (Figure 13.8a) and in addition to that define an internal compression ratio due to their geometry. With regard to the low viscosity of the gas, the deflection of the screws is smaller. That makes this compressor design the most preferable of the rotating positive displacement machines and it reaches – with cooling – pressures of more than 1.5 MPa per machine and up to 5 MPa (big machines).

Remark: Without "installed" compression rate and in case of an almost leakproof conveying chamber, even shortly before the opening of the discharge side the pressure inside the chamber would be almost the same as the suction pressure.

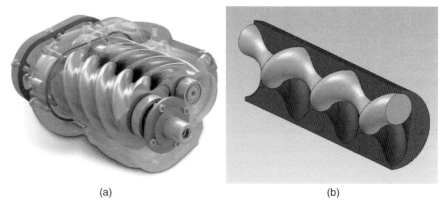

(a) (b)

Figure 13.8 Screw machines: (a) screw compressor (*source*: Kaeser Company, Germany); (b) progressing cavity pump with two complete stages (*source*: Institute of Process Machinery and Systems Engineering, University of Erlangen-Nuremberg, Germany).

If the chamber on the discharge side was opened, the gas would stream back into the conveying chamber and would all at once compress the gas at discharge pressure level. This would frequently involve loud noise but also certain harmful thermal effects.

13.6.4
Progressing Cavity Pump

This kind of machine is part of the group of screw pumps but has only one screw that rotates eccentrically inside a purpose-built stator. This shows that the screw's pitch is twice that of the stator. Therefore, one fluid chamber (stage) is formed with each revolution of the stator and in a way the rotor screws it along the stator toward the pressure side. In doing so, the rotor moves in an eccentric circle (Figure 13.8b).

The seal against pressure is possible by a gap, and the higher the pressure, the narrower the gap. Because of the comparably sophisticated component designs, however, narrow gaps are difficult to manufacture. And additionally such tight fits cause a slightly higher sensibility to wear. For this reason, elastomer stators are preferably used that are slightly oversized and thus bring about a compressed sliding seal effect. The pressures to be reached by this design range from 0.4 to 0.6 MPa per step. As especially with this type of pump, very long screws with many steps are easily feasible well up to 15 MPa (24 stages). In principle, the author, however, does see a chance to reach even higher pressures by using high-precision and nonelastic stators.

The typical range of application for such machines in high-pressure use is found in process engineering as well as in secondary oil recovery. The tube-like construction is ideal for lowering it into deep drilling holes for pumping crude oil from there up to the Earth's surface. This, however, means multiphase conveyance with abrasive contents, which the pump can handle almost without any problem because stator elastomers react elastically to particles imported between rotor and stator and thus keep the wear within bounds. Basically, it is possible to convey gas in higher pressure ranges with nonelastic stators, but this takes an utmost precision in production as a basis and nevertheless requires additional lubrication in order to avoid wear. But also the fluids pumped with elastic stators (elastomers) need to have a certain lubricity. At the same time, they are thermally sensitive. Therefore, the elastomers can easily be damaged by compression heating of the gases and lack of lubrication. This is why progressive cavity pumps have not been used as compressors so far.

13.7
Reciprocating Positive Displacement Machines

Reciprocating positive displacement machines exist in versions such as piston pumps, membrane pumps, piston compressors, and membrane compressors. These machines achieve the highest pressures and the pumps are additionally able to reach a high conveying precision. The main reason for these features is the leakproof discharge chambers (operating areas of the positive displacement machines).

Especially for high-pressure technology, however, certain additional constructional aspects have to be observed. Stroke adjustable as well as only speed adjustable pump versions are available.

13.7.1
Drive Technology for Reciprocating Positive Displacement Machines

Each technology, regardless of whether mechanic, electric, or hydraulic, which is able to generate reciprocating movements, is basically suited for reciprocating positive displacement machines. However, the generation of high pressures requires operating principles that can create large forces. Therefore, most of the machines are equipped with a crankshaft drive (Figure 13.9a), spring-cam drive (c), or magnetic drive, and only a few are equipped with a hydraulic drive (b).

Hydraulic linear drives, impelled by hydraulic aggregates, are basically suited for generating high and even very high forces. But due to shifting procedures and ensuing fluid dynamics (strokes) and fluid effects (gas cavitation in the beginning of the suction process (see below)), they are to a large extent limited in the frequency of shifting procedures per time unit and also in the feasible stroke frequency (flow = $f(n)$). As a consequence, these machines are only rarely used and if so preferably with very high pressures (up to 2200 MPa, extreme value). Magnet drives instead can generate only small forces and are therefore preferred for low-pressure applications. With the spring-cam drive, on the other hand, the force of the piston onto the cam is localized in one point or on a small space and creates high tensions there. The bigger the machine (higher force), the higher the stress and it soon reaches inadmissible extents or requires elaborate designs. Therefore, this machine technology is a typical and economical solution for small dosing pumps (up to 40 MPa) with the advantage of providing for an utmost dosing accuracy. As the piston is always

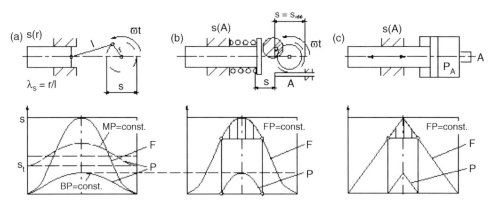

Figure 13.9 Drive technologies for reciprocating positive displacement machines: (a) crankshaft with stroke adjustment with the eccentric radius $s(r)$; (b) spring-cam drive with stroke adjustment (A); (c) linear drive; $s(A)$: individual stroke length; UT: back end point; MT: midpoint piston stroke; VT: front end point (source: Institute of Process Machinery and Systems Engineering, University of Erlangen-Nuremberg, Germany).

pressed onto the cam without clearance, the piston stroke remains almost unaffected by any clearance on the driving side (e.g., slide bearing tolerance).

Nearly every type of drive unit can be equipped with or without stroke adjustment.

13.7.2
Flow Behavior of Reciprocating Positive Displacement Machines

The reciprocating movement inside the common machines describes an exact (spring cam) or approximately exact sinusoidal curve (crankshaft). Linear drives, on the other hand, may also deviate from the linear movement due to certain standard measures. When the piston moves forward, it performs the pressure stroke, and when it moves backward, it performs the suction stroke. If the fluid is conveyed, it is at first sucked into the operating chamber at suction pressure level, compressed between 1 and 2 (Figure 13.10, right), and at the same time the pump components are strained by the arising pressure and afterward between 2 and 3 discharged toward the discharging side. During the suction stroke, the fluid that is still present in the operating chamber and the pump components is released from 3 to 4 and after the suction pressure (4–1) has been reached fresh fluid will be sucked in. Having completed the suction stroke (1), the whole process as described starts all over again. When the process – as just described in chronological sequence – is presented as function $p(V_S)$, a cycle diagram is created that among experts is called indicator diagram. Figure 13.10 shows such indicator diagrams for a piston pump (right) and a piston compressor configuration.

The gradient of the compression curve (dp/dV) depends on the total fluid volume and its compressibility, which is contained in the operating chamber. Compared to

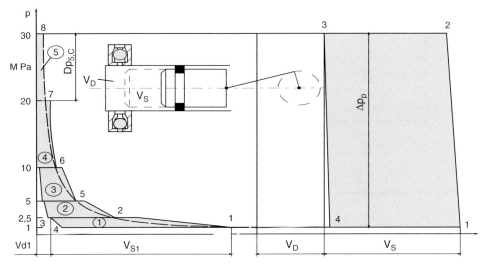

Figure 13.10 Indicator diagrams of a piston pump and a piston compressor configuration for a realistic application toward 30 MPa; V_D: detrimental space; V_S: stroke volume (*source*: Institute of Process Machinery and Systems Engineering, University of Erlangen-Nuremberg, Germany).

that, the decompression curve, unless it proceeds absolutely vertical, is bound to relax the balance volume in the operating chamber, which is not located in those areas of the operating chamber that are reached by the pistons.

The effective suction stroke is reduced by the stroke ratio, which is necessary for alleviation of tension. Among experts this remaining volume is therefore called detrimental space as it brings loss to the machine's efficiency. This is, however, not a total loss of energy. It is just the amount of friction energy used for the "unproductive part" of the stroke that is to be judged as a loss of energy. But the positive displacement volume of the machine has to be bigger than it would be without the detrimental space.

As can be seen in Figure 13.10, the pump is able to reach the 30 MPa pressure in one step, whereas the compressor needs five steps in total. The reason for this is compression heating, which was limited to 200 °C within the compressor due to the risk of oil ignition. As the compressor worked at 500 strokes/min, there was no effective possibility for cooling during the compression procedure. The compression therefore proceeds from adiabatic to polytropic (Eq. (13.2)). Additional friction effects generate additional heat. In order to provide the same thermal entry criteria for the next compression step as in the previous step, the gas is submitted to an interim cooling. If compression entry of the first step is connected to the end points of the interim cooling stages, this curve is equivalent to an isothermal compression (dashed line). The pressure ratio on the compressor outlet and inlet is called pressure stage ratio π_K. If several steps are needed, it is intended to get the same end temperatures for each step. This is reached as soon as unchanging step pressure ratios are realized. Therewith it is possible to calculate from a total pressure ratio (π, Eq. (13.5)) with a known number of steps K the single pressure step ratios and from that the step pressures.

$$\pi_K = \sqrt[K]{\pi} \tag{13.5}$$

The higher the pressure, the bigger the part of the stroke movement that is used up for the compression. Assuming a sinusoidal movement of the piston, this compression as part of the piston stroke is not used for the conveying procedure (Figure 13.11). The consequence is shown in the figure of the piston speed $v(t)$. The fluid speed in the tube on the pressure side between 1 and 1a equals zero. In 1a, the pressure valve opens and the liquid column inside the tube on the discharge side is rapidly accelerated. This signifies a water hammer, approximately according to Eq. (13.6) (Joukowsky shock), and the longer the connected tube (L), the higher the density, and the bigger the speed jump Δv from 1a to 1b, the bigger the water hammer.

$$\Delta p_J = \varrho L \Delta v \tag{13.6}$$

It is therefore obvious that reciprocating high-pressure pumps always show a pulsating conveying behavior due to the sinusoidal conveying movement with interruption during the suction stroke, which is additionally interfered by water hammers. It is important, however, that the amplitude of the water hammer is not a function of the system pressure. The higher the system pressure, the less dangerous the water hammers are for the affected components. However, this discontinuous

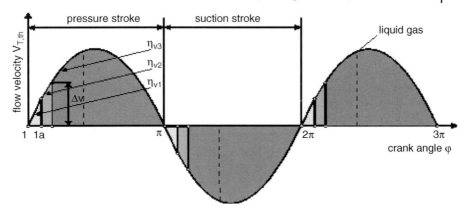

Figure 13.11 Course of the flow speed in the pipeline on the discharge side with different fluid compressibilities with related phase steps. Δv: speed jumps; η_{vi}: volumetric efficiency, $i = 1 = 100\%$ (complete sinus curve) (source: Institute of Process Machinery and Systems Engineering, University of Erlangen-Nuremberg, Germany).

behavior always has to be taken into account with sophisticated conveying processes (dosing, sensitive measuring instruments, etc.) and has possibly even to be dampened by special methods (see below).

Often, especially with big flows, multiple pumps connected in parallel (N machine units) are used to reduce the pulsation. The kinematics of each machine thereby act with a phase shift of $360°/N$ to the neighboring machine. But nevertheless a remaining pulsation will occur.

13.7.3
Pulsation Damping

Fluid conveyance that pulsates and is afflicted by pressure surges is able to create disturbances beyond the above-mentioned flow variations, to induce vibrations in the system or even to cause damages in the pump or the compressor by repercussions from the system. For all these reasons, the dampening of those effects is required. This can be accomplished by absorption dampers (for liquids) or resonators (for liquids and gases) (Figure 13.12). Whereas the former are able to reduce low-frequency pressure and flow variations, resonators are able to dampen fluid acoustic vibrations as, for example, caused by Joukowsky shocks. For the well functioning of absorptions, storage of a downstream pressure loss $\Delta p(v)$ is necessary. If the flow speed v and in consequence Δp is high, a part of the flow is led into the dampener where it compresses the gas. If Δp and v decrease, this volume is led back into the piping. The rule of thumb for construction is $V_{Gas} = 8-10 V_S$. Usually, liquid resonators have a distinctly bigger volume (100–200 V_S) and therefore also have absorbent effects. Moreover, their shape causes a reflection of the acoustic vibrations with the development of interferences right up to its extinction. Gases have a similar effect, but the resonator is distinctly smaller in size due to the gas compressibility.

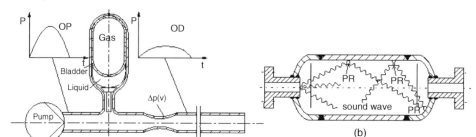

Figure 13.12 Absorption damper (a) and resonator (b). OP: output pump; OD: output dampener; PR: points of wave reflection (*source*: Institute of Process Machinery and Systems Engineering, University of Erlangen-Nuremberg, Germany).

With very high pressures, even the flexibility of the piping can have a dampening effect. The best results can be reached here, if the piping comes through several winding passages into a coil. At the pressure peak, the coil is slightly bent up and thus enlarges its volume content.

13.7.4
Design Versions

The constructive rules for high-pressure components and especially for those under dynamic stress are as follows: as few notches as possible (cross bores, steps, and edges), smooth surface, ductile materials, and pressure-loaded chambers as small as possible. The figures as described in the following show examples of high-pressure machines that were put into practice.

13.7.4.1 Vertical Pump Head for 70 MPa

Figure 13.13a shows a pump head for 70 MPa, which usually is run in vertical direction. Both pressure valve and suction valve are arranged in the piston direction on top of each other directly above the piston. All pressure-loaded edges are rounded. The suction valve (8, valve closing body) is designed as an annular disk valve. This means that the fluid is sucked in through the channel system in component 6 from its surrounding chamber. Then the fluid is conveyed concentrically through component 6 toward and through the discharge valve (5, valve closing body) to the ring channel around component 3 and then left into the pressure line. The complete pump head internals are clamped over component 1 and screws 9 and sealed with static, very well trapped plastomer seals (e.g., 13 and 14). The piston sealing is realized as a stuffing box with about 10 fabric rings and support rings of bronze (because of good emergency running properties). The detrimental space reaches from the top edge of the piston to the discharge valve and is definitely significant. In most cases, this design is realized as triplex monoblock technology. This means that three of such designs are placed next to each other in block part 7. The drives operate with a 120 phase angle offset and thus accomplish a relatively steady conveyance.

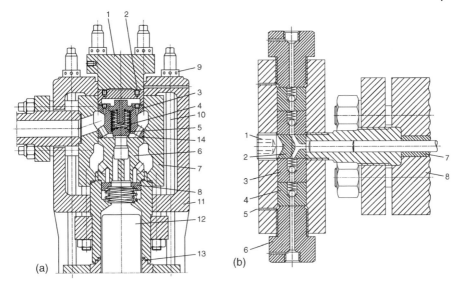

Figure 13.13 High-pressure piston pump heads: (a) vertical monoblock pump head (according to Hammelmann); (b) piston pump head with Y-piece (*source*: Institute of Process Machinery and Systems Engineering, University of Erlangen-Nuremberg, Germany).

13.7.4.2 Horizontal Pump Head with Y-Piece for 300 MPa

Figure 13.13b also shows a design where the valve parts (2, 3) are braced as a pile with the same diameter over screw connections (6) on the discharge side and on the suction side. The major difference, however, is that there have been invariably used metallic seals and a central piece (Y-piece, 2) serves as bifurcation of the bore. This component bears a high fatigue risk due to the tensions around the bore interface. Therefore, the bores are burnished and the material has to be absolutely flawless and ductile. Furthermore, with such pressures ball valves are used without exception and the piston seals are usually customized. As presented in the example, a gap seal is implemented. Other options are up to about 50 MPa stuffing packings, and above that individually manufactured plastomer non-ferrous metal combinations.

13.7.4.3 Diaphragm Pump Heads

If aggressive or toxic materials have to be conveyed, hermetically sealed constructions are necessary. Diaphragm pumps meet this requirement because the never absolutely leakproof piston seal (dynamic or grinding seals are never absolutely leakproof) has been replaced by the diaphragm's static seals at the diaphragm clamp.

Figure 13.14a displays a hydraulic diaphragm pump with a plastomer diaphragm (high deformation capacity) for 35 MPa and Figure 13.14b shows a version with a metal diaphragm (low deformation capacity) for 100 MPa. Between piston and diaphragm there is a hydraulic fluid, which serves as transmitter of the piston movement via the diaphragm on the fluid. Whereas Figure 13.14a uses two piston rings as piston seal, in Figure 13.14b a very narrow gap seal (approximately 10 μm gap

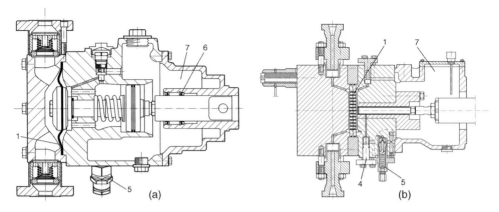

Figure 13.14 Diaphragm pump heads: (a) with PTFE diaphragm for 40 MPa; (b) with metal diaphragm for 100 MPa. 1: Diaphragms; 4: replenishing valve; 5: pressure relief valve; 7: oil reservoir (*source*: LEWA Company, Germany).

width) is realized. Both pumps are equipped with replenishing valves, which enable – as they are vacuum operated – the replacement of the oil volumes that were lost through the piston seal. In addition, pressure relief valves are installed, which in case of overpressure discharge hydraulic oil into the reservoir. It is important for the conveying behavior that the hydraulic oil volume has to be regarded as detrimental space, which distinctly increases the phase control compared to piston pumps and reduces the volumetric efficiency. Besides the component breathing is bigger as well due to the bigger components (cf. piston diameter to pump head diameter).

The diaphragm clamp in Figure 13.14a is geared, but its performance is limited to maximum pressures of 40 MPa. Beyond that there are designs with elastomer diaphragms up to 80 MPa for special applications. For higher pressures, that is, up to 300 MPa, metal diaphragms are used without exception. The constructional design follows the guidelines mentioned above.

13.7.4.4 Piston Compressor for 30 MPa at the Maximum

The only differences to the piston pumps are the valve positions and their constructional layout. On pumps the suction valve is always at the bottom and the pressure valve is always at the top, because gases in the working chamber can be discharged much easily in an upward direction. With compressors, though, such a requirement does not exist. The position of the valves can therefore be chosen freely in order to create as little detrimental space as possible. Furthermore, gas as the material to be transported is definitely less viscous and therefore causes less pressure loss at the cross-section steps. This allows for the use of valves with multichannel plates (Figure 13.15). As a result, the closing bodies and in consequence the closing energies become noticeably smaller, clearing the way for higher stroke frequencies.

Because of the bad lubricity of gases, piston compressors have to be either supplied with lubricating oil or fit for dry running. As the extraction of the oil out of the gas stream is essential after the compressor step, oil traps are arranged subsequently.

13.7 Reciprocating Positive Displacement Machines | 331

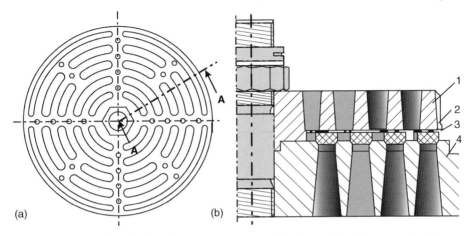

Figure 13.15 Multichannel valves for piston compressor: (a) valve on sight; (b) cross-sectional drawing. 1: Upper part; 2: spring; 3: closing parts; 4: valve seat (*source*: Institute of Process Machinery and Systems Engineering, University of Erlangen-Nuremberg, Germany).

Unfortunately, the separation is not entirely successful. An oil residue dissolved in the gas always remains within the gas flow. But this is not acceptable for many processes. Therefore, the demand for oil-free processes is increasing. This, however, means dry running. Dry runners in less critical applications work with plastic piston rings in absolutely varying shapes (Figure 13.16). If, in addition, there are conveyed

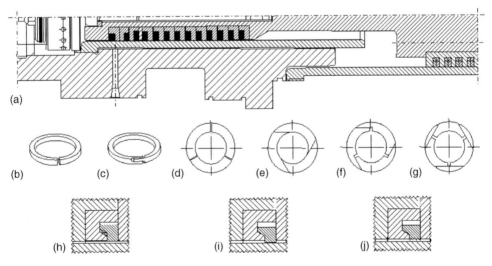

Figure 13.16 Compression piston with piston ring seals for a compression stage up to 30 MPa and different versions of piston rings.
(a) Compression stage in half-section;
(b) normal split PR – much leakage, leakage increases with wear; (b) PR with cover – less leakage, increases with wear; (d–g) PR with compensation for wear (*source*: Institute of Process Machinery and Systems Engineering, University of Erlangen-Nuremberg, Germany).

reactive gases such as oxygen, this involves prohibiting the use of any inflammable material. A labyrinth seal remains the only possible solution in this case.

For critical gases, there are also diaphragm compressors. These are – like the pumps – hydraulically driven and with regard to the compression heating they are equipped with metal diaphragms without an exception. A special feature of this technology is that the leakage loss through the piston seal toward the end of the pressure stroke is being replenished by a little high-pressure pump. The discharge volume is designed in such a manner that the diaphragm is pressed against the front wall free of voids leaving just the channels to the valves as detrimental spaces. With the help of this method, compression ratios of up to $\pi_K = 20$ can be realized.

13.7.4.5 Compressor for 300 MPa

If gas is compressed, its density approaches more and more the one of liquids. From a density of about $700\,\text{kg/m}^3$, it is called liquid gas. So the difference between a compressor and a pump becomes increasingly blurred. Figure 13.17 shows a compressor for carbon dioxide and 300 MPa. The housing is accomplished as a shrinked design and equipped with cooling channels within the compression area. The valves are put together as cartridges, engineered as ball valves and metallic sealing. The detrimental space reduction by component 5 is remarkable as it ends exactly before the top dead center. The piston seal is made of steel support rings 7 and 8 and the high pressure seals 9. These components consist of a filled PTFE, FEP, or

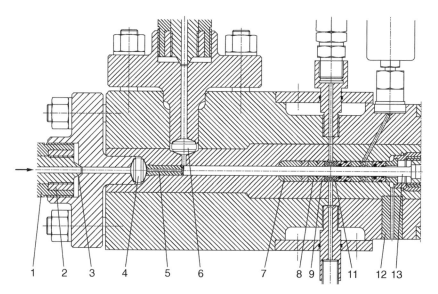

Figure 13.17 Piston compressor for 300 MPa for carbon dioxide (according to Hofer) (*source*: Institute of Process Machinery and Systems Engineering, University of Erlangen-Nuremberg, Germany).

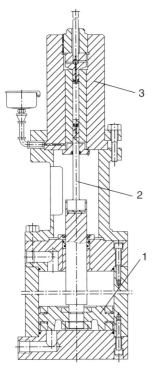

Figure 13.18 Compressor for nitrogen up to 1400 MPa with a hydraulic linear drive (according to Harwood) (*source*: Institute of Process Machinery and Systems Engineering, University of Erlangen-Nuremberg, Germany).

ECTFE with good sliding properties. This section is followed by a lantern ring that is used for purging and therefore, in a way, for lubrication and cooling of the piston. Next to this is placed a series of low-compression seals. The piston is hardened and characteristically lean for the accomplished pressure level in order to keep the driving forces low.

13.7.4.6 Piston Compressor for 1400 MPa

Figure 13.18 shows a compressor for 1400 MPa for nitrogen. The pressure-loaded area consists only of strait channels and as far as possible no notches. The detrimental space is close to zero and the seals are based on lens geometries from high-strength steel. The head (3) is shrinked and the piston (2) is driven by a hydraulic cylinder (1) acting as a pressure converter. At such a pressure, the difference in density between gas and liquids is close to negligible and also the compressibility is very small. But the lubrication ability of gas is still bad. Due to that, the piston seal, which is traveling with the piston, is the most critical part. It consists of only one special shaped ring made out of some wear-resistant polymer.

References

1 Schlücker, E., Ellermeier, J., and Depmeier, L. (2002) Fatigue behaviour of high pressure parts from different steel types in different chemicals. 4th International Symposium on High Pressure Technology and Chemical Engineering, Venice, Italy.

2 Vetter, G. and Plappert, S. (2001) Improve durability and safety of thick-walled components of reciprocating fluid machines by autofrettage. World Congress Chemical Engineering, Melbourne.

3 Mughrabi, H. (2006) Specific features and mechanisms of fatigue in the ultrahigh-cycle regime. *Int. J. Fatigue*, **28** (11), 1501–1508.

14
High-Pressure Measuring Devices and Test Equipment
Arne Pietsch

14.1
Introduction

The need to know system parameters arose from the beginning of technical high-pressure processing and naturally system pressure and temperature were the first to be observed. Pressure manometers were in use not with the very first steam engines but invented shortly and at the end of the nineteenth century complex engineering handbooks did not even see the necessity to explain them in detail anymore [1]. Their principle has not changed since then: the mechanical displacement of Bourdon[1] tubes, diaphragms, or bellows is transferred to an indicating pointer. Siphons filled with oil were also used for early reaction vessels, keeping process media from spoiling the manometer [2]. Scientists applied a surprising variety of methods to accurately determine pressures a century ago – details on Bourdon gauges, mercury columns, free piston (pressure balance) and manganin resistance gauges, Petaval gauges for rapid pressure changes, and early systems using the piezoelectric effect are well explained by Newitt [3] in 1940.

Today, sensors with electric signals and computing dominate the industrial application. Common sensor types are offered by more than one manufacturer and advice on application is readily available (Figure 14.1). This chapter will not focus on the functionality inside sensors but rather give an overview on available principles and their benefits as well as critical aspects of usage to keep in mind. Most information will be given in tabular style only for easy access and overview; "high pressure" is defined as a pressure above 10 MPa (100 bar, 98.6 atm, or 1450 psi). The following abbreviations are used in the text: "p" for pressure, "T" for temperature, "HP" for high pressure, and "MAWP" for the maximum allowable working pressure.

---

1) Eugène Bourdon was a watchmaker and engineer who in 1849 invented the Bourdon gauge, a pressure measuring instrument still in use today. It could measure pressures up to 100,000 psi" – quoted from Wikipedia.com.

Industrial High Pressure Applications: Processes, Equipment and Safety, First Edition. Edited by Rudolf Eggers.
© 2012 Wiley-VCH Verlag GmbH & Co. KGaA. Published 2012 by Wiley-VCH Verlag GmbH & Co. KGaA.

Figure 14.1 Instrumentation of high-pressure vessel (MAWP 32.5 MPa).

Listed ratings are typical ones; extremer ones may be available, though. Legal aspects have to be considered according to the local laws and are not part of this chapter. Technology is advancing rapidly, so please do not expect an omniscient monograph; the author welcomes comments and suggestions from readers for future use.

14.2
Process Data Measuring – Online

Online measuring of pressure and temperature is of crucial importance for high-pressure processes in order to avoid bursting of components due to exceeding material tensile strength. While the change from mechanical p or T indicators to electric systems has taken place in most applications, the online detection of additional parameters is not widely applied yet, except for flow measurements; most others are still restricted to very specific needs.

14.2.1
Sensor Choice and Installation

To get started, a basic checklist can help to take relevant aspects into account when choosing a sensor type (Table 14.1). Besides the main function, other aspects should be considered in process plants at an early stage, latest when potential suppliers are contacted.

Self-evident but sometimes ignored is the fact that the maximum pressure for a complete plant or system is governed by the component with the lowest admissible operating pressure. Therefore, sensor choice needs prior specification of the MAWP of the complete plant. Important are also conformity declarations according to requirements regarding design approval (boiler codes), explosion-proof technology, and foreign bodies protection/ingress of water protection.

Table 14.1 Checklist – general aspects to consider about sensor choice for high-pressure applications.

- Is the sensitive part of the sensor really where you want to measure?
- How fast will the sensor react in the application? What speed is necessary?
- Is the connection a high-pressure design and leakproof? Is this certified by the supplier? Advisable is to use a standardized connection/seal type that is generally used in the plant
- Ensure that the MAWP is not lower than the rating of the plant/system
- Can the sensor be blocked/retarded or fail by particles/dirt in the system? What would be the consequence? Need measures to be taken?
- Are there special requirements about hygienic needs or a cleaning performance? Design for CIP processes (other fluids with other corrosion potency) or easy disassembling needed?
- Is it necessary to repair or maintain the sensor without depressurization?
- Check if there are requirements like design approval [2],[3],[4]
- Needed electric safety – protection against touching and water contact, for example, check IP rating [5],[6]. For example, IP69K stands for "6" dust tight and "9" protection against high-pressure and steam cleaning
- Is there an explosion risk? For example, check EU ATEX classification [7]
- Does the sensor type match electrically the process and data bus standards (programmable logic controller)?
- Cost (cost of planning, commissioning, maintenance, and disposal)

---

2) Pressure Equipment Directive (PED) 97/23/EC.
3) ASME Boiler and Pressure Vessel Code – 2007 edition, American Society of Mechanical Engineers, New York, USA.
4) Japanese regulatory codes, for example, High pressure gas safety law and Gas industry law.
5) International protection classes according to DIN EN 60529/IEC 529/VDE 047 T1.
6) International protection classes according to IEC 60529.
7) ATEX Directive 94/9/EC and ATEX Directive 1999/92/EC.

Table 14.2 Overview on industrial high-pressure sensors.

Sensor type	P_{max}	T_{max}	Other parameters	Typical use
Bourdon gauge (mechanical)	700 MPa	60 °C (brass) 300 °C (ss)	Robust	Vessel pressure Pump pressure
Pressure transducer	1400 MPa	120 °C; higher possible		Process pressure controlling
Calibration only				
Comparison pumps	250 MPa and higher	Atmospheric	Compare two manometers	Lab and calibration only
Dead-weight tester	250 MPa and higher	Atmospheric	Very accurate, 0.05% f.s.	Lab and calibration only

14.2.2
Pressure and Differential Pressure

Pressure measurements are essential for high-pressure processing. Although often neglected in high-pressure business, it is advisable to distinguish between reference levels *atmospheric* or *gauge* (units *psig* and *barg*) and *zero* (units *psia* and *bara*). The unit *psid* relates to a differential pressure reading within a process.

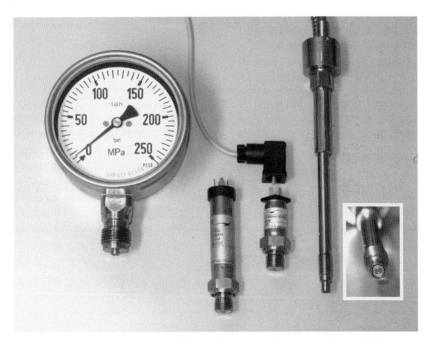

Figure 14.2 Pressure sensors. From left to right: gauge, 250 MPa; transducer, 60 MPa; transducer, 40 MPa; transducer with diaphragm seal on tip, 50 MPa/315 °C; and details of its diaphragm.

Classical pressure sensors are the purely mechanical pressure gauges (Table 14.2). Reasons for their wide use are robustness, low price, simple installation, and independence from electric power supplies. In the so-called Bourdon gauge, a dead-end tube in helical form with several turns expands due to elastic deformation when pressure is increased. This movement is transferred mechanically to a mostly circular scale. Although these instruments can be overloaded, it is common practice to operate them at no more than three-quarters of their full scale. These gauges are manufactured by a large number of companies and can be obtained in various qualities; Table 14.3 gives advice on what to consider while choosing.

Electric pressure sensors allow signal transformation to other places, for example, a control room and its electronic/PC controlling structures in unlimited complexity. Therefore, these sensors are much more useful and adaptable than mechanical gauges, especially when applied in control systems and data acquisition. Important are a good linearity of the sensor output, a high degree of temperature independence, and robustness. Already in 1960, electric sensors for very high pressures were available: manganin resistance coils for up to 3000 MPa bar and the Harwood bulk modulus cell for up to 1400 MPa [4]. Today, the predominant principle is usage of a micromachined silicon diaphragm with a piezoresistive strain gauge diffused into it and a supporting silicon or glass backplate. Standard maximum pressure rating is around 700 MPa, but specialty sensors are available. Pressure transducers are those sensors that are built into a housing with process connection. If, in addition, full signal conditioning is integrated into the housing of the sensor, it is a pressure transmitter. For further insight into the sensor technologies, see Refs. [5, 6].

For high-pressure applications, keep in mind that lifetime of the instrument can be restricted by a maximum number of load cycles. This number depends on the course of the pressure (height of changes, rise time, and pressure drop time) in the realized application. Close to the maximum number of load cycles, leakages due to hairline cracks may occur and cause a safety risk.

A widely used combination in high-pressure technology is the mechatronic pressure measuring instrument (Figure 14.2). This is a mechanical pressure gauge fitted with electrical components or contacts. Two types are available: pressure gauges with electric output signals and gauges with switch contacts. Advantageous is the ongoing indication during electric power outages.

Differential pressure sensors are special pressure sensors with two chambers separated by a membrane. The pressure difference between those two chambers is detected. This setup allows determination of pressure differences significantly smaller than either one of the applied pressures with high accuracy. Fields of application are the pressure losses over certain unit operations such as separation columns or the use of differential pressure to monitor two other process parameters: flow and level. The latter applications are discussed in the subsequent chapter. In high-pressure technology, two aspects have to be observed closely. First, the maximum allowable differential pressure of the transducer can easily be exceeded due to high system pressures, thus destroying or offsetting the sensor. Second, fluid

---

8) Directive DIN EN 837-1, 1997.

Table 14.3 Options when choosing pressure sensors.

- Instrument style and size
- MAWP, measuring accuracy, and certificates!
- Process connection – avoid sequences of high-pressure adapters by thorough order
- Throttle screw to dampen pulsations – note the safety risk: easy blocking with particles and dirt
- Corrosion resistance of body and medium wetted parts; also observe maximum temperature
- Consider high viscosity and abrasive properties of process fluids if given
- Oil-free calibration at manufacturer site when used for oil-free applications, for example, oxygen
- Safety glass and a blowout back are self-evident in all high-pressure applications
- Overrange protector in case excessive overpressurization can occur. These devices protect the gauge for very high overloads, especially exceeding a typical value of 130% MAWP [8]. Note the risk of blocking and malfunctioning with dirt
- Diaphragm seals. For highly corrosive fluids, an additional diaphragm seal in connection line can be used. This device separates the process media from the delicate membrane of the sensor by means of a pressure-transmitting fluid and a more robust membrane contacting the process fluid. Typically, the whole unit is delivered filled and calibrated. The liquid should be chosen so that there is no risk with process or product in case of leakage. Mercury can still be found today but should be avoided
- Diaphragm seals at senor tip. Also called "flush diaphragm." In case dead volume between sensor and process needs to be extremely small, a diaphragm seal at sensor tip can be used. The sensor then does not have any internal cavities that are to be filled by the process fluid by using a membrane directly at the tip. This variation is less sensitive but easy cleanable. It is used for viscous or easily solidifying fluids, for example, in polymer processing or processes that need precision cleaning. In cases of higher temperatures, the thermal expansion of the inclusion liquid needs to be taken into account; some models allow sensor adjustment when at process conditions
- Pressure gauge cocks/valves with purge ports are rarely used in high-pressure applications for safety reasons
- Siphons even in trumpet form will not avoid entry of liquid media into the gauge in case of large pressure alterations; therefore, they are used for steam applications only
- Number of load cycles can be limited; check with manufacturer for your individual application

Gauge only

- Instrument mounting style, for example, classical case or panel mounting
- Measuring accuracy, scale accuracy, dial units (parallel up to three possible)
- "Red mark" for process MAWP if needed, for example, for pressure vessels
- Liquid filling (glycerin or silicone oil) to dampen pulsations, for example, for piston pumps
- Additional electric signaling – for example, to set an maximum alarm – is often available (mechatronic)
- Customer logo is often available

Transmitters only

- Case version and size, for example, field case for outdoors installation
- Output signal – choose according to specification control system
- Choose if seals at the sensor membrane are acceptable or welded versions are needed. Elastomers need to withstand the fluids

---

temperature in the usually colder connection tubes leading to the sensor can interfere with the measurement due to fluid density changes. Hydrostatic pressure of condensed liquids or just denser fluid phases can grossly falsify the measurement. Diaphragm seals at the process line and connection tubing to the sensor filled with a hydraulic liquid can help to avoid these effects.

14.2.3
Temperature

Common temperature sensors types are thermocouples (types J and K) and resistance temperature detectors (RTDs) and can often be used for high-pressure processes as long as three critical aspects are looked after: first the pressure resistance of the sensor sheath itself, second the sealing, and third the correct interaction with the pressurized fluids. Helpful to meet the first two requirements is a cylindrical or rod-type shape of the sensor that is generally the case anyway. Apparently, this is also the reason why self-made solutions with standard temperature sensors can be found in high-pressure applications quite often. This is in no way recommendable since the user is fully liable for malfunctions or even accidents; the hazard risk potential of a sensor shooting from high-pressure equipment should not be underestimated. Regarding pressure resistance of the sensor sheath, it is advantageous that sensor diameter is small. More critical is design of the sensor tips; welded versions are preferable. But as stated above, suppliers should certify material quality and yield strength of the sensors. If – in very special cases – a conformity declaration is not obtainable, the user has to qualify the sensor himself according to safe engineering practice by design calculations based on material certificates and pressure testing. It should be noted that used small-sized tubing for standard sensor manufacturing is not in all cases seamless and high-pressure proof. If the sensor needs to be bent, this has to be taken into account too.

The second major aspect of high-pressure sealing touches specific manufacturer know-how. The sensor itself – most commonly a thermocouple because of its smaller diameter – needs to be connected with a high-pressure metal fitting, for example, with a thread and defined seal or any other shape that can be properly installed. General methods to do this are listed in Table 14.4 and Figure 14.3 depicts such a sensor installed into a high-pressure T-piece. Technical details of the sealing methods are often manufacturer know-how and the listed parameter values are for orientation only.

Besides pressure rating, the durability and reliability need to match the intended use. It goes without saying that a vessel made of high-performance alloys needs to be equipped with sensors of the same or better grade material. Additional wear resistance is realizable with additional coating of the tip, for example, carbide-coated tips for use in low-density polyethylene (LDPE) plants.

In research applications, specialty measuring methods are known. For example, temperature measurement through high-pressure sapphire view glasses is possible using dispersed thermochromic liquid crystals (see Chapter 9).

Even with good and approved temperature sensors (ratings listed in Table 14.5; options in Table 14.6), erroneous readings can occur in high-pressure as well as in standard setups due to wrong positioning, too long response time, aging of sensors (especially at high temperatures), fouling, corrosion, and last but not least heat transfer effects to the sensor. To mention the extremes, turbulent flow of a liquid ensures a good heat transfer to the sensor while static gas charges are less simple to measure. In the latter case, it is advisable to use more than one sensor and check if temperature homogeneity in the system is given. If not, additional convective flows can occur and their effects on the process should be taken into account. Thick vessel

Table 14.4 Mounting methods of temperature sensors to high-pressure fittings.

Method	Manufacturing details	Limitations
Gluing	Glue the sensor into a lightly larger drill hole	• For tiny diameters only • Lowest pressure rating • High-risk interaction with fluid media • Maximum T depending on glue, 200 °C • No official certification • Not recommendable for high pressure!
Soft soldering	Solder sensor to fitting	• For tiny diameters only • Low pressure only, for example, 5 MPa • Risk of corrosion • Maximum temperature approximately 150 °C
Brazing	Braze sensor to fitting	• Up to 30 MPa and 500 °C • Risk of sensor damage during brazing
Welding	Weld sensor to fitting	• Up to 100 MPa and 1100 °C • Risk of sensor damage during welding
Clamping (see Figure 14.4)	Fix the sensor with a ferrule like a piece of tubing. If the diameter is too small, liner reinforcement rings can be welded onto the sensors	• Standard up to 50 MPa and 600 °C • Easy mounting but risk of sensor "shooting out," if not mounted properly or non-approved technology!

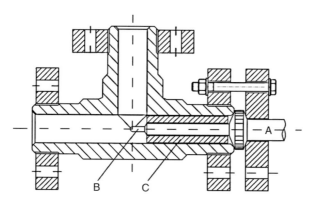

Figure 14.3 Industrial T-sensor installation (standard HP T-piece; A: turning workpiece with lens joint as sealing; B: thermocouple welded to A; C: filling bushing to reduce stagnant volume). Figure according to a drawing in Ref. [7].

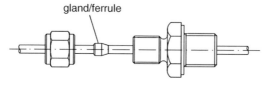

Figure 14.4 Philips Thermolok® gland seal (MAWP 50 MPa, 700 °C)[9].

9) Figure according to a technical drawing from Thermocoax GmbH, Stapelfeld, Germany.

Table 14.5 Rating of HP temperature sensors.

Sensor type	MAWP catalogue examples	T_{max} sensor limitation	Typical use
Resistance temperature detector	210 MPa @ 20 °C 400 MPa @ 500 °C	Standard type PT100 750 °C	Advantage: robust calibration
Thermocouple	280 MPa @ 315 °C 515 MPa @ 20 °C 700 MPa @ 500 °C	Depending upon types J (Fe-CuNi): 760 °C K (Ni-CrNi): 1372 °C S (Pt-RhPt): 1600 °C	Advantages: small diameter, quick response

walls as used in high-pressure processes can influence the temperature of the sensor when wall and fluid temperatures are not identical: heat conduction through the metals can override the temperature equilibrium that should form between the fluid and the sensor tip only. This effect of course depends upon the specific design and allocation of the components. Short and thick sensors close to the wall and with intense thermal contact to the vessel can lead to significant mal readings and this occurs even more during alteration of process parameters (Tables 14.5 and 14.6).

Table 14.6 Options when choosing thermocouples for HP temperature measurement.

- Design approved by supplier for required pressure rating!
- Maximum temperature – today sensors for more higher than 1500 °C are available
- Check measuring accuracy and response time
- Output signal depending upon type, for example, "K" – check compatibility with data acquisition
- Check needed corrosion resistance and wear resistance
- Request needed conformity certificates

14.2.4
Flow

Flow quantification is essential in the vast majority of all processing plants and distribution systems of gases and liquids (Fig. 14.7). Processes operating at high pressure are no exception and therefore a variety of robust and field-proven sensor technologies are offered today. All large-scale chemical reaction plants involving catalytic synthesis or polymerization, liquefaction, and separation of gases and all other processes such as those dealt with in this book require feed flow control. In some cases, feed flows can be monitored before pressurization with conventional sensors, but there are also measuring points requiring high-pressure sensors. This is especially true for the oil and gas business. Offshore production and subsea exploitation, transmission, liquefaction, and processing are designed to operate at elevated and high pressures for enhanced capacity and energetic reasons. In Tables 14.7 and 14.8, an overview on available sensor types and their principles is given, while Table 14.9

Table 14.7 HP flow detection sensors without moving parts – overview.

Sensor type and range	Principle	MAWP (MPa)	T (°C)	Advantages/limitations	Fields of application
Coriolis (0.03 kg/min to 12 t/min)	Mass flow detection. Detection of an oscillating system: phase changes induced by Coriolis forces of a flow	90	−200 to +400	• High pressures • Also for aggressive and contaminated media • No pressure effects • No moving parts/long lifetime • Observe orientation • Firm mounting needed • Single phase only	• All, if not too costly • Determination of compressed gas and liquid flows without influence of p, T, density • Often with additional online density (e.g., quality control oils) • Some units with additional online viscosity (e.g., spraying processes)
Thermal mass flow sensor – gas (0.2–11 000 N m³/h)	Mass flow detection. Part of gas flow through a bypass. Bypass is precisely heated and rise of temperature detected	70	−10 to +100	• Low cost especially in combined versions as "mass flow controller" • High precision • No moving parts • Calibration necessary • No aggressive gases	• Determination of *pure* gas flows, for example, N_2, pressurized air without influence of p, T, density • Mixing, charging, research • Pure overheated steam (T_{max} 200 °C)

Thermal mass flow sensor – liquid (1 mg/h to 20 kg/h)	Mass flow detection. Method as above	40 (100)	−10 to +70	As above, for example, • High precision	• Pure liquids, for example, oils, water without influence of p, T, density • Mixing, charging, research
Differential pressure	Velocity detection. Measuring of pressure loss on orifice restrictors or Pitot tubes	40	+1000	• No moving parts • Maintenance free • No drift after calibration	• Steam • For cryogenic liquids, for example, CO_2 • Monitoring oil loops in transformers
Vortex	Capacitive detection of vortex eddies caused by bluff body	25	+400	• Extremely robust • Dirt resistant • Minimum Re 20 000	• Steam saturated and overheated • Thermo oil (solar power plant)
Electromagnetic	Fluid velocity	35	+200	• Influence of p, T, fluid • Calibration necessary	• Conductive liquids only, for example, water • Usually only low pressure

(*Continued*)

Table 14.7 (Continued)

Sensor type and range	Principle	MAWP (MPa)	T (°C)	Advantages/limitations	Fields of application
Ultrasonic, clamp on	Ultrasonic measurement from outside of tube	Limit by maximum tube wall size	+230	• No pressure drop • No medium contact • Maximum steel wall ~70 mm • Less good with polymer tubes • Thick tube walls require high fluid densities	• Liquids only • Tube sizes 20 mm and more • Pressurized steam • Hydraulic oil
Nucleonic	Monitor movement of sharp pulse of radioactive tracer	50 MPa and higher		• Special application for subsea pipelines • Sensor moves along pipeline or multiple sensors	• Diagnostics for pipeline performance, find stuck pigging tool or solid buildup/blockages

Table 14.8 HP flow detection sensors with moving parts – overview.

Sensor type and range	Principle	MAWP (MPa)	T (°C)	Advantages/limitations	Fields of application
Turbine	Mediate volume transmitter. Stainless steel turbine in sapphire cups, Hall sensor	400	+125	• Very high pressures • High accuracy • Corrosion resistance • Small size • Easy sterilization • Limited instrument span	• Light fuel oil, solvents, liquefied gases • Dosing accurately in lab and processing • Jet cutting • Subsea: modifiers for oil wells, submarines
Gear wheel	Positive displacement. Plastic coupled gear wheels, Hall sensor	350	+180	• Independent of viscosity • Very high pressures	• Liquids • Viscous liquids, for example, wax, polymers • Dosing accurately, for example, print colors
Helical flow meter	Positive displacement. Helical coil axial, Hall sensor	40	+200	• High accuracy • Corrosion resistance	• Liquids • High-viscosity liquids, for example, polymers, oil, glue, heavy fuel oil in power plants • Spray PVC, underbody coating • Dosing accurately, for example, colors
Microflow meter (double ring piston)	Positive displacement using double ring piston	40	+180	• Extremely low flow volumes	• Low-viscosity liquids, for example, additives, water, pharmaceuticals • Injection odorant into natural gas • Batching and filling applications

(*Continued*)

Table 14.8 (Continued)

Sensor type and range	Principle	MAWP (MPa)	T (°C)	Advantages/limitations	Fields of application
Piston system (switch only)	Spring-supported piston with magnetic coupling to adjustable microswitch	25	+350	• Robust • Good repeatability	• For example, alarm high or low • Monitoring hydraulic systems, for example, in machine tools
Paddle (switch only)	Spring-supported paddle in the flow. Magnetic coupling to adjustable microswitch	25	+350	• Robust, dirt resistant • Low pressure losses	• For example, alarm high or low

Table 14.9 Flow detection sensors for pressures 10 MPa and higher – choice with regards to process fluid (" + " suitable).

Type of flow meter	Gas or supercritical fluid	Steam	Liquid, conductive	Liquid, nonconductive	Bidirectional metering (forward/reverse)
Coriolis	++	+	++	++	+
Thermal mass flow	++	+	+	+	
Differential pressure	++	++	++	++	+
Vortex	+	++	+	+	
Electromagnetic			+		+
Ultrasonic, clamp on			+	+	+
Turbine, gear wheel			++	++	
Flow switch			+	+	

gives initial advice for sensor choice with regard to the respective media. Some are discussed in more detail in the text.

Coriolis-type mass flow sensors have found wide acceptance in the field of high-pressure technology. Especially with fluids and gases at varying process conditions (density, viscosity, p, and T), their striking advantage is independence from these parameters and direct flow determination without the need to know the actual gas properties. In addition, these instruments concurrently supply data on fluid temperature and density. Wetted material can be chosen from a wide range including Hastelloy, Monel, tantalum, and other exotic materials to meet specific material compatibility requirements. Flow accuracy of 0.1% and repeatability better than 0.05% are standard. Different ranges, sizes, and styles are available: the traditional shape is the omega tube type but also parallel tubes and other shapes are used. Figure 14.5 displays the inside of an omega-type Coriolis flow meter to explain the nontrivial principle. The meter uses an oscillating device with drive coils mounted to a mass bar ensuring controlled amplitude of the system. A corresponding Coriolis force is formed when liquid or gas flows through the oscillating omega-shaped tubes and the resulting phase shift is detected with pickup coils.

Firm mounting is necessary to eliminate the influence of other oscillating plant parts. And installation position is also important to prohibit buildup of interfering second phases; for example, an instrument for detection of a gas flow should be oriented as shown in Figure 14.5 so that liquid droplets will not collect in the curved tubes. In industrial processes, especially cyclic high-pressure processes, a decision has to be made considering optional positions with different parameters (temperature, pressure rating, risk of contamination, or two-phase flow).

Thermal mass flow sensors have also found wide application for pure media, such as metering of N_2, compressed air, or oils because of their compact design, precise detection, and lower cost than comparable Coriolis meters. Their handicap is the

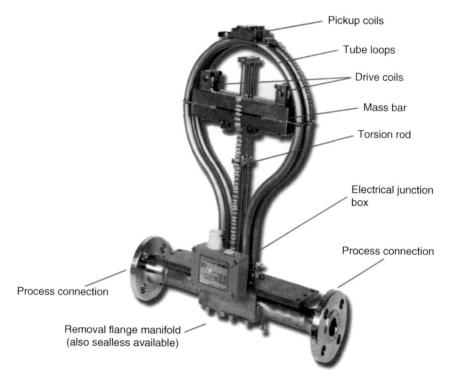

Figure 14.5 Inside of Coriolis-type flow sensor [10].

limitation to pure media. A gear wheel flow meter is a positive displacement meter, similar to a gear pump. The measuring fluid rotates two gears, which are engaged with minimum play. The flow is in proportion with the rpm of the gear. Hall sensors integrated into the housing pick up the rpm of the gears. Differential pressure detection is simple and robust and common for flow detection in well-defined systems, for example, transportation and delivery of pure gases. A system installed on a gas tank truck can be seen in Figure 14.6. Ultrasonic inline detectors, visual rotors, or variable area meters (float in a transparent tube) are not used in high-pressure technology; they are limited to applications below 5 MPa (Figure 14.7).

14.2.5
Fluid Level

Different methods for level detection are in use in today's processing plants (Figure 14.8). Fluid level determination in high-pressure processes is mainly

10) Figure from brochure "Inside of Coriolis flow meter Rheonik," Schwing Verfahrenstechnik, Neukirchen-Vluyn, Germany.

Figure 14.6 Flow determination with differential pressure detection, horizontal tube with orifice indicated.

needed to monitor vessel contents and control the bottoms receiver in separation and distillation columns as well as any receiver after a condensation step. The variety of sensors suitable for high pressure is limited; especially some of the pure mechanical devices are not available or fail easily. An overview is given in Table 14.10 and examples are depicted in Figure 14.8 and 14.9.

14.2.6
Density

Density data are of interest in various fields; methods and some applications are compiled in Table 14.11.

14.2.7
Viscosity

Online determination of viscosity is mainly implemented to monitor quality of products, correct turbine-type mass flow readings, or control processes with wetting procedures (coating, painting, and spray processes). In research and plant design, HP viscosity values are frequently needed for various flow calculations. Examples are transport processes for natural oil and gas and – more scientific – calculation of film thicknesses and traction effects in elastohydrodynamic lubrication using organic liquids. At present, small-scale viscosity sensors are developed that will operate in the pressure range up to 100 MPa attached to the drilling rods in oil exploration (Table 14.12).

Figure 14.7 Gas mixing. Upper unit: Coriolis-type sensor with subsequent regulation valve. Lower unit: thermal mass flow controller (MAWP 20 MPa).

14.2.8
Concentration – Solute in High-Pressure Gases and Fluids

Process photometry is the dominating method to detect and quantify substances dissolved in high-pressure fluids. Photometers use the interaction of light (absorption or scattering) with the process fluids and can either detect at a fixed wavelength (photometer) or evaluate a spectrum (spectrometer). Depending upon the appropriate range of wavelengths, the detectors can be differentiated as ultraviolet (UV), visual light (Vis), near-infrared (NIR), or infrared (IR). For application in high-pressure processes, the light beam needs to pass the high-pressure gas or fluid and therefore pressure-resistant sight glasses are needed, usually two opposing ones

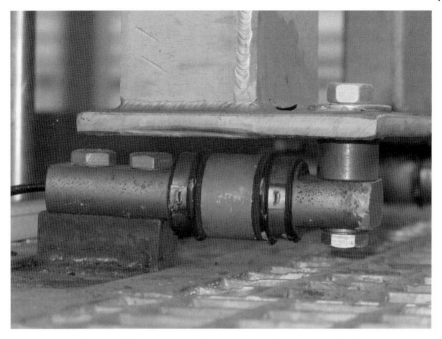

Figure 14.8 Robust level determination in high-pressure tank using a load cell.

Figure 14.9 Gamma densitometer for level determination and density in high-pressure vessel (50 l, MAWP 30 MPa) using a ^{137}Cs isotope.

Table 14.10 High-pressure level gauges.

Sensor type	Principle	MAWP (MPa)	$T_{min/max}$ (°C)	Remarks
(a) Continuous level detection				
Float	Level indication. Float in vertical bypass tube	30	−196 +400	• Fluid density >0.5 • Risk of failure with density changes (gases), dirt, turbulences
Radar	Reflection of microwave impulses	16	−200 +450	• Storage vessels • Sensor inside vessel top
Nucleonic/radiometric	Density change detected with gamma densitometer (see Figure 14.9)	>30	−20 +125	• High reliability
		Limit by wall thickness versus needed accuracy		• Nearly maintenance free • Multiphase possible • Densities 0–3000 kg/m³ • Strict safety regulations • High cost • Used in nuclear industry, research, and subsea
Capacity	Capacity detection. Level switch or continuous	6	−50 +200	• Limited to certain liquids • Liquefied gases only when dry (e.g., fresh gas tank)
(b) Level switch only				
Optic sensors	Refractive index	50	200	• Bottoms of column
Vibrating tuning fork	Frequency change of piezo-driven vibrating tuning fork	10	−50 +280	• All liquids • Without adjustment • For example, level between compressed gas and water

Table 14.11 Overview on HP density determination, $p > 10$ MPa.

Sensor type	Principle	Range (kg/m³)	MAWP (MPa)	T_{max} (°C)	Remarks
(a) Online					
Tuning fork		0–3000	100	200	• Oils • Dry and clean gas, for example, natural gas, ethylene • Supercritical CO_2 in refrigeration plants
Coriolis	Multifunctional sensor. Mass flow sensor with additional pickup of resonance frequency	0.5–6000	40	−200 to +350	• High pressures • Also for aggressive and contaminated media • No pressure effects • No moving parts/long lifetime • Single phase only
Nucleonic/radiometric	Gamma densitometer	0–3000	>30 Limit by wall thickness versus needed accuracy	>150 Limit by wall material	• High reliability • Nearly maintenance free • Multiphase possible • Strict safety regulations • High cost • Used in nuclear industry, research, and subsea
(b) Additional for off-line/lab					
Pycnometer	Weighing fluid filled cell with precisely known volume		70 and higher	+175	• Limited accuracy (volume versus pressure correction needed) • PVT studies

Table 14.12 Overview on HP viscosity determination, $p > 10$ MPa.

Sensor type	Principle	Range (mPa s) (cP)	MAWP (MPa)	T_{max} (°C)	Remarks
(a) Online					
Coriolis	Multifunctional sensor. Mass flow sensor with additional pickup of resonance frequency	n.s.	90	−200 to +400	• High pressures • No pressure effects • No moving parts/long lifetime • Single phase only • Also for aggressive and contaminated media
Electromagnetic oscillating piston	Speed of electromagnetically driven piston detected by circuitry analysis [11]	0.02–10 000	260 and above	+190	• With T sensor • Small fluid quantity • Oil and gas exploration, PVT
Tuning fork	Frequency change of vibrating tuning fork	1–20 000	21	+200	• Newtonian behavior
Rotational	Force to rotate cylinder in fluid	1–720 000	30	+350	• Non-Newtonian behavior, for example, emulsions
Quartz	Torsional oscillation of a piezocrystal	1–10 000	1000	+250	• Highest pressures • Detection if Newtonian fluid
(b) Additional for off-line/lab					
Vibrating rod	Frequency change of vibrating rod	1–1 000 000	70	+200	• Heavy oil viscometer

		Limit of cell (vertical visual field)	Limit of view cell	Limit of view cell	
Falling sphere or rising bubble	Sphere sinks or bubble rises in fluid in HP view cell [8]				• Ensure terminal velocity • No commercial units
Falling needle	Needle sinks in fluid, velocity with Hall sensors	0.7–20 000	420	−40 to +150	• Non-Newtonian behavior
Rolling ball	Travel time of rolling ball, different detectors	0.3–1000	70 and above, for example, 1000 [9]	+175	• Newtonian behavior
Rotational	Force to rotate cylinder in fluid	1–100 000	70	+280	• For non-Newtonian behavior
Capillary	Differential pressure of a precisely controlled capillary flow	0.02–0.3 (gas) 0.3–10 000 (liquid)	500	−10 +150	• For non-Newtonian behavior • Complex technology

11) Cambridge Viscosity Brochure, Cambridge, Medford, MA, USA.

mounted as an optical cell unit. Sapphire glasses or other high-quality glass materials are necessary. It shall be noted that adsorption spectra in different fluids are not identical, a fact that is sometimes ignored even in scientific studies. To give an example, the absorption maximum for β-carotene (all-E) in hexane is at 450 nm while in pressurized CO_2 it is found at 434 nm (20 MPa, 35 °C). In addition, the maximum in a dense gas, for example, CO_2, can shift significantly with changing pressure and temperature. It is obvious that erroneous readings due to inaccurate choice of wavelengths can occur.

Maximum pressures are mainly limited by the mechanical rating of the view cell system allowing extreme research applications above 1 GPa. Industrial continuous high-pressure inline measurement is applied today in the pressure range up to 30 MPa. Robust realization is possible using process photo- or spectrometers as long as the optical glasses can be prevented from staining. Advantageous are highly polished glass surfaces and a high fluid turbulence around those glasses. Often the view cells are designed in such a way that the glasses are not aligned with the tube wall but rather protrude into the fluid flow. Deposition of dirt is more likely in gas flows with their limited density compared to liquids and therefore it is not surprising that most industrial applications have sensors for liquids. Color and concentration measurements can be used to monitor feed flow (catalyst dosing), product quality (color formation by overheating or corrosion, impurities, oil traces), or phase separation (level control).

Detection of water in high-pressure fluids can be accomplished with NIR technology but is limited to the lower percentage region. Systems for pressures up to 100 MPa and 180 °C are available[12]. Other moisture sensors use capacity changes of a polymer or aluminum oxide. They are rated up to 35 MPa and commonly used at high pressure to monitor moisture in gases (7–10 MPa), like natural gas after dehydration with molecular sieve driers. The requirement is to keep moisture content low to prevent hydrate forming that can block the flow of natural gas during transmission. Quite similar is the application in CO_2 pipelines.

14.2.9
Concentration – Gas Traces Dissolved in Liquids

Detection of gases dissolved in liquids at high pressure is in focus, for example, at marine systems and will gain increased attention with future research on new energy resources (methane hydrate) and climate research (marine CO_2). To give two examples, methane can leak from the seafloor and hydrogen is emitted in hydrothermal plumes. Hydrostatic pressure at these sites can be as high as 50 MPa and requires special sensors. Online on-site monitoring of, for example, methane, hydrogen, or CO_2 in a water phase can be accomplished with specialized pressure-proof sensors consisting of a silicone membrane contacting the water phase and a suitable detector in the sensor interior [10]. Diverse detector types are used

---

[12] Mobile NIR in-line device for monitoring of supercritical CO_2 based processes, Brochure SITEC-Sieber Engineering, Zurich, Switzerland.

14.3 Lab Determination – Additional Offline Test Equipment

14.3.1 Phase Equilibrium

Phase equilibrium (PE) studies have been performed since the very beginning of high-pressure technology. General principles are listed in Table 14.13. Steam condensation, condensation of gases and their mixtures, and behavior of reactants and products in reaction processes were early fields of research to understand and develop new processes. The basic method is – optimized and refined – still in use today: a pressure vessel is filled with the substances and fluids to be investigated, pressurized, and time given to equilibrate. Subsequent samples are taken from different levels of the vessel and analyzed after decompression. This method is named *analytical* and requires significant density differences between the involved phases and furthermore sufficient volume of each phase to be sampled and analyzed. *Synthetic* methods use a vessel filled with a mixture of precisely known composition. System behavior and detection of phase boundaries is determined by changing of parameters, for example, T or p. Description of these methods is, for example, given

Table 14.13 General principles for phase equilibrium measurement with at least one fluid.

	Synthetic	Analytical	Analytical dynamic
Principle	Known content in vessel, observation by change of parameters (e.g., p, T)	Samples from vessels after equilibration	Continuous flow of a fluid through a second fluid or bed of solids
Example	Observe phase change in optical view cell	Stirred tank, sampling, gravimetric, GC, and so on	Flow saturator, detection of solute at outlet
Advantages	• In critical region • Often easy and quick	• For complex systems • For low solubility	• For large samples • For extremely low solubility
Limitations	• In multiphase systems composition of phases not known • Not for low solubility	• Poor at critical region • Not for isopycnic phases • Not for small phase volumes • No PVT data	• Uncertainty if equilibrium established • Maximum solubility <10%

by Stahl et al. [11] with focus on systems with dense gases but validity for all high-pressure fluid equilibria measurements and most recently with an abundance of examples and details by Dohrn et al. [12].

Exemplary flow schemes of setups for analytical equilibrium determination with high-pressure CO_2 depicted in Figures 14.10–14.12 are taken from the author's work [13]. While the stirred vessel allows rapid equilibration, sample volume is restricted to low quantities in order to avoid pressure loss of the system. Note that the gas-phase sampling line is equipped with a rinse port to flush precipitated substances into the sample receiver using a suitable solvent. The setup using a high-pressure column with a floating piston allows larger sample volumes by maintaining the pressure level constant with the use of a hydraulic system actuating the piston. Equilibrium is attained with an isobaric circulation pump conveying the denser phase from the vessel bottom back to the top, thus intensifying phase contact. With the large vessel volumes displayed, the isothermal state of the complete system needs to be looked after thoroughly. Measurements following the analytical dynamic principle can be, for example, made in a plant as sketched in Figure 14.12. A grained solid substrate is placed into the vessel (A) and solubility at p and T can be calculated from the amount received in the separator versus fluid flow. Of course, complete solute precipitation in the separator is required and sometimes a sequence of separators or total depressurization to atmospheric conditions is necessary, obviously depending upon the properties of the investigated solid. For rough qualitative studies with viscous/sticky materials or extremely low solubility, the same setup is applicable using a horizontal vessel (B) with inserted trough holding the substrate and weight determination of the trough before and after a suitable operation time.

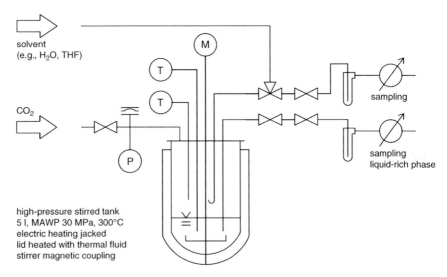

Figure 14.10 Analytical PE measurement stirred vessel.

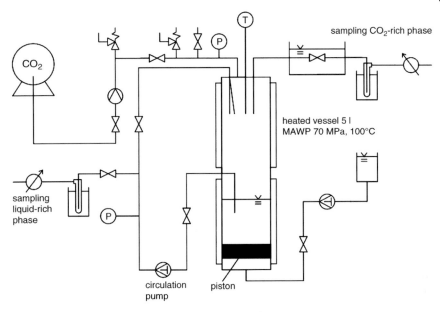

Figure 14.11 Analytical PE measurement column with floating piston.

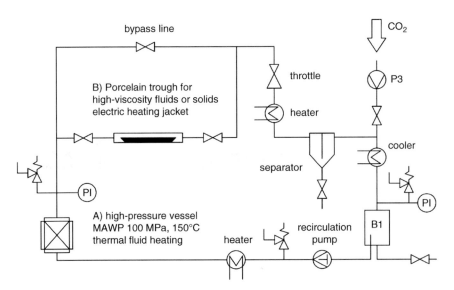

Figure 14.12 Analytical dynamic PE measurement up to 100 MPa.

14.3.2
Magnetic Sorption Balance

In the past 15 years, magnetic coupled sorption balances have found wide application in high-pressure research. The basic principle is shown in Figure 14.13 and allows online weighing of samples inside a high-pressure vessel. These samples can be solids as well as powders or liquids in a small receptacle. The samples are contacted with a compressed fluid and subsequent mass changes can be recorded. This method allows to measure precise HP densities and solubilities (e.g., a dense gas in a polymer) when equilibrium conditions are attained. Following transient data can allow calculation of permeability and diffusion coefficients. If the measuring cell is equipped with additional view glasses, swelling behavior can also be monitored. Magnetic coupled balances can operate up to 200 MPa and – though not at as high pressures – up to 2000 °C and can provide the skilled researcher with various insights.

14.3.3
Interfacial Tension and Wetting

Interfacial effects are gaining increasing attention today. Although significantly less known than viscosity, the interfacial tension can be a crucial parameter in technical processes. Interfacial tension and contact angles can be determined with the pendant

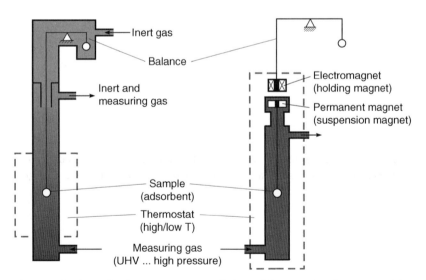

Figure 14.13 Principle of a magnetic sorption balance (left: conventional apparatus; right: magnetic suspension balance).

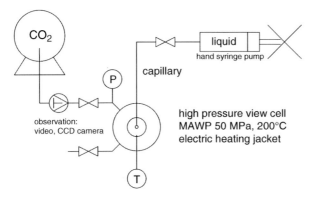

Figure 14.14 Principle of an interfacial tension measurement using a view cell.

drop method using a high-pressure view cell (see Figure 14.14), CCD image recording, and software evaluation [14].

14.3.4
Gas Hydrates

Gas hydrates are crystalline solids composed of water and gas and can occur at low temperatures. The necessary amount of water is surprisingly low and in pressurized processes or environments hydrates form also at ambient or even higher temperatures, for example, ethylene hydrate at 30 °C/80 MPa. The formation of hydrates in exploitation and production of natural gas and oil is of significant economical relevance; for example, natural gas hydrates can block gas transmission lines at temperatures above the ice point easily (for more information see Chapter 7). Therefore, hydrate research is very intense with specialized laboratories and more than 4000 publications in this field in the last decade. Typical research parameters are 270–305 K and 0.1–100 MPa. A comprehensive list of applied research methods can be, for example, found in the monograph of Sloan *et al.* [15] Properties such as phase equilibrium as well as kinetics and morphology need to be known and understood. Measuring technology is mainly based on three lines. The first is visual observation to get evidence of hydrate existence. Therefore, HP view cells are often used. Second, indirect means are used, such as a hysteresis pressure course while changing temperature (isochoric) or gas take-up determination while cooling (isobaric method). Third, highly specialized methods using X-ray, NMR, neutron, or Raman spectroscopy and others are used to investigate complex hydrate chemistry and morphology. Traditional equilibrium apparatuses use stirring or other means of agitation to distribute the involved phases evenly and speed up equilibration. The quartz crystal microbalance (QCM) works with one drop of water positioned on a mass sensor and therefore allows significant faster equilibration. For further insights on hydrate measuring, see the related scientific literature.

Table 14.14 Industrial specialty sensors to be used at high pressures.

Parameter	Principle	p_{max} (MPa)	T_{max} (°C)	Application
Acoustic velocity	Ultrasonic sound	10	120	• Concentration and phase properties • Oils in refrigeration plants
DSC	Differential scanning calorimetry	100	300	• Heat capacities • Kinetics/rate of reaction, for example, gas hydrates
Admittance	Admittance detection	20	150	• Gas content in liquid or v.v. • Foaming in transportation processes, columns • Oil–water systems
Gas bubble	Capacity	10	200	• Gas bubbles in liquid gas, for example, butane, propane

14.3.5
Other Properties Online

Besides equilibrium data, transient effects can be of interest for high-pressure processes too. For example, design optimization of high-pressure reactors requires data on heat and mass transfer in pressurized fluids, diffusion coefficients, and rates of reaction. In most cases, custom-made research units using one or more principles as explained above will be needed to obtain experimental data. The setting of the zero point poses a challenge for these experiments because most setups do not allow prompt final pressure conditions at time zero but rather pressure adjustment takes some time while material interactions have already started. In Table 14.14, some further available sensors for application in high-pressure systems are listed. Often these sensors are combined with others. For example, ultrasonic sound signals are linked to the sought-after concentrations in a complex way and pressure and temperature changes produce nonlinear deviations. Therefore, ultrasonic sound sensors are often complemented with pressure and temperature detection.

14.4
Safety Aspects

Although appropriate sensors are built to withstand the applied pressures, it is advisable to consider additional protective measures to avert potential danger resulting from smaller parts being ejected in the case of faults. This can be achieved

Figure 14.15 Boiler explosion about 1850 – vessel instrumentation scarcely in use at that time [13].

by a spatial orientation away from users as well as with protective devices that should be designed in the way that they cannot be removed without using tools.

As a matter of course, reliable pressure detection is of crucial importance for the general safety of high-pressure processes. Fortunately, today's sensor technology and sophisticated safety concepts have massively cut down vessel explosions since early days of pressure technology (Figure 14.15). Although not issue of this chapter, mandatory safety valves for each process vessel featuring sufficient flow rating shall be at least mentioned at this point. Observe that all sensors with safety function need to have design approval for the intended use and often duplicate installation is advisable if not mandatory. Although freedom from defects/malfunction of electronic devices is continuously upgraded, wiring of sensors with safety function should be conventional and irrespective of electronic and computer systems. A separate cabinet is advisable too and general observation of safety rules regarding automation and electric installations[14]. Keeping the appropriate degree of electric shock protection as well as explosion-proof rating has been mentioned before.

Regarding pressure sensors, keep in mind that gauges allow readings even during power outings and are therefore indispensable for all pressure vessels. Furthermore, the maximum working pressure of the respective vessels needs to be marked on the scale of these gauges so that MAWP exceedances are readily identifiable for everybody. Clearance for opening procedures of vessels – especially of those equipped

---

13) Picture of ruptured steam engine boiler, public domain, taken from www.wikipedia.org.
14) Directive IEC/EN 62061: Safety of machinery – Functional safety of safety-related electrical, electronic, and programmable electronic control systems.

Figure 14.16 Example safety instrumentation of a clamp-closure HP vessel.

with quick closures – requires reliable detection of the nonpressurized state. Since pressure sensors show low accuracy close to atmospheric conditions, the sole use of a standard pressure sensor is not safe. Additional measures need to be taken, for example, dead times after opening of relief valves. Usually, vessel lids and clamps need to be equipped with sensors to detect their correct working position. These can be, for example, mechanical contacts, inductive proximity switches, and electric locking pins (Figure 14.16). Mechanical locking devices activated by vessel pressure are more favorable than exclusive electric solutions.

With the knowledge that material tensile strength declines with rising temperature (especially in the range above 200 °C), it is obvious that reliable temperature detection is also a safety issue in high-pressure technology.

Summing up high-pressure processing needs close observation of all compulsory national regulations. Moreover, additional technical measures can be necessary depending on the given process and its risks. Process knowledge from the user needs to amend official requirements. Note that surveillance authorities do not have exhaustive knowledge in all cases due to the high degree of specialization high-pressure processing requires. Equipment manufacturers interacting with the users have to assume the responsibility for safety.

14.5
Future

The necessity to monitor high-pressure processes for design optimization, safety, and controlling will most probably not vanish in the future; on the contrary, the need to enhance efficiencies will lead to heighten the need for sensors. Maintenance intervals are increasing while on-site manpower is consequently reduced, thus enforcing augmented data monitoring. New or changing high-pressure technologies can be expected at the subsea level: declining reservoir capacities of natural gas and crude oil demand sophisticated technology at extremer conditions, thus asking for higher MAWP ratings. The advocated idea of CCS (carbon dioxide capture and storage from combustion processes) and activities regarding hydrates are other potential fields. These may lead to an increasing use of high-pressure sensors, for example, flow detectors at subsea level. On the other hand, large-scale high-pressure processes are costly and constantly under review if replaceable by alterative technologies. The launch of new HP processes is for the same reason generally stringy but new technologies are constantly under review. They will require various new data and not all of them will be accessible by pure theoretical calculation and thus enforce experimental determination with sophisticated high-pressure sensor technology – at least in the lab. Sensor sizes will decrease following the needs for application in limited space (e.g., automotive) and minimizing plant technology (microsystem technology). Manifold high-pressure sensors can be of further help in this direction.

References

1 Verein Hütte (1883) *Ingenieurs Taschenbuch*, 12 Auflage, Verlag von Ernst & Korn, Berlin, p. 612.
2 Goodwin, H. (1925) *Autoclaves and High Pressure Work*, Ernest Benn Ltd, London, p. 109f.
3 Newitt, D.M. (1940) *The Design of High Pressure Plant and the Properties of Fluids at High Pressures*, Oxford University Press.
4 Newhall, D.H. and Abbot, L.H. (1961) High-pressure measurement, in *Instruments & Control Systems* (ed. R. Rimbach), vol. 34, Chilton Company, p. 232ff.
5 Wilson, J.S. et al. (2009) *Test and Measurement: Know It All*, Elsevier.
6 Wilson, J.S. (ed.) (2005) *Sensor Technology Handbook*, Elsevier.
7 Buchter, H.H. (ed.) (1967) *Apparate und Armaturen der Chemischen Hochdrucktechnik: Konstruktion, Berechnung und Herstellung*, Springer, Heidelberg.
8 Schlunder, E.-U. and Lockemann, C.A. (1995) Liquid-phase viscosities of the binary systems carbon dioxide–oleic acid, carbon dioxide–methyl myristate, and carbon dioxide–methyl palmitate at high pressure. *Chem. Eng. Process.*, **34**, 487–493.
9 Izuchi, M. and Nishibata, K. (1986) A high pressure rolling-ball viscometer up to 1GPa. *Jpn. J. Appl. Phys.*, **25**, 1091–1096.
10 Garcia, M.L. and Masson, M. (2004) Environmental and geologic application of solid state methane sensors. *Environ. Geol.*, **46**, 1059–1063.
11 Stahl, E., Gerard, D., and Quirin, W. (1988) *Dense Gases for Extraction and Refining*, Springer, Berlin.
12 Dohrn, R., Peper, S., and Fonseca, J. (2010) High-pressure fluid-phase equilibria: experimental methods and systems investigated (2002–2004). *Fluid Phase Equilib.*, **288**, 1–54.
13 Pietsch, A. (2000) Die Gleichstromversprühung mit überkritischem Kohlendioxid an den Beispielen Hochdruckentcoffeinierung und Carotenoidaufkonzentrierung. Doctoral thesis, TU Hamburg-Harburg.
14 Jaeger, P. and Pietsch, A. (2009) Characterization of reservoir systems at elevated pressure. *J. Petroleum Sci.*, **64**, 20–24.
15 Sloan, E.D. and Koh, C.A. (2008) *Clathrate Hydrates of Natural Gases*, CRC Press, Boca Raton, FL.

15
Sizing of High-Pressure Safety Valves for Gas Service
Jürgen Schmidt

High-pressure safety valves are presently sized according to ISO 4126-7. Equations for ideal gases are presented there, but no indications are given as to how the real gas factor and the isentropic exponent for real gases should be calculated. For this reason, a sizing method comprising real gas effects was derived and compared with the ideal gas model according to EN-ISO 4126-7.

15.1
Standard Valve Sizing Procedure

The predominant numbers of safety valves are used at set pressures below 5 MPa (50 bar). For gas service, they are sized according to EN-ISO 4126-7 [1] or national regulations derived from it [2].

In general, the seat cross-sectional area of a safety valve A_{seat} must at least be large enough to discharge the minimum required mass flow rate $Q_{m,OUT}$ according to the sizing scenario for a pressurized system. For simplification, the dischargeable mass flow rate Q_m through the valve is calculated by means of an equivalent nozzle flow model for ideal gases corrected by an experimentally determined discharge coefficient $K_{d,g}$. With an appropriate nozzle flow model, all property data at the operating conditions are covered; that is, only the geometric effects are included in the discharge coefficient. The more accurate the nozzle flow model reproduces the actual mass flow rate, the less the discharge coefficient is influenced by property data. This leads to the sizing criteria for safety valves:

$$Q_m = K_{d,g} Q_{m,nozzle} \geq Q_{m,OUT} \tag{15.1}$$

With the sizing coefficient C_g for a nozzle with a throat area equal to the seat area of the safety valve, the mass flow rate is defined in a dimensionless form:

$$C_g = \frac{Q_{m,nozzle}}{A_{seat}\sqrt{2p_0/v_0}} \tag{15.2}$$

Industrial High Pressure Applications: Processes, Equipment and Safety, First Edition. Edited by Rudolf Eggers.
© 2012 Wiley-VCH Verlag GmbH & Co. KGaA. Published 2012 by Wiley-VCH Verlag GmbH & Co. KGaA.

Equations (15.1) and (15.2) are definitions that are valid for both ideal gas and real gas flow.

The sizing coefficient for steady-state conditions of an ideal gas through the nozzle without friction and heat exchange with the wall is written as (isentropic flow of an ideal gas)

$$C_{g,ISO} = \sqrt{\frac{\kappa_0}{\kappa_0 - 1} \eta^{2/\kappa_0} [1 - \eta^{(\kappa_0 - 1)/\kappa_0}]}, \qquad \eta = \frac{p}{p_0} \qquad (15.3)$$

Equation (15.3) was developed for plenum flow with negligible inlet velocity. If the inlet velocity may not be neglected, a (fictitious) stagnation pressure – the actual inlet pressure increased by the pressure difference computed by an isentropic deceleration of flow to the velocity of zero – must be introduced in Eq. (15.3). This equation is not valid if gases condense or chemical reactions occur.

The critical pressure ratio is the lowest pressure ratio that can be established under these flow conditions in the narrowest flow cross section of the valve:

$$\eta_{crit,ISO} = \left(\frac{2}{\kappa_0 + 1}\right)^{\kappa_0/(\kappa_0 - 1)} \qquad (15.4)$$

According to EN-ISO 4126-7, the isentropic exponent κ is to be related to the entry conditions ($\kappa = \kappa_0$). Since no additional information has been provided for the calculation, it is obvious to compute the isentropic exponent as well as the sizing coefficient for an ideal gas:

$$\kappa_0 = \frac{c_{p,0}}{c_{p,0} - R} \qquad (15.5)$$

The specific volume at stagnation conditions in Eq. (15.2) should be determined not for ideal gases but rather for a real gas. No method is recommended in EN-ISO 4126-7 for computing the real gas factor.

In Appendix 15.A, an example is given how the sizing coefficient is calculated in accordance with the standard sizing method EN-ISO 4126-7. The sizing coefficient $C_{g,ISO}$ is calculated for a discharge of ethylene from 100 MPa (1000 bar) and 300 K to the atmosphere. The real gas factor defining the specific volume of the ethylene is deduced from an equation of state (see below). Specific heat capacities are taken from IUPAC[1] to calculate the isentropic coefficient by definition for ideal gases. The sizing coefficient is 0.458. By means of Eq. (15.2), it leads to a dischargeable mass flux of $Q_{m,nozzle}/A_{seat} = 2.44 \times 10^5$ kg/(m² s). To get the mass flux through the safety valve, it must be corrected by the discharge coefficient of the safety valve.

The discharge coefficient is in general determined experimentally by the valve manufacturer for a fixed nominal lift. It corresponds to the average value of the ratio from mass flows measured and calculated with the nozzle flow model [1]. At least

---

1) www.iupa.org.

three measurements are performed at each of three set pressures on two or three sizes of the same valve type or three significantly different springs in the same valve.

$$K_{d,g} = \frac{1}{n}\sum_{n}\left(\frac{Q_{m,exp}}{Q_{m,nozzle}}\right) \tag{15.6}$$

This testing of safety valves at a fixed lift presumes that the valves will open at least to the nominal lift in case of an emergency relief.

Overall, the sizing method according to EN-ISO 4126-7 is based on very simple equations mainly for ideal gases. In the literature, the validity of this sizing method was shown by numerical calculations up to pressures of about 10 MPa (100 bar) [3, 4]. No limits of application are formulated in the standards for high-pressure valves. In addition, the discharge coefficients for safety valves are generally measured at pressures between 0.1 and 5 MPa – the application limit of current test facilities amounts to about 25 MPa (250 bar). The question arises whether the application of the current standard beyond a pressure of more than 10 MPa is allowed.

15.2
Limits of the Standard Valve Sizing Procedure

During the flow of a fluid through safety valves, a very complex shaped flow channel is formed (Figure 15.1), in which the fluid entering the valves is first strongly accelerated and subsequently strongly deflected under the valve disk. The exact contour of the flow channel is determined by the inlet conditions, the type of gas, and the geometry of the valve. In the valve seat (point 1 in Figure 15.1) or under the valve disk (points 2 and 3 in Figure 15.1), as a rule, a critical pressure ratio is established – there the gas reaches the velocity of sound. Immediately thereafter, the pressure decreases very strongly, in some cases significantly below ambient pressure, and a region of supersonic velocity is formed in the valve housing. At the periphery of this region, the flow is extremely decelerated (compression shock) and the pressure decreases to the valve outlet flange pressure. Typically, at very high set pressures, the

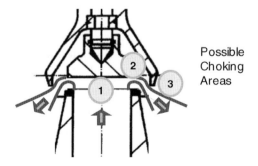

Figure 15.1 Example of the contour of flow in a safety valve with the places where the velocity of sound may be developed.

pressure in the outlet flange is again decreased by a second compression shock down to almost the back pressure of the valve. Any change of conditions in the inlet may result in a change of the position and size of the narrowest cross section of flow and the location of pressure shocks in the valve and the outlet line.

Despite the complex flow phenomena, the technical rules and standards give simple nozzle equations without application limits for the sizing of safety devices, and the valve manufacturers report only two discharge coefficients as constant values for all gases/vapors and liquids in their catalogs. Physically, this is only possible if the flows in the valves behave similarly. The location of the sound velocity must be essentially the same, and the narrowest flow cross section in the valve may not vary significantly (Reynolds and Mach number similarity).

Near the thermodynamic critical point, conversely, gases no longer behave ideally but as real gases. As a result of this, the stagnation temperature in the nozzle varies; therefore, the gas generally becomes cooler in the presence of moderate inlet pressures (negative Joule–Thomson effect); however, it heats up considerably if the inlet pressure is very high (positive Joule–Thomson effect). This real expansion behavior is displayed by many gases even at very low pressures (e.g., methane). It influences the location and size of the narrowest flow cross section and thereby the flow characteristics of the valve. It can also be changed by wall friction and the transport of heat to the wall if the gas densities are so high that these effects are no longer negligible. In addition, the similarity of the flow may also be nullified by geometric factors. Safety valves of the same type are frequently not geometrically similar at different nominal widths for economic and engineering reasons. Since these effects are not accounted for at all in the simple nozzle flow model, they would have to be included in the discharge coefficients of the valves. As a consequence, the coefficients cannot be constant. For physical reasons, the discharge coefficients for gases and liquids at the thermodynamic critical point should even be the same, because the densities of both phases are identical there. The discharge coefficients, however, are not measured under the above-noted boundary conditions.

It is expected that the mass flow rate through a safety valve and possibly also the discharge coefficients for gases/vapors are influenced by an entire series of parameters (gas type, valve geometry, and property data) that are presently not fully accounted for either in the nozzle model per EN-ISO 4126-7 or in the testing procedure.

15.3
Development of a Sizing Method for Real Gas Applications

In engineering practice, it is often argued that at very high pressures or low temperatures, gases usually behave in a distinctly real manner. But in fact, the proximity to the thermodynamic critical point is most important. As an example, nitrogen behaves as an ideal gas under ambient conditions and also at around 100 MPa (1000 bar). In-between, nitrogen is a real gas; that is, intermolecular forces are no longer negligible. The caloric and thermal properties of the gas depend on the temperature *and* on the pressure. In addition, friction and heat exchange with the wall increase considerably if the density of the gas is on the order of liquid densities. As a consequence, the mass flow rate through a nozzle or even a safety valve can no longer be determined by simple relations for ideal gases.

In the case of flow of a real gas through an adiabatic nozzle, the stagnation enthalpy between the entrance and the narrowest flow cross section, the nozzle throat, remains constant (h_t = constant). The change in enthalpy corresponds to the change in kinetic energy and the irreversible frictional pressure loss is expressed by the resistance coefficient ς:

$$-\int_0^{\text{throat}} dh = \frac{w^2}{2}(1+\varsigma), \qquad w_0 \ll w \tag{15.7}$$

$$Q_{m,\text{nozzle}} = \frac{A}{v}\sqrt{\frac{-2}{1+\varsigma}\int_0^{\text{throat}} dh} = C_g A \sqrt{2\frac{p_0}{v_0}} \tag{15.8}$$

The mass flow rate is obtained by a progressive variation of the state conditions in the nozzle throat at a constant stagnation enthalpy – or approximately in the case of isentropic flow (no heat transfer, $\varsigma = 0$) – until the back pressure or a maximum of the mass flow rate (critical pressure ratio) is reached.

This method is suitable when the state diagram of the gas is available or can be plotted. It applies not only for pure substances but also for mixtures of gases. If graphical methods are not efficient, the mass flow rate through a nozzle can also be determined with an equation for real gases, which can be derived from the first law of thermodynamics:

$$dh = c_p\, dT + \left[v - T\left(\frac{\partial v}{\partial T}\right)_p\right] dp \tag{15.9}$$

The second term corresponds to the Joule–Thomson effect, which includes the variation of the stagnation temperature during the expansion of the gas. This term is equal to zero for ideal gases.

The temperature dependence of the specific volume at constant pressure in Eq. 15.9 may be derived from the second law of thermodynamics:

$$ds = c_p \frac{dT}{T} - \left(\frac{\partial v}{\partial T}\right)_p dp \tag{15.10}$$

By means of Eqs. (15.9) and (15.10), the derivation of an analytical equation for the mass flow rate through the nozzle is quite simple if the change of state between the entry and the nozzle throat is assumed to be isentropic ($ds = 0$) and the frictional losses neglected at this time as well as the heat transfer with the wall is accounted by the velocity coefficient ϕ; otherwise, the equations cannot be analytically integrated.

$$\phi = \frac{w}{w_{is}} \cong 0.9 - 1 \tag{15.11}$$

Herewith, the solution of the integral equation (15.7) is traced back to the calculation of an isentropic change of state of a real gas:

$$C_g = \phi\left(\frac{v_0}{v}\right)\sqrt{\frac{-\int_0^{\text{throat}} dh_{is}}{v_0 p_0}}, \quad \text{where } dh_{is} = v\, dp \tag{15.12}$$

The change of enthalpy within the isentropic nozzle flow can be calculated by using the first and second laws of thermodynamics, if the equation of state for real gases

$$v = \frac{ZRT}{p} \tag{15.13}$$

including the real gas factor Z and the isentropic exponent κ (see definition in Eq. (15.28)) is introduced:

$$dh_{is} = ZR\frac{\kappa[1 + K_T]}{\kappa[1 - K_p] - 1} dT = \frac{ZR}{\Pi} dT \tag{15.14}$$

$$\Pi = \frac{\kappa[1 - K_p] - 1}{\kappa[1 + K_T]} \tag{15.15}$$

where K_T and K_p are the gradients of the real gas factor, which correspond to a value of zero for ideal gases:

$$K_T = \left(\frac{T}{Z}\frac{\partial Z}{\partial T}\right)_p, \quad K_p = \left(\frac{p}{Z}\cdot\frac{\partial Z}{\partial p}\right)_T \tag{15.16}$$

The isentropic exponent and the specific heat capacity for ideal and real gases are given in Table 15.1.

Equation (15.12) must be integrated numerically. For convenience, this should be done in three steps: (1) isothermal integration at nozzle stagnation entry temperature T_0 from the starting stagnation pressure p_0 to zero pressure, where ideal gas assumptions are valid for any gas; (2) isobaric integration at zero pressure along

Table 15.1 Property data of ideal and real gases.

Ideal gas	Real gas
$\kappa = \dfrac{c_p}{c_p - R}$	$\kappa = \dfrac{c_p}{c_p[1 - K_p] - ZR[1 + K_T]^2}$
$c_p = \dfrac{\kappa R}{\kappa - 1}$	$c_p = ZR\dfrac{\kappa[1 + K_T]^2}{\kappa[1 - K_p] - 1}$
$c_v = \dfrac{R}{\kappa - 1}$	$c_v = ZR\dfrac{[1 + K_T]^2}{\kappa[1 - K_p]^2 - [1 - K_p]}$

an ideal gas path from stagnation temperatures T_0 to T; and (3) a second isothermal integration at stagnation nozzle throat temperature T from zero pressure to the stagnation nozzle throat pressure p. Ideal gas specific heat properties, which are well known for almost any gas, instead of temperature- and pressure-dependent properties, can be used for the three-step integration procedure [14, chapter 5].

As an alternative, Eq. (15.12) can also be integrated if the average values of the parameter Z and Π, $\langle Z \rangle$ and $\langle \Pi \rangle$, in each case averaged between the stagnation conditions in the entry and the nozzle throat conditions, are introduced. The velocity in the nozzle throat in the case of an isentropic change of state is obtained from this:

$$C_g = \phi \frac{v_0}{v} \sqrt{\frac{1}{\langle \Pi \rangle} \frac{\langle Z \rangle}{Z_0} \left(1 - \frac{T}{T_0}\right)} \tag{15.17}$$

$$\langle \Pi \rangle = \left(\frac{\kappa[1 - K_p] - 1}{\kappa[1 + K_T]}\right)\Bigg|_{T_0}^{T} = \frac{\langle \kappa \rangle [1 - \langle K_p \rangle] - 1}{\langle \kappa \rangle [1 + \langle K_T \rangle]} \tag{15.18}$$

With the isentropic relationships for real gases,

$$\frac{T}{T_0} = \left(\frac{p}{p_0}\right)^{\langle \Pi \rangle} = \eta^{\langle \Pi \rangle} \tag{15.19}$$

$$\frac{v_0}{v} = \left(\frac{p}{p_0}\right)^{1/\langle \kappa \rangle} \tag{15.20}$$

and the velocity coefficient ϕ, the sizing coefficient for a nozzle in case of a frictional flow of a real gas is obtained:

$$C_g = \phi \sqrt{\frac{1}{\langle \Pi \rangle} \frac{\langle Z \rangle}{Z_0} \eta^{2/\langle \kappa \rangle} (1 - \eta^{\langle \Pi \rangle})} \tag{15.21}$$

For an ideal gas and a frictionless flow through an adiabatic nozzle, where $\langle Z \rangle = Z = 1 \rightarrow K_T = K_p = 0$ and $\phi = 1$ is valid, Eq. (15.21) merges into Eq. (15.3).

Typical values for the velocity coefficient in a frictional nozzle flow are 0.9–1 – Sigloch ([5], S. 392) gives an average value of 0.97. Numerical calculations on a high-pressure safety valve lead to the conclusion that the velocity coefficient is close to 1, [14, chapter 5]. The real gas factor is generally determined with a suitable equation of state.

15.3.1
Equation of State and Real Gas Factor

In the industry, predominantly cubic equations of state are used, because a large number of coefficients are tabulated (Table 15.2). They can be reduced back to a basic equation:

$$p = \frac{RT}{v - b} - \frac{\Theta}{v^2 + \delta v + \varepsilon} \frac{v}{RT} \tag{15.22}$$

Table 15.2 Coefficients of the cubic equations of state.

Equation of state	Θ	δ	ε
Redlich–Kwong [7]	$a/\sqrt{T}$	b	0
Soave–Redlich–Kwong [8]	$a\alpha(T)$	b	0
Peng–Robinson [9]	$a\alpha(T)$	$2b$	$-b^2$

Here, a and b are material-specific parameters that are adapted to measured data or can be calculated from the thermodynamic critical data of the gas:

$$a = \Omega_1 \frac{R^2 T_c^2}{p_c}, \quad b = \Omega_2 \frac{RT_c}{p_c} \tag{15.23}$$

$$\alpha(T) = \left[1 + S\left(1 - \left(\frac{T}{T_c}\right)^{0.5}\right)\right]^2 \tag{15.24}$$

$$\omega = -\log\left(\frac{p}{p_c}\right)_{T/T_c=0.7} - 1 \tag{15.25}$$

Significant differences between the equations of state may arise when the coefficients a and b are adapted directly to the measured data. Generally, with the Redlich–Kwong equation of state, the properties of almost spherical and nonpolar molecules can be described (e.g., of inert gases Ar, Kr, and Xe). The Soave–Redlich–Kwong and Peng–Robinson equations are clearly better suited for polar and larger molecules. The equations have not been developed for very high pressures of 1000 bar and above. Here there are special equations of state for individual gases.[1] In this pressure range, a comparison with measured data is always recommended. In the case of gas mixtures, the coefficients a and b (see Eq. (15.23) and Table 15.3) can be determined with the aid of mixing rules [10, 11].

The real gas factor Z is calculated from the equation of state by adjustment of Eq. (15.22) with the specific volume $v = ZRT/p$. For example, for the Soave–Redlich–Kwong equation, we have

$$\frac{ZRT}{p(ZR(T/p) - b)} - \frac{a\alpha}{RT(ZR(T/p) + b)} - Z = 0 \tag{15.26}$$

Table 15.3 Coefficients of the cubic equations of state.

Equation	Ω_1	Ω_2	S
Redlich–Kwong	0.42748	0.08664	0
Soave–Redlich–Kwong	0.42748	0.08664	$0.48 + 1.574\omega - 0.176\omega^2$
Peng–Robinson	0.45724	0.0778	$0.37474 + 1.54226\omega - 0.26992\omega^2$

15.3 Development of a Sizing Method for Real Gas Applications

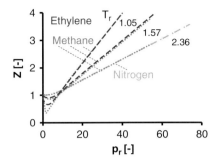

Figure 15.2 Real gas factor Z for ethylene, methane, and nitrogen as functions of reduced stagnation pressure and reduced stagnation temperature.

The equation for computing the real gas factor Z and the derivatives K_T und K_p can be derived analytically. Figure 15.2 shows the real gas factor calculated by the Soave–Redlich–Kwong equation of state for methane in the pressure range between 0.01 and 250 MPa at temperatures of 200, 300, and 450 K, for nitrogen at a temperature of 300 K, and for ethylene at 296.4 and 443.1 K. Pressure and temperature are based in all cases on the critical data of the fluid as reduced pressure p_r and reduced temperature T_r:

$$p_r = \frac{p}{p_c}, \qquad T_r = \frac{T}{T_c} \qquad (15.27)$$

The thermodynamic critical data of certain gases are reproduced in Table 15.4.

Nearby the thermodynamic critical temperature ($T_r = 1.05$), the real gas factor drops off, at first very strongly, reaches a minimum at a reduced pressure of somewhat over 1, and then increases again. The further away the temperature of the gas is from the thermodynamic critical point, the less strongly pronounced the minimum is.

The real behavior of a gas essentially depends on how far away the actual pressure and temperature are from the thermodynamic critical point and not on the absolute values of pressure or temperature of the gas. The assumption that a gas behaves ideally ($Z = 1$) may lead to significant errors in the sizing of safety valves. Basically, the required cross-sectional area of the valve seat is rather underestimated if a too small real gas factor is assumed.

The gradients of the real gas factor, K_p and K_T, are given in Figures 15.3 and 15.4 for ethylene, methane, and nitrogen as functions of the reduced pressure. According to it,

Table 15.4 Thermodynamic critical pressure, temperature, and acentric factor for different gases.

Gas	p_c (MPa)	T_c (K)	ω
Nitrogen	3.3978	126.2	0.04
Methane	4.596	190.564	0.008
Ethylene	5.04	282.3	0.089

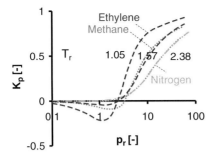

Figure 15.3 Pressure gradient of the real gas factor for ethylene, methane, and nitrogen as functions of reduced stagnation pressure and reduced stagnation temperature.

the pressure gradient K_p is approximately negligible up to a reduced pressure of about 2 as long as the temperature of the gas is not very close to the thermodynamic critical temperature. Neglecting the temperature gradient, K_T may, on the other hand, lead to significant errors already at a moderate reduced pressure.

15.3.2
Isentropic Exponent

The isentropic exponent κ is defined as ratio of pressure fluctuations and corresponding fluctuations of the density at isentropic conditions and may be calculated, for example, by measurements of the velocity of sound [12, 13]:

$$\kappa = -\frac{v}{p}\left(\frac{\partial p}{\partial v}\right)_s = \frac{\varrho w_{\text{crit}}^2}{p} \quad \text{with } w_{\text{crit}} = \sqrt{\frac{1}{-(\partial v/\partial p)_{\text{is}}}} \qquad (15.28)$$

For an ideal gas, the isentropic exponent according to Eq. (15.28) can be transferred into the ratio of specific heat capacities at constant pressure and volume (see Table 15.1). There are various calculation procedures in the literature for approximate

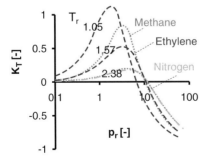

Figure 15.4 Temperature gradient of the real gas factor for ethylene, methane, and nitrogen as functions of reduced stagnation pressure and reduced stagnation temperature.

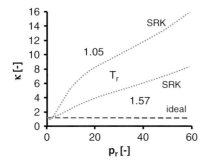

Figure 15.5 Isentropic exponent for ethylene according to EN-ISO 4126-7 and a cubic equation of state (SRK) as a function of reduced pressure and reduced temperature at stagnation inlet conditions.

solutions, for example, isentropic exponent for ideal gases (cf. Table 15.1), and the calculation with an equation of state.

The results of the approximation equations are given in Figure 15.5 for ethylene at temperatures of 296.4 and 443.2 K and pressures up to 270 MPa (2700 bar). Depending on the calculation model, highly differing values for the isentropic exponents are obtained, calculation with the equation of state being the most accurate method. Inaccuracies should be expected there if the coefficients of the equation of state are determined only with the thermodynamic critical values of the gas and not adapted to the measured data.

In general, the isentropic exponents take very high values at large reduced pressures; the description of this factor on the assumption of an ideal gas behavior is then no longer permissible. The calculation by means of measured specific heat capacities according to the isentropic exponent defined in Table 15.1 leads to almost equal values even at very high reduced pressures.

15.3.3
Critical Pressure Ratio

The velocity in the narrowest cross section in a safety valve may reach the sound velocity at maximum. Simultaneously, the pressure decreases to its minimum value, the fluid dynamic critical pressure, and the mass flow rate is maximized.

The fluid dynamic critical pressure and the critical pressure ratio $\eta_{crit} = p_{crit}/p_0$ in a nozzle throat may be calculated by differentiating the sizing coefficient, Eq. (15.21), with respect to the pressure ratio and setting the result equal to zero. Further mathematical transformations lead to the critical pressure ratio of a real gas:

$$\eta_{crit} = \left[\frac{2}{2 + \langle \Pi \rangle \langle \kappa \rangle}\right]^{1/\langle \Pi \rangle} \tag{15.29}$$

An estimation of the critical pressure ratio is necessary to calculate the average values of $\langle \Pi \rangle$ and $\langle \kappa \rangle$. The critical pressure ratio of an ideal gas is a good first estimate for the iteration ($K_T = K_p = 0 \to \langle \Pi \rangle = \langle \kappa \rangle - 1/\langle \kappa \rangle$), whereas the isentropic coefficient should be related to the entrance stagnation condition to avoid any iteration of the start value ($\langle \kappa \rangle = \kappa_0$):

$$\eta_{crit,id} = \left[\frac{2}{\kappa_0 + 1}\right]^{\kappa_0/(\kappa_0-1)} \tag{15.30}$$

The critical pressure ratio according to Eq. (15.30) is given in EN-ISO 4126-7.

The critical pressure ratio η for calculating the mass flow rate, cf. Eqs. (15.2) and (15.3) (ideal gas) and Eqs. (15.2) and (15.21) (real gas), corresponds to the maximum of either the back pressure ratio $\eta = \eta_b = p_b/p_0$ or the critical pressure ratio $\eta = \eta_{crit} = p_{crit}/p_0$.

15.4
Sizing of Safety Valves for Real Gas Flow

The most precise method for calculating the critical mass flux through a safety valve accounting for real gas effects is a numerical integration of Eq. (15.12), defining the sizing coefficient for a nozzle for real gas service. For the integration along the nozzle path, from the inlet to the throat, either the enthalpy or the specific volume of the real gas is needed depending on temperature *and* pressure.

For most gases in practice, the specific heat capacity c_p is not available depending on temperature and pressure. But the specific heat capacity for the ideal gas is known – it depends only on temperature. Hence, the integration of Eq. (15.12) should follow the following integration path [14, chapter 5]:

1) Isothermal change of state at inlet stagnation temperature between the inlet stagnation pressure and zero pressure, where the gas behaves as ideal.
2) Isobaric change of state of an ideal gas at zero pressure from inlet stagnation temperature to the nozzle throat temperature.
3) Another isothermal change of state at nozzle throat temperature between the zero pressure and the pressure in the nozzle throat.

To follow this integration path, the first and second laws of thermodynamics should be rewritten as given in Appendix 15.A.

In Appendix 15.A, an example is given how the sizing coefficient is calculated if real gas effects are accounted for. The real gas specific volume was used for an integration of Eq. (15.12). The sizing coefficient C_g is again calculated for a discharge of ethylene from 250 MPa (2500 bar) and 373 K (100 °C) to the atmosphere. The real gas factor defining the specific volume of the ethylene is deduced from the Soave–Redlich–Kwong equation of state. Specific heat capacities are taken from IUPAC[1] to calculate the isentropic coefficient by definition for real gases (see Table 15.1). The sizing coefficient is 0.746. By means of Eq. (15.2), it leads to a dischargeable mass flux of

$Q_{m,nozzle}/A_{seat} = 3.98 \times 10^5$ kg/(m² s). It is, therefore, 63% larger than the value in accordance with EN-ISO 4126-7. If the same discharge coefficient of the valve is used, the minimum required area of the safety valve would be 63% lower.

In engineering practice, it is often more convenient to solve the integral of Eq. (15.12) by an approximation with a stepwise sum of the specific volume times the pressure interval. The solution can be performed in a spreadsheet.

Regarding the example in Appendix 15.A, the use of EN-ISO 4126-7 would be overconservative when sizing a safety valve. But the discharged mass flow rate – generally necessary for sizing downstream appliances – is highly underestimated. Unfortunately, close to the thermodynamic critical point, the standard valve sizing procedure would lead to smaller required valve areas than calculated with a nozzle flow model accounting for real gas effects.

As an alternative, the analytic solution based on Eq. (15.17) may be used for sizing a valve. Arithmetic averages between the inlet and the throat of the nozzle for the isentropic exponent, the real gas factor, and its gradients can only be recommended if the change of the isentropic coefficient and the real gas factor is almost linear; that is, pressure and temperature are far from the thermodynamic critical condition.

The results for a discharge of ethylene from stagnation conditions of 250 MPa (2500 bar) and 373 K (100 °C) are shown in Appendix 15.A. The discharge coefficient of an ideal nozzle amounts to 0.704 and the mass flux equals 3.744×10^5 kg/(m² s). It is 5.6% lower than the result of the numerical calculation.

The calculated sizing coefficients for ethylene are reproduced in Figure 15.6. According to EN-ISO 4126-7 with an assumption of an ideal behavior of the gas the sizing coefficient varies only slightly. Considering real gas effects, considerably different values can be found: sizing coefficients about 74% lower are calculated with EN-ISO 4126-7 at an inlet pressure of 2700 bar. Overall, the type of gas

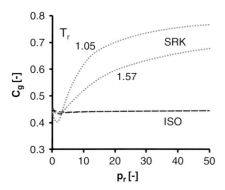

Figure 15.6 Sizing coefficient versus reduced inlet stagnation pressure calculated according to EN-ISO 4126-7 and a nozzle flow model including real gas effects for ethylene at inlet stagnation temperatures of 300 and 443 K.

generally has a smaller effect on the sizing coefficient than the state conditions at the entrance.

In general, the sizing coefficients presented herein are valid, if no condensation occurs up to the nozzle throat (the narrowest flow cross section of the valve). Therefore, pressures and temperatures at the inlet and the nozzle throat should be above the saturation line of the fluid considered.

The REAL nozzle flow model was validated with measurements performed on a high-pressure nozzle with nitrogen at a test facility of BASF. The measurements were conducted at inlet pressures of up to 1000 bar [15]. In addition, the calculations agree very accurately with the values in DIN EN ISO 9300 [13], based on many measurements on nozzles performed worldwide at pressures below 200 bar. Moreover, Beune [6] has intensively studied the flow behavior through high-pressure safety valves by numerical simulations with ANSYS CFX. He underlined the precision of the real gas nozzle flow model. In addition, he verified experimentally for a certain valve type that the discharge coefficient of the safety valve in combination with the REAL nozzle flow model can be regarded as constant as is presently stated in EN-ISO 4126-7.

In practice, often an ideal behavior of gases is assumed at moderate pressures when sizing a safety valve for gas service. Real gas behavior is only assumed at a very high pressure, for example, at a pressure of more than 100 bar. In general, the real gas behavior is rather determined from the proximity of the thermodynamic critical point. With the reduced thermodynamic pressure and the reduced thermodynamic temperature, the deviation from ideal behavior can be described much better than with the absolute values of pressure and temperature. If the reduced pressure and the reduced temperatures at the entrance of the nozzle exceed $p/p_c > 0.5$ or $T/T_c > 0.9$, the deviations from the ideal behavior are usually no longer tolerable.

Applying the REAL nozzle flow model, care must be taken for averaging the parameter. Close to the thermodynamic critical point, it might be necessary to stepwise average the parameter or to solve the equation numerically.

15.5
Summary

Currently for sizing safety valves, the only methods available have been those developed for applications in which no strong real gas effects are expected, cf., for example, EN-ISO 4126-7 [1, 2]. For high-pressure applications, the calculation of the gas properties is not adequately described in the technical standards. The application of equations according to EN-ISO 4126-7 and AD 2000 is limited. If real gas effects are considered, the sizing coefficients for ideal nozzle flow result in much larger values by trend. For the sizing of a safety valve a lower sizing coefficient is conservative, whereas for the proper operation and the sizing of downstream appliances it may not.

Appendix 15.A Calculation of Sizing Coefficient According to EN-ISO 4126-7 and a Real Gas Nozzle Flow Model

15.A.1
Inlet Stagnation Conditions

$T_0 := 373.15$ K   inlet temperature

$p_0 := 2500 \times 10^5$ Pa   inlet pressure

15.A.2
Property Data and Coefficients for Ethylene

Real gas coefficient $Z(p, T)$ as a function of pressure p and temperature T according to Soave–Redlich–Kwong equation of state calculated elsewhere:

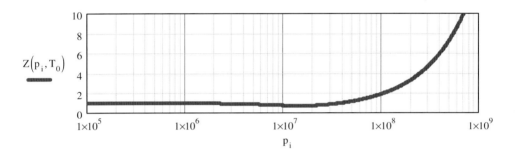

$$K_p(p, T) := \frac{p}{Z(p, T)} \frac{d}{dp} Z(p, T), \quad K_p(p_0, T_0) = 0.851$$

$$K_T(p, T) := \frac{T}{Z(p, T)} \frac{d}{dp} Z(p, T), \quad K_T(p_0, T_0) = -0.759$$

$$v(p, T) := \frac{Z(p, T)RT}{p}, \quad v(p_0, T_0) = 1.767 \times 10^{-3} \text{m}^3/\text{kg}$$

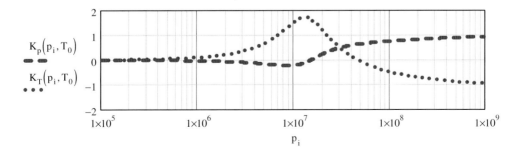

$v(p_i, T_0)$

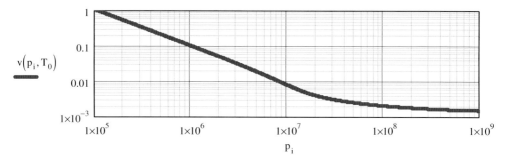

15.A.3
Calculation of Flow Coefficient According to EN-ISO 4126-7

$$\kappa_{ISO} := \frac{c_{p,id}(T_0)}{c_{p,id}(T_0) - R} = 1.196, \quad \eta_{crit,ISO} := \left(\frac{2}{\kappa_{ISO} + 1}\right)^{\kappa_{ISO}/(\kappa_{ISO}-1)} = 0.565$$

$C_{ISO}\left(\dfrac{p_i}{p_0}, \phi\right)$

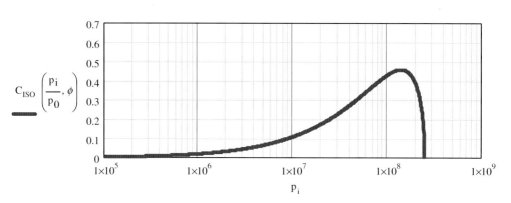

Fluid dynamic critical pressure and temperature in the throat area of the nozzle and the narrowest flow cross section of the safety valve:

$$T_{crit,ISO} := \left(\frac{2}{\kappa_{ISO} + 1}\right) T_0 = 339.797 \text{ K}$$

$$p_{crit,ISO} := \eta_{crit,ISO} p_0 = 1.413 \times 10^3 \text{ bar}$$

$$C_{g,ISO} := C_{ISO}(\eta_{crit,ISO}, \phi) = 0.458$$

$$\eta_{crit,ISO} = 0.565$$

$$m_{ISO} := C_{g,ISO} \sqrt{2 \frac{p_0}{v(p_0, T_0)}} = 2.437 \times 10^5 \text{ kg}/(\text{m}^2 \text{ s})$$

15.A.4
Calculation of Flow Coefficient Accounting for Real Gas Effects

Iterative procedure to calculate the fluid dynamic critical pressure and temperature. Estimated values to start the iteration: critical conditions according to EN-ISO 4126-7.

$$\kappa_{\text{ideal}}(T) := \frac{c_{p,\text{id}}(T)}{c_{p,\text{id}}(T) - R} \qquad \text{isentropic coefficient for ideal gases}$$

$$T_{s,\text{id}}(p_0, p, T_0, T) := T_0 \left(\frac{p}{p_0}\right)^{(\kappa_{\text{id}}(T)-1)/\kappa_{\text{id}}(T)} \qquad \text{temperature at pressure } p \text{ for an ideal gas}$$

Second law of thermodynamics (isentropic change of condition):

Vorgabe

$$\ln\left(\frac{p_0}{p}\right) + \int_{T_0}^{T} \frac{c_{p,\text{id}}(T_x)}{RT_x} dT_x + \int_{p_0}^{0} -[Z(p_x, T_0)(1 + K_T(p_x, T_0)) - 1]\frac{1}{p_x} dp_x$$

$$+ \int_{0}^{p} -[Z(p_x, T)(1 + K_T(p_x, T)) - 1]\frac{1}{p_x} dp_x = 0$$

$$T_s(p_0, p, T_0, T) := \text{Suchen}(T) \qquad \text{isentropic temperature at pressure } p \text{ for a real gas}$$

First law of thermodynamics (integral dh):

$$\text{INT } dh(p_0, p, T_0, T) := \int_{T_0}^{T} c_{p,\text{id}}(T_x) dT_x$$

$$+ \int_{0}^{p_0} Z(p_x, T_0) RT_0 K_T(p_x, T_0) \frac{1}{p_x} dp_x$$

$$+ \int_{0}^{p} -Z(p_x, T) RT K_T(p_x, T) \frac{1}{p_x} dp_x$$

Sizing coefficient for adiabatic frictionless flow of real gases through nozzles:

$$C_g(p_0, p, T_0, T, \phi) := \phi \sqrt{\frac{v(p_0, T_0)}{2p_0}} \frac{\sqrt{-2\text{INT } dh(p_0, p, T_0, T)}}{v(p, T)}$$

Derivative of sizing coefficient is set to zero for maximum search and solved numerically:

Vorgabe
$$T_{is}(p_x) := T_s(p_0, p_x, T_0, T_x) \qquad \text{real gas isentropic temperature}$$

$$\frac{d}{dp_x} C_g(p_0, p_x, T_0, T_{is}(p_x), \phi) = 0 \quad \text{maximum flow coefficient at critical pressure}$$

$$p_{crit} := \text{Suchen}(p_{xx})$$

Fluid dynamic critical pressure and temperature in the throat area of the nozzle and the narrowest flow cross section of the safety valve for real gas flow:

$$p_{crit} = 4.946 \times 10^7 \text{ Pa}, \quad \eta_{crit} := \frac{p_{crit}}{p_0} = 0.198$$

$$T_{crit} := T_s(p_0, p_{crit}, T_0, T_x) = 309.773 \text{ K}$$

$$C_{g,crit} := C_g(p_0, p_{crit}, T_0, T_{crit}, \phi) = 0.74565$$

Mass flux of real gas through the nozzle:

$$m_{real,crit} := C_{g,crit} \sqrt{2 \frac{p_0}{v(p_0, T_0)}} = 3.967 \times 10^5 \text{ kg}/(\text{m}^2 \text{ s})$$

15.A.5
Approximation of Mass Flux by an Analytical Method (Averaging Method)

Estimated pressure in the nozzle throat: $p_1 := 1068$ bar
Isentropic temperature at estimated pressure:

$$T_1 := T_s(p_0, p_1, T_0, T_0) = 337.142 \text{ K}$$

Real gas coefficient, adiabatic exponent, gradients of real gas coefficients, and averaged values:

$$Z_0 := Z(p_0, T_0) = 3.99413, \qquad Z_1 := Z(p_1, p_1) = 2.11479$$

$$\kappa_{real,0} := \kappa_{real}(p_0, T_0) = 8.33531, \qquad \kappa_{real,1} := \kappa_{real}(p_1, T_1) = 7.01172$$

$$K_{T,0} := K_T(p_0, T_0) = -0.7589, \qquad K_{T,1} := K_T(p_1, T_1) = -0.5656$$

$$K_{p,0} := K_p(p_0, T_0) = 0.8507, \qquad K_{p,1} := K_p(p_1, T_1) = 0.80711$$

$$Z_m := \frac{Z_0 + Z_1}{2} = 3.05446$$

$$\kappa_{real,m} := \frac{\kappa_{real,0} + \kappa_{real,1}}{2} = 7.67352$$

$$K_{p,m} := \frac{K_{p,0} + K_{p,1}}{2} = 0.82891$$

$$K_{T,m} := \frac{K_{T,0} + K_{T,1}}{2} = -0.66225$$

$$\Pi_m := \left[\frac{\kappa_{\text{real},m}(1 - K_{p,m}) - 1}{\kappa_{\text{real},m}(1 + K_{T,m})}\right] = 0.12072$$

Flow coefficient and mass flux according to averaging method, Eq. (15.21):

$$C_{g,m} := \phi \sqrt{\frac{1}{\Pi_m} \frac{Z_m}{Z_0} \left(\frac{p_1}{p_0}\right)^{2/\kappa_{\text{real},m}} \left[1 - \left(\frac{p_1}{p_0}\right)^{\Pi_m}\right]} = 0.704$$

$$m_{m,\text{crit}} := C_{g,m} \sqrt{2 \frac{p_0}{v(p_0, T_0)}} = 3.744 \times 10^5 \text{ kg}/(\text{m}^2 \text{ s})$$

Pressure in the nozzle throat (check of estimated pressure):

$$\eta_{\text{crit},m} := \left(\frac{2}{2 + \Pi_m \kappa_{\text{real},m}}\right)^{1/\Pi_m} = 0.04273$$

$$p := \eta_{\text{crit},m} p_0 = 1.068 \times 10^7 \text{ Pa}$$

Appendix 15.B List of Symbols

a	Variable of the equation of state of real gases (m^5/(kg s^2))
A	Narrowest flow cross section in the nozzle (m^2)
A_{seat}	Valve seat cross section (m^2)
b	Variable of the equation of state of real gases, covolumes (m^3/kg)
c_p	Specific heat capacity at a constant pressure (J/(kg K))
c_v	Specific heat capacity at a constant volume (J/(kg K))
C_g	Sizing coefficient for gases including real gas effects
$C_{g,\text{ISO}}$	Sizing coefficient for gases according to EN-ISO 4126-7
h	Specific enthalpy (J/kg)
h_t	Specific stagnation enthalpy (J/kg)
H	Enthalpy (J)
$K_{d,g}$	Discharge coefficient of a safety valve for gases/vapors
K_p	Pressure gradient of the real gas factor
$\langle K_p \rangle$	Mean pressure gradient of the real gas factor (average between inlet and nozzle throat)
K_T	Temperature gradient of the compressibility number
$\langle K_T \rangle$	Mean temperature gradient of the real gas factor (average between inlet and nozzle throat)
M	Molar mass (kg/kmol)
n	Number of measurement data
p	Pressure (bar)

p_b	Counter-pressure (bar)
p_r	Reduced pressure
Q_m	Dischargeable mass flow rate through a safety valve (kg/s)
$Q_{m,OUT}$	Mass flow to be discharged from a pressurized system (kg/s)
$Q_{m,nozzle}$	Dischargeable mass flow through an ideal frictionless nozzle (kg/s)
R	Specific gas constant (J/(kg K))
S	Parameter in the equation of state (J/(kg K))
s	Specific entropy (J/(kg K))
T	Temperature (K)
T_r	Reduced temperature
v	Specific volume (m³/kg)
V	Volume (m³)
w	Flow velocity (m/s)
z	Height (m)
Z	Real gas factor
$\langle Z \rangle$	Mean value of the real gas factor (average between inlet and nozzle throat)
$a(T)$	Variable for computing the equation of state of real gases
χ	Parameter of the equation of state of real gases
δ	Parameter of the equation of state of real gases
ε	Parameter of the equation of state of real gases
γ	Parameter of the equation of state of real gases
η	Pressure ratio
ϕ	Velocity coefficient
κ	Isentropic exponent
$\langle \kappa \rangle$	Mean adiabatic exponent (average between inlet and nozzle throat)
λ	Parameter of the equation of state of real gases
μ	Dynamic viscosity (Pa s)
Π	Exponent for real gases
$\langle \Pi \rangle$	Mean value of exponent for real gases (average between inlet and nozzle throat)
ϱ	Density (kg/m³)
ω	Acentric factor

Subscripts

none	Static condition
0	Inlet stagnation condition
c	Critical (thermodynamic)
crit	Critical (fluid dynamic), state in narrowest cross section of the flow channel
CFX	Calculated with the program ANSYS CFX
exp	Experimental
g	Gas
is	Isentropic (loss-free and adiabatic)

num	Numerical
r	Reduced magnitude
SV	Safety valve
t	Total

References

1 EN-ISO 4126 (2004) *Safety Devices for Protection Against Excessive Pressure. Part 7, Common data*, Beuth Verlag GmbH, Berlin. Until 2012: EN-ISO 4126 Part 1, Safety Valves (Content has been shiftet from Part 1 to Part 7 during a revision in 2012).

2 Arbeitsgemeinschaft Druckbehälter (AD) 2000 Merkblatt A2 (2001) *Sicherheitseinrichtungen gegen Drucküberschreitung – Sicherheitsventile*, Beuth Verlag GmbH, Berlin.

3 Johnson, R.C. (1964) Calculations of real gas effects in flow through critical flow nozzles. *J. Basic Eng.*, **86**, 519–526.

4 Baurfeind, K. and Friedel, L. (2003) Berechnung der dissipationsbehafteten kritischen Düsenströmung realer Gase. *Forsch. Ingenieurwes.*, **67**, 227–235.

5 Redlich, O. and Kwong, J.N.S. (1949) On the thermodynamics of solutions. V. An equation of state. Fugacities of gaseous solutions. *Chem. Rev.*, **44**, 233–244.

6 Beune, A. (2009) Analysis of high pressure safety valves. Ph.D. thesis, TU Eindhoven, The Netherlands.

7 Sigloch, H. (2007) *Technische Fluidmechanik*, 6 Auflage, Springer.

8 Soave, G. (1972) Equilibrium constants from a modified Redlich–Kwong equation of state. *Chem. Eng. Sci.*, **27**, 1197–1203.

9 Peng, D.Y. and Robinson, D.B. (1976) A new two-constant equation of state. *Ind. Eng. Chem. Fundam.*, **15**, 59–64.

10 Dohrn, R. (1994) *Berechnung von Phasengleichgewichten (Calculation of Phase Equilibria)*, Vieweg Verlag, Braunschweig.

11 Gmehling, J. and Kolbe, B. (1992) *Thermodynamik*, 2 Auflage, VCH Verlagsgesellschaft, Weinheim.

12 Baehr, H.-D. (1967) Der Isentropenexponent der Gase H_2, N_2, O_2, CH_4, CO_2, NH_3 und Luft für Drücke bis 300 bar (The isentropic exponent for the gases H_2, N_2, O_2, CH_4, CO_2, NH_3 and air for pressures up to 300 bar). *BWK Brennst.-Wärme-Kraft*, **19** (2), 65–68.

13 DIN EN ISO 9300 (2003) *Durchflussmessung von Gasen mit Venturidüsen bei Kritischer Strömung*, Beuth Verlag GmbH, Berlin.

14 Schmidt, Jürgen (ed.) *Process and Plant Safety Applying Computational Fluid Dynamics* 1. Edition - 2012.

15 Schmidt, J., Peschel, W., and Beune, A. (2009) Experimental and theoretical studies on high pressure safety valves: sizing and design supported by numerical calculations (CFD). *Chem. Eng. Technol.*, **32** (2), 252–262.

Appendix: International Codes and Standards for High-Pressure Vessels

Introduction

This appendix is a compilation of codes and standards for high-pressure vessels. The center of interest is the dimensioning and the lifetime evaluation of high-pressure vessels. Several codes and standards are evaluated detailed with a view to the applicability for high-pressure vessels. Primarily are described the respective formula for dimensioning of cylindrical shells under internal pressure. The limits of validity are a quite good indication for the application of the codes for high-pressure vessels.

Abbreviations

d	Inner diameter of pressure vessel
D	Outer diameter of pressure vessel
P	Internal pressure
S	Allowable stress
t_{min}	Minimum required wall thickness, including mechanical and corrosion allowances
Y	Ratio of outer diameter and inner diameter of pressure vessel
S_y	Yield strength

Corresponding International Codes and Standards for Unfired Pressure Vessels

Table A.1 shows a collection of international codes and standards for unfired pressure vessels. The column "Codes for construction of pressure vessels" contains the codes that are usually used for normal pressure vessels. The codes contained in the next column "Alternative rules" have the possibilities to evaluate the results of an FE analysis and also give guidelines to assess the protection against failure from cyclic loading. The codes to be normally used for high-pressure vessels are included in the last column.

Industrial High Pressure Applications: Processes, Equipment and Safety, First Edition. Edited by Rudolf Eggers.
© 2012 Wiley-VCH Verlag GmbH & Co. KGaA. Published 2012 by Wiley-VCH Verlag GmbH & Co. KGaA.

Table A.1 International codes and standards for unfired pressure vessels.

	Codes for construction of pressure vessels	Codes for construction of pressure vessels: alternative rules	Codes for construction of pressure vessels: alternative rules for high pressure
China	GB 150 [7]	JB 4732 [7]	JB 4732 [7]
Europe	EN 13445 [8]	EN 13445 [8]	
France	CODAP 2005 Division 1 [12]	CODAP 2005 Division 2 [13]	
Germany	AD2000 [9]	AD2000 [9]	
Great Britain	PD 5500 [11]	PD 5500 [11]	
India	IS 2825 [16]		
Japan	JIS B 8265 [14]	JIS B 8266 [15]	HPIS C106-2005 [4]
Korea	KEPIC MG [6]		
The Netherlands	RTOD [10]	RTOD [10]	
Russia	GOST R 52857 [5]	GOST R 52857 [5]	
United States	ASME VIII-1 [1]	ASME VIII-2 [2]	ASME VIII-3 [3]

The summary of the codes for high-pressure vessels represents the current state of the codes (2009). Some of the statements of the codes are not based directly on the codes but on publications about them.

Several codes and standards are evaluated detailed with a view to the applicability for high-pressure vessels.

United States – ASME Section VIII-2 [2]

Code	Validity	Minimum required wall thickness
ASME Section VIII-2 [2] "Alternative rules" Part 4.3: Design rules for shells under internal pressure		Cylindrical shell under internal pressure (4.3.3): $t_{min} = \dfrac{d}{2}\left(\exp\left[\dfrac{P}{SE}\right] - 1\right)$

Notes

Section VIII Division 2 [2] has requirements for materials, design by rule, design by analysis, fabrication, inspection and examination, pressure testing, and overpressure protection. In addition, this division provides design rules for layered vessels.

There is the possibility to evaluate the results of an FE analysis based on Part 5 "Design by analysis requirements." Part 5 describes the elastic stress analysis method, limit load analysis method, and the elastic–plastic stress analysis method.

For "thick-walled" components, the plastic analysis methods (limit load analysis method and the elastic–plastic stress analysis method) are more adequate. Part 5 also gives a guideline to assess the protection against failure from cyclic loading.

The rules of Section VIII-2 [2] do not specify a pressure limitation but are applicable to all types of high-pressure vessel constructions. Therefore, some additional considerations to these rules may be necessary to meet the design principles and construction practices essential to very high-pressure vessels. As an alternative to Division 2 [2], Division 3 should be considered for the construction of vessels intended for operating pressures exceeding 68.95 MPa (see the following section).

United States – ASME Section VIII-3 [3]

Code	Validity	Minimum required wall thickness
ASME Section VIII-3 [3] "Alternative rules for construction of high-pressure vessels" KD-221.1: Cylindrical monobloc shells	Closed-end cylindrical shell and open-end cylindrical shell for $Y \leq 2.85$	Cylindrical shell under internal pressure (KD-221.1): $t_{min} = \dfrac{d}{2}\left(\left[\dfrac{P}{2.5856 S_y} + 1\right]^{1/0.268} - 1\right)$
	$Y > 2.85$	$t_{min} = \dfrac{d}{2}\left(\exp\left[\dfrac{5P}{4 S_y}\right] - 1\right)$

Notes

The rules of Division 3 [3] describe the design, construction, inspection, and overpressure protection of metallic pressure vessels with design pressures generally above 68.95 MPa.

Applications include hot and cold isostatic pressing, food sterilization, quartz crystal growth, polyethylene production, oil and gas production, hydrogen transport and storage, and research and development.

Division 3 is a design-by-analysis code, although a few detailed design rules are provided. The types of constructions specifically considered are conventional welded vessels, forged nonwelded vessels, forged layered vessels assembled by shrink fitting, concentrically wrapped welded layered vessels, welded layered vessels assembled by shrink fitting, wire wound vessels, vessels with fiber-reinforced polymer composite hoop wrapping, and vessels for impulsive (explosive) loading.

Methodology and calculations for the following failure modes are provided: through the thickness and local yielding, leak due to fatigue cracks, fast fracture due to unstable crack growth, buckling, and ratcheting.

In this division, there is the possibility to estimate the results of an FE analysis based on the KD-230 "Elastic–plastic analysis" and KD-240 "Linear elastic analysis." KD-3 "Fatigue evaluation" and KD-4 "Fracture mechanics evaluation" also give a guideline to assess the protection against failure from cyclic loading.

The minimum required thickness can be determined by using closed-form equations for cylindrical and spherical shells, blind ends, threaded closures, and clamp connections. The elastic–plastic analysis can be used in most cases and for all D/d ratios. The linear elastic analysis is permitted only if $D/d < 1.25$ [17].

Fatigue analysis in Division 3 can be done using the "traditional SN method" or the "structural stress method" (limited to the analysis of welds) only if leak-before-burst behavior can be demonstrated. Otherwise, the fracture mechanics method must be used.

The fracture mechanics method is the most robust method provided in Division 3 to determine a service inspection interval [17].

The material tables in Chapter KM-4 list materials suitable for welded constructions with yield strengths up to 760 MPa. Materials with yield strengths up to 965 MPa are listed for nonwelded primary pressure boundary construction. Materials with a minimum specified yield strength greater than 835 MPa can be used for confined liners where leak-before-burst criteria can be met.

Europe – EN 13445 [8]

Code	Validity	Minimum required wall thickness
EN 13445 [8] "Unfired pressure vessels"	This rule is valid for cylindrical shells of pressure vessels with the limitations $t_{min}/D \leq 0.16$	Cylindrical shell under internal pressure (7.4.2):
Chapter 7: Shells under internal pressure		$t_{min} = \dfrac{dP}{2SE-P}$

Notes

The EN 13445 Section 1 [8] excepts layered vessels, autofrettage vessels, and prestressed vessels (p.e. wire wound vessel).

In EN 13445 Section 3 [8], there is the possibility to evaluate the results of an FE analysis based on Appendix C "Procedure of stresses categories for the dimension based on analysis methods." Appendix C describes essentially the elastic stress analysis method. Stresses are determined using an elastic analysis, classified into categories, and limited to allowable values that have been conservatively established so that a plastic collapse will not occur. For "thick-walled" components, the plastic analysis methods in Appendix B "Directly dimension with analysis methods" are more adequate. This appendix also contains a guideline to perform a plastic analysis based on the upper bound limit load method, a ratcheting analysis, and a buckling analysis.

Chapter 18 in EN 13445 Section 3 [8] "Detailed calculation of the cyclic life" gives a guideline to calculate the cyclic design life of pressure vessels. But EN 13445 Section 3 is not applicable for extremely high-pressure vessels, because the design fatigue curves are limited by a material tensile strength of 1000 MPa.

Germany – AD2000 [9]

Code	Validity	Minimum required wall thickness
AD2000 [9] "Pressure vessel"	This rule is valid for cylindrical shells of pressure vessels within the limitation $1.2 < D/d \leq 1.5$ provided that the shell sustains the full axial stress and the material of the shell shows ductile behavior	Cylindrical shell under internal pressure (AD2000-B10, 6.1.1):
Chapter B10: Thick-walled cylindrical shells under internal pressure		$t_{min} = \dfrac{dP}{2.3S - 3P}$

Notes

The AD2000 code [9] is generally used for "thin-walled" unfired pressure vessels ("thin walled" means $D/d \leq 1.5$). There are only a few options to dimension high-pressure vessels. An exception is Chapter B10 for "thick-walled" cylindrical shells within the limitation $1.2 < D/d \leq 1.5$. Beyond this limitation, AD2000-B10 refers to the technical book "Apparate und Armaturen der Chemischen Hochdrucktechnik" by H.H. Buchter (Springer Verlag, 1967). This book contains many options to size different types of high-pressure vessels and design guidelines. But since it is not a code, the book only provides few proposals for safety factors.

In AD2000 code [9], there is the possibility to evaluate the results of an FE analysis based on Chapter S4 "Estimation of stresses based on computed and experimental strength analysis." Chapter S4 describes basically the elastic stress analysis method: Stresses are determined using an elastic analysis, classified into categories, and limited to allowable values that have been conservatively established so that a plastic collapse will not occur; for "thick-walled" components, the plastic analysis methods are more adequate. In Chapter S4 [9], there is no guideline for such an analysis, but the application is not forbidden.

Chapter S2 [9] "Evaluation of cyclic loading" gives a guideline to calculate the cyclic design life of pressure vessels. However, Chapter S2 is not applicable for extremely high-pressure vessels, because the design fatigue curves are limited by the material tensile strength of 1000 MPa.

In addition, Chapter HP 801/39 "Pressure vessel of Isostat presses" [9] provides special requirements for the in-service inspections for this type of high-pressure vessels.

Special Aspects for Test Pressure Definition for High-Pressure Vessels

There are special considerations to define the test pressure for prestressed pressure vessels. The normal reason of a pressure test is to demonstrate the integrity of the vessel and to induce compression residual stresses in the area of notches. Prestressed high-pressure vessels have mostly a simple design without notches. The test pressure should be limited, in order not to change the prestressed status of the pressure vessel. The determinations in /3/ for test pressure definition are practicable. Additional nondestructive test procedure should be avoided.

References

1. ASME Boiler and Pressure Vessel Code, Section VIII, Division 1, 2009 edition.
2. ASME Boiler and Pressure Vessel Code, Section VIII, Division 2, 2009 edition.
3. ASME Boiler and Pressure Vessel Code, Section VIII, Division 3, 2009 edition.
4. Susumu Terada, Development of alternate methods for establishing design margins for ASME Section VIII Division 3 (parts 1 and 2), ASME PVP 2009.
5. Boris Volfson, New Russian national standards on pressure vessel and apparatus design and strength calculation, ASME PVP 2009.
6. Hoon-Seok Byun, The present and the future of the Korea electric power industry code, ASME PVP 2008.
7. Shou Binan, Recent development of the pressure vessel codes and standards in China, ASME PVP 2008.
8. EN 13445: Unfired pressure vessels, 2002 edition.
9. AD2000 – Regelwerk: Pressure vessel, 2009 edition.
10. RTOD: Regel voor Toestellen Onder Druk, 2005 edition.
11. PD 5500: Specification for unfired fusion welded pressure vessels, 2009 edition.
12. CODAP Division 1: France design rules for unfired pressure vessel, 2005 edition.
13. CODAP Division 2: France design rules for unfired pressure vessel, 2005 edition.
14. JIS B 8265: Construction of pressure vessel – general principles, 2008 edition.
15. JIS B 8266: Alternative standard for construction of pressure vessels, 2003 edition.
16. IS 2825: Code for unfired pressure vessels, 1969 edition.
17. J. Robert Sims, ASME Section VIII, Division 3: Alternative rules for construction of high pressure vessels, PVPD-60.

Index

a

absorption 19, 35, 128, 173, 174, 195, 206, 327
adiabatic nozzle 373
advanced process control (APC) 94, 95
– MFI and MFR trend 95
– utility 95
aerogels, supercritical drying 202, 203
– process, flow sheet 203
ammonia synthesis process 55
– ammonia content in equilibrium 57
– basics and principles 56
– chemical and physical hydrogen attack 64
– development, of process and pressure 58–63
– disadvantage of turbo compressors 63
– history, ammonia process 3–6, 57, 58
– material selection 63
– metal dusting 63, 64
– nitriding 64
– stress corrosion cracking 64
– world ammonia production 56
angle control valve 304
anthracite 125
antisolvent 174
Archimedes number 34
Austenitic–ferritic duplex steels 288
autoclave reactor 84–85
– safety requirements 88–89
autofrettage 5, 259, 260, 262, 264, 273, 298, 299, 314, 317
– to improve fatigue lifetime 260–264

b

Bauschinger effect 298
benzaldehyde 77
binary diffusion coefficient 22, 28
biodiesel 235
bioethanol 235
biogas 235

biogenic waste 236
biomass 235, 236
– fuel conversion, technologies 236
– hydrothermal conversion 236–237
biomatter, influence of pressure 212–215
bolted cover 296
Bourdon gauge 339

c

Cahn–Hillliard theory 14
capillary pressure 15, 16
carbonates 146
carbon capture storage technology (CCS) 30, 123
carbon dioxide, as miscible fluid 148
Carnot's efficiency 127
cellulose, hydrothermal decomposition 254
centrifugal pumps 311
chemical equilibrium 51, 55, 212
chemical polymerization 77
cinematic viscosity 34, 35
coal-fired steam power plant 123, 125
– evolution of steam parameters 124
– net efficiency 124
– power plant efficiency 125–127
– thermal efficiency of Rankine cycle 126
– thermodynamics 125–127
coal for power generation 123
CO_2 emissions 123
cold bending process 307
cold pasteurization 211
combustion 5, 6, 20, 127, 156, 160, 246, 247, 367
compression shock 370
compressors 3, 6, 68, 82, 155, 184, 260, 311
– centrifugal 63
– diaphragm 332

– influence of fluid on selection and design 313, 314
– piston-type 63, 323, 330
– screw 319, 322
– turbo 63, 314
condensation processes 43–45
– average heat transfer coefficient 44
– deviation of real velocity profile 44
– dispatching of condensate film 45
– enhancement factor 45
– impact of high pressures on 45
– in inert gases 45
– shear stress, evaluation of 44, 45
conical pipe connection 307
core–shell nanoparticles (CSN) 112, 113
corrosion 64, 71, 72, 111, 136, 191, 287, 288, 294, 298, 340, 342, 358
critical heat flux q_{max} 38
cutting devices 273–275

d

dairy processes 107–109
density profile of CO_2 13, 21
diffusion coefficients 22, 35
– calculation of 36
– of CO_2 in oil 35
double pipe heat exchanger 302
drag coefficients, of droplets in high-pressure atmosphere 20
droplet
– agglomeration 107
– coalescence 104–107
– disruption 103, 104

e

elastic stress 291
elastomer O-rings 300
emulsification, and process functions 103–107
– droplet disruption 103, 104
– homogenization valves
– – droplet agglomeration in 107
– – droplet coalescence in 104–107
enhanced oil recovery (EOR) 148, 150, 155–157, 162
– carbon dioxide capture and storage (CCS) in 160
– chemical injection 158
– gas injection 159, 160
– thermal recovery 158, 159
– water flooding 157, 158
enthalpy 17, 22, 125
entropies 127
enzyme activity 212, 215, 216, 222, 224

enzyme inactivation kinetics, influence of pressure 215–218
– adiabatic process 219
– Arrhenius equation 217
– biphasic/multiphasic behavior 216, 217
– comparison of temperature fields 221
– down-flow velocity 219
– Eyring equation 217
– factors 218
– first-order kinetics 216, 222
– heterogeneous temperature field in autoclave 221
– at higher temperatures 223
– industrial inactivation 219
– inhomogeneous inactivation 225
– lipoxygenase (LOX), activity of 223
– particle tracks of enzymes 224
– polyphenoloxidase (PPO), activity of 223
– size of autoclaves, significant effect on 224
– technological aspects 218–226
– temperature stratification in autoclave 222
– thermal heterogeneities
– – effect of packaging on 225
– – quality indicators, need of 226
EOR. See enhanced oil recovery (EOR)
ethylene
– approximation equations 379
– coefficients for 383
– compression 77
– copolymers 82
– decomposition 88
– hydrate 363
– isentropic exponent 379
– polymerization 79
– real gas factor 377
– total ethylene flow 86
evaporation process 17, 37–43
– burn out point 38
– convective boiling 40
– critical heat flux 39
– – dependent on 43
– dry out point 40
– film boiling (upflow of water in a tube) 42
– forced convection boiling 40
– heat transfer
– – in case of flow boiling 38
– – regime of nucleate boiling 39
– limits of validity 42
– nucleation regime 38
– qualitative course of boiling curve 38
– regime of nucleate boiling 41

f

film thickness 18
– laminar 18
– turbulent 18, 19
film velocity 18, 19
flange pressure 370
flow coefficient calculation 384
fluid dynamic critical pressure 379, 384, 386
fluid saturation in reservoir rocks 145
forced convective mass transport 35
forged vessels 294
forgings 294
fossil fuels 123
free convective transfer 30, 35, 37
– heat and mass transfer, general equation 24
– at high pressures in CO_2 and N_2 34
frettage 298
Froude number 197, 198

g

gas-assisted high-pressure extraction 205, 206
– gas-assisted pressing of oilseed 205
gas, caloric and thermal properties 372, 373
gas extraction effect, industrial application 169
gas hydrates 363
gasification 235, 240–242
– entrained flow gasifiers 244–248
– fixed bed gasifier 242, 243
– fluidized bed gasifiers 243, 244
gasket 295
geometries 26
Gibbs energy 53
graphite 303
Grashof number 23, 25, 30, 34
Grayloc® clamped pipe connection 306

h

Haber–Bosch process 55, 63
Hagen-Poiseuille equation 320
heat capacity 26
heat conductivity 21
heat exchangers
– double pipe 302
– high-pressure 302, 303
– vessels/piping elements 301
heat flux 37–40, 42
heat transfer
– coefficients, for liquid flow 30
– correlations applicable for high-pressure processes 31, 32
– enhancement 29
– in near-critical pressure 30

heavy oil recovery 161, 162
high-pressure bends, fabrication 307
high-pressure centrifugal pumps 317–319
high-pressure components 285
– austenitic–ferritic duplex steels 288
– austenitic stainless steels 288
– chromium–molybdenum hydrogen-resistant steels 288, 289
– heat exchangers 301–303
– high-strength high-alloy steels 287, 288
– high-strength low-alloy steel 287
– high-strength steels
– – fatigue and fracture properties 289
– materials 285
– piping 304–309
– pressure vessels 290–301
– – leak before burst 292
– – nonwelded pressure vessels 294–298
– – prestressing techniques 298–300
– – sealing systems 300, 301
– – welded pressure vessels 292–294
– steel selection criteria 286
– steel selection guide 286
– valves 303–304
– weldable fine-grain/high-temperature structural steels 287
high-pressure (HP) equipment 72, 163
– bolting 74
– design standards 314–316
– flexible tube, design 164
– gaskets 74
– influence of fluid
– – on selection and design 313, 314
– materials testing 316, 317
– multilayered/multiwall vessels 73
– – external frame-supported end closures 297
– – externally clamped end closure 296
– pipes 164
– pumps 163
– seals 164–165
– selection of materials for 316, 317
– separators 165–166
– vessel design, recommendations 73, 74
– – external frame-supported end closures 297
– – externally clamped end closure 296
high-pressure fuel injection 6
high-pressure–high-temperature plants 123
high-pressure homogenizers (HPHs) 97
– design 98
– disruption systems 98
– – nozzles 99, 100

– – orifices 99, 100
– – valves 98, 99
– flow conditions 100
– – in disruption system 100, 101
– – homogenization valves on emulsion droplets, effect on 101
– – simultaneous emulsification and mixing (SEM) system 101, 102
high-pressure machines. *See* high-pressure (HP) equipment
high-pressure phase equilibrium 8
– data collection 10
– experimental methods 10
– measuring devices 10–12
– software 10
– thermodynamic aspects 9
high-pressure processes
– milestones 3–5
– realization 285
– working pressures 5
high-pressure pumps 3, 269–272, 311
– influence of fluid on selection and design 313, 314
high-pressure safety valves sizing
– real gas applications
– – critical pressure ratio 379, 380
– – equation of state 375–378
– – isentropic exponent 378, 379
– – real gas flow, safety valves sizing 380–382
– – sizing method, development 372–375
– standard valve sizing procedure 369–371
– – limits of 371, 372
high-pressure steam turbines 138
– design features 139–142
– longitudinal section of 140
– modern steam turbines, configuration 138, 139
– three-dimensional view 140
high-pressure turbocompressors 317–319
high-pressure vessels 4, 5, 30, 170, 200, 202, 288, 290, 297, 299, 362
– advantages 73
– instrumentation of 336
highstrength low-alloy (HSLA) steel 287, 288
homogenization processes, using SEM-type valves 107
– dairy processes 107–109
– emulsion droplets as templates 112–114
– formulation of nanoporous carriers for bioactives 116, 117
– melt homogenization 111, 112
– nanoparticle deagglomeration 116, 117
– pickering emulsions 109–111

– submicron emulsion droplets as nanoreactors 114–116
homogenization valves
– droplet agglomeration in 107
– droplet coalescence in 104–107
hot bending processes 309
HPHs. *See* high-pressure homogenizers (HPHs)
HTC. *See* hydrothermal carbonization (HTC)
hydrates in oil recovery 162, 163
– phase diagram 163
hydraulic valve 305
hydroforming 257, 258
hydrostatic pressures 15
hydrothermal carbonization (HTC) 248–250
– char and coke obtained after 251
hydrothermal gasification 253
– catalytic 253
– supercritical 254–256
hydrothermal liquefaction 251–252
hydrothermal processes 235, 248–250
hydroxymethylfurfural (HMF) 251

i

ideal/real gases, property data 374
interfaces in high-pressure processes 12
interfacial tension 15, 16, 17, 362, 363
– principle of measurement 363
– of system CO_2–H_2O 15
internal thread end closure 295
isentropic coefficient for ideal gases 385
isentropic exponent 370, 378, 379
isostatic pressing 258, 259

j

Joule–Thomson effect 372, 373

k

Kelvin equation 16, 17

l

law of chemical affinity 51
law of mass action 51, 52
LDPE. *See* low-density polyethylene (LDPE)
leak before burst (LBB) 292
le Chatelier–Braun principle 52
lens ring gasket 300
lens ring seal 300
lignin 251
lignite 125
linear low-density polyethylene (LLDPE) 78
lipoxygenase (LOX) 223
liquefaction 5, 21, 238–240, 248, 343
– hydrothermal 250

liquid natural gas (LNG)   6
liquids processing, with supercritical fluids   194
LLDPE. *See* linear low-density polyethylene (LLDPE)
low-density polyethylene (LDPE)   77
– applications   89
– – blow molding   90, 91
– – in blown film   89, 90
– – copolymers   91
– – extrusion coating   90
– – injection molding   90
– – wire and cable   90
– chain transfer   80–81
– free radical polymerization process for   78
– high-pressure process
– – historical background   77
– – latest developments   78
– – polyethylene high-pressure processes   78
– – polyethylene resin   77
– – reaction kinetics and thermodynamics   78–81
– homopolymer resins   89
– – applications   89–91
– initiation   79
– markets   77
– overall process description   82–84
– – flow sheet, industrial process   82
– – steps/process units   82
– plant
– – off-line applications   91–93
– – online application   93–95
– – training simulator image for   93
– propagation   79–80
– properties   77
– reaction
– – kinetics   56, 78–81
– – steps   78
– reactor (*See* autoclave reactor; tubular reactor)
– termination   81
– thermodynamics   78–81
lower heating value (LHV)   123
Lurgi multipurpose gasification (MPG) technology   248

m

Mach number   372
magnetic coupled sorption balances   362
– principle   362
mass flow rate   373
mass flux approximation   386
mass transfer   20
– correlations relevant for free and forced convection   33

material properties   20, 21
melt homogenization   111–112
MFR soft sensors   94
microorganisms, treatment of   203
– action of CO_2 on microorganisms   204
– hydrostatic high-pressure process   203, 204
minimum miscibility pressure (MMP)   147
– determination, methods for   147
– efficiency   148
– equipment   148
miscibility at elevated pressures   147–148
molecule distances   8

n

near-critical cinematic viscosity data   26
near-critical dynamic viscosity data   25
near-critical heat conductivities   24
nonpseudocritical pressure range, for turbulent flow   28
nozzle flow model   369, 370
nozzle throat   373, 375, 387
– pressure   375
Nusselt number   24, 30

o

off-line applications   91
– dynamic simulation of process   92, 93
– flow sheet simulations   91, 92
– steady-state simulation of tubular reactor   92
Ohnesorge number   19
oil reservoir stimulation   161
online application   93
– advanced process control   94, 95
– soft sensors   93–94
online measurements   336
– density determination   351, 355
– differential pressure measurements   338–340
– flow quantification   343–350
– – Coriolis-type flow sensor   350, 352
– – HP flow detection sensors   344–349
– fluid level detection   350, 351, 353
– – gamma densitometer   353
– – using load cell   353
– gas traces dissolved in liquids   358, 359
– high-pressure level gauges for   354
– parameters   337
– pressure measurements   338–340
– – sensors selection   338, 340
– safety aspects   364–366
– sensor choice and installation   337
– – checklist   337

– solute concentration measurements 352, 358
– temperature detection 341
–– industrial T-sensor installation 342
–– mounting methods of temperature sensors 342
–– rating of HP temperature sensors 343
–– resistance temperature detectors (RTDs) 341
–– seal 342
–– thermocouples for HP temperature measurement 343
– viscosity determination 351, 356, 357
original oil in place (OOIP) 146, 156, 157

p

particle-stabilized emulsions (PSE) 109
PCSAFT equation 9
PE. See phase equilibrium (PE)
Peng-Robinson equations 376
petaval gauges 335
petroleum
– factors influencing migration 146
– reservoir 145
phase equilibrium (PE) 55, 359
– analytical
–– dynamic measurement up to 100 MPa 361
–– measurement column with floating piston 361
–– measurement stirred vessel. 360
– principles for measurement with one fluid 359
phase inversion temperature (PIT) method 97
pickering emulsions 109–111
piezoelectric effect 335
pipe connection, with lens ring gasket 306
piping system 304–309
plate layered vessels
– disadvantages 294
– limitations 294
plug flow reactor (PFR) 85
polyethylene high-pressure processes 78
polymer properties 77
poly(methylmethacrylate) (PMMA) 113
polyphenoloxidase (PPO) 223
power generation 123
Prandtl numbers 25, 28, 30
– for CO_2 28
pressure vessels
– expansion energy 293
– external frame-supported end closures 297
– externally clamped end closure 296
– fabrication 292

pressure–volume–temperature (PVT) test 147
prestressing techniques 299
propylene (P) 164
pseudocritical temperature 21
pyrolysis 235, 237, 238

q

quartz crystal microbalance (QCM) 363

r

reaction kinetics 53–55
real gas behavior 382
real gas factor 27, 370, 374, 377, 378
real gas nozzle flow model, sizing coefficient 382, 383
– EN-ISO 4126-7, flow coefficient 384
– ethylene, property data/coefficients 383, 384
– flow coefficient 385–386
– inlet stagnation conditions 383
– mass flux by analytical method 386, 387
reciprocating positive displacement machines 323, 324
– design versions 328
– diaphragm pump heads 329, 330
– drive technology 324, 325
– flow behavior 325–327
– high-pressure piston pump heads 329
– horizontal pump head 329
– piston compressors 330–333
– pulsation damping 327, 328
– vertical pump head 328, 329
Redlich–Kwong equation of state 376
reservoirs
– conditions 145–147
– hydrocarbon 146
– petroleum 145
– PVT tests for reservoir fluid 147
reservoir systems, at elevated pressures
– physical chemical properties 148
–– density 148–150
–– diffusivity 153, 154
–– interfacial tension 151
–– permeability 154, 155
–– rheology 150, 151
–– wetting 151–153
Reynolds number 19, 25, 30, 372
rising bubble apparatus (RBA) 147
rotating positive displacement machines 319
– discharge rate 319, 320
– gear pumps 320, 321
– progressing cavity pump 323
– screw pumps 321–323

rotor–stator machine (RSM) 106
rupture disk assembly 305

S

safety instrumentation, of clamp-closure HP vessel 366
safety valves 304
– contour of flow 371
– narrowest flow cross section of 384
– sizing criteria 369
sandstones 146
Schmidt number 25
sealing element 300
sealing systems 300, 301
seat cross-sectional area of a safety valve A_{seat} 369
self-diffusion coefficient 36
SEM valves 108, 109
– geometric modification 110
– operational modes for production of o/w emulsions in 102
sensors 335, 338
– choice for high-pressure applications 337
– – checklist 337
– for differential pressure 339
– industrial specialty sensors 364
– safety aspects 364–366
shear stress 18
Sherwood number 24
simulation tools 91
sizing coefficient
– for ethylene 381
– real gas service, nozzle 380
– *vs.* reduced inlet stagnation pressure 381
sludge-to-oil reactor system (STORS) 252
Soave–Redlich–Kwong equations 376, 380, 383
solid hydrocarbon 162
solids–like polymers 35
Spray formation 19
static wetting angle 17
steam-assisted gravity drainage (ES-SAGD) 160
steam generator 130
– cross-sectional view 131
– design 130–133
– development of materials used in 133, 134
– final superheater heating surface 135, 136
– final superheater outlet header
– – and live steam piping 136–138
– HP system 131
– membrane wall 134, 135
– tube configuration 132
steam power plants

– configuration 127–129
steel, strength-toughness 289
Stokes–Einstein equation 35
supercritical extraction of solids, applications of 184
– cleaning and decontamination of cereals like rice 189, 190
– cleaning of cork 193
– decaffeination of green coffee beans 184
– economics 193, 194
– extraction of essential oils 186, 188
– extraction of γ-linolenic acid 189
– extraction of spices and herbs 186, 187
– impregnation of wood and polymers 190–192
– production
– – high-value fatty oils 189
– – hops extract 184, 185
– – natural antioxidants 188
– yield in carnosolic acid 188
supercritical fluids 169
– critical conditions 170
– deacidification of vegetable oil, pilot plant 196
– efficiency, in high-pressure separation process 198
– Froude number and Reynolds number of liquid phase 198
– for generation of renewable energy 204
– high-pressure column processes 195
– – designing 195
– high-pressure nozzle extraction 200
– high-pressure spray processes using 198, 199
– – advantages 198, 200
– hydrodynamic behavior of countercurrent flow 197
– – equation, prediction of flooding point 197
– interfacial tension at high pressures 198
– problem in calculating rate of mass transfer 197
– processing of liquids with 194–202
– thin film extraction (TFE) process 200
– – application 202
– – design of prototype pilot plant 200, 201
– – TFE pilot unit 201
– viscous materials 200
– – for continuous processing 201, 202
supercritical processing 169, 170
– classification of 171
– compressor process 183, 184
– design
– – criteria 177
– – with use of basket 178, 179

– energy optimization 182
– hydrodynamics 182
– mass transfer 179–182
– mass transfer kinetics 171
– phase equilibrium data 171
– pressure range 172
– pretreatment of raw materials 176, 177
– processing of solid material 172–174
– pump process 182, 183
– separation of dissolved substances 173
– – extractable substances 175
– – isobaric process 174
– – selective extraction 176
– – single/cascade operation 174, 175
– – total extraction 176
– solubility in supercritical CO_2 173
– thermodynamic conditions 179
– typical trend of extraction lines 172
surface tension 14, 203

t

tangential stress 290
tetrafluoroethylene (TFE) 164
TFEP (AFLAS®) copolymer 164
thermal expansion coefficient 30, 34
thermodynamic critical point 381
thermodynamic critical pressure, gases 377
thermodynamics, second law 385
thermophysical data at high pressures, correlations for 27
transport
– coefficients 21
– data 20, 21
– phenomena 55
– processes 20
tubular reactor 85–88, 86
– configurations
– – multiple feed reactor 86
– – single ethylene feed/S-Reactor 86
– multiple cold ethylene feed points 86
– multiple feed tubular reactor
– – temperature profile 87
– product properties, affected by 88
– safety requirements 88–89
– steady-state simulation of 92
turbocompressors 311
turbomachines 311, 313

– in multistage design 313
twister separator, design 165

u

urea synthesis 64
– basics and principles 65–67
– history of urea process 67–71
– integration with ammonia processes 71
– material selection 71, 72

v

vanishing interfacial tension (VIT) 147
vapor pressure 16, 17
vessel wall 295
viscosity 19, 22, 27, 30, 45, 101, 107, 112, 156, 159, 197, 221, 313, 320, 322, 349, 362
volumetric pumping behavior 312
von Mises criterion 291

w

water alternating gas injection (WAG) 160
waterjet cutting technology 259, 260
– advantages 265
– with 6000 bar 272, 273
– cutting process, and parameters 267–269
– – abrasive, influence on cutting results 268
– – material surface after AIWJ cutting 269
– – successive phases of stock removal during 268
– generation of waterjets 265–267
– new trends 276
– – abrasive water suspension jet 276
– – medical applications 277, 278
– – microcutting 276, 277
Weber number 19
welded pressure vessels 293
– heating and cooling 294
– steels, economical importance 287
welding 73, 74, 134, 257, 287, 292, 294, 302, 304, 342
wetting angle
– pressure dependence of 18
– solid surfaces 16
– water and oil on flat steel surface 16, 17

y

Young equation 16, 17
Young–Laplace equation 15